Advanced Research in Combustion Energy: Optimization, Applications, and Analysis

Advanced Research in Combustion Energy: Optimization, Applications, and Analysis

Guest Editor

Zekai Hong

MDPI

Basel • Beijing • Wuhan • Barcelona • Belgrade • Novi Sad • Cluj • Manchester

Guest Editor
Zekai Hong
Aerospace Research Center
National Research Council of Canada
Ottawa, ON
Canada

Editorial Office
MDPI AG
Grosspeteranlage 5
4052 Basel, Switzerland

This is a reprint of the Special Issue, published open access by the journal *Energies* (ISSN 1996-1073), freely accessible at: https://www.mdpi.com/journal/energies/special_issues/M1MI1R5AQI.

For citation purposes, cite each article independently as indicated on the article page online and as indicated below:

Lastname, A.A.; Lastname, B.B. Article Title. *Journal Name* **Year**, *Volume Number*, Page Range.

ISBN 978-3-7258-5009-9 (Hbk)
ISBN 978-3-7258-5010-5 (PDF)
https://doi.org/10.3390/books978-3-7258-5010-5

Contents

About the Editor

Zekai Hong

Zekai Hong received his Ph.D. in Mechanical Engineering from Stanford University in 2010. He began his career at GE Global Research in Niskayuna, NY, where he worked as a combustion research engineer on GE's next-generation turbofan engines. In 2017, he joined the National Research Council Canada (NRC) in Ottawa, where he currently serves as a Principal Research Officer. He also holds an adjunct professorship at the University of Ottawa. His research interests include combustion, laser diagnostics, chemical kinetics, combustion instability, and shock waves. Dr. Hong received the Early Career Award from the Global Power and Propulsion Society (GPPS) in 2019 and was named an Associate Fellow of the American Institute of Aeronautics and Astronautics (AIAA) in 2021.

energies

MDPI

Review

Biofuels in Aviation: Exploring the Impact of Sustainable Aviation Fuels in Aircraft Engines

Ramozon Khujamberdiev and Haeng Muk Cho *

Department of Mechanical Engineering, Kongju National University, Cheonan 31080, Republic of Korea; khujamberdievramozon@gmail.com
* Correspondence: hmcho@kongju.ac.kr; Tel.: +82-(10)-87113252

Abstract: This comprehensive review examines the role of sustainable aviation fuels (SAFs) in promoting a more environmentally responsible aviation industry. This study explores various types of biofuels, including hydroprocessed esters and fatty acids (HEFAs), Fischer–Tropsch (FT) fuels, alcohol-to-jet (ATJ) fuels, and oil derived from algae. Technological advancements in production and processing have enabled SAF to offer significant reductions in greenhouse gas emissions and other pollutants, contributing to a cleaner environment and better air quality. The review addresses the environmental, economic, and technical benefits of SAF, as well as the challenges associated with their adoption. Lifecycle analyses are used to assess the net environmental benefits of SAF, with a focus on feedstock sustainability, energy efficiency, and potential impacts on biodiversity and land use. Challenges such as economic viability, scalability, and regulatory compliance are discussed, with emphasis on the need for supportive policies and international collaboration to ensure the long-term sustainability of SAF. This study also explores current applications of SAF in commercial airlines and military settings, highlighting successful case studies and regional differences driven by policy frameworks and government incentives. By promoting technological innovation and addressing regulatory and economic barriers, SAF has the potential to play a crucial role in the aviation industry's transition toward sustainability.

Keywords: sustainable aviation fuels; aircraft engine; hydroprocessed esters and fatty acids; synthetic iso-paraffin; Fischer–Tropsch fuels

1. Introduction

1.1. Background on Aviation and Environmental Concerns

Aviation is a significant contributor to global carbon emissions, posing substantial environmental challenges that exacerbate climate change. Aviation emissions, particularly carbon dioxide (CO_2), nitrogen oxides (NOx), and water vapor, directly contribute to global warming and climate change through the greenhouse effect and cloud formation enhancements [1,2]. The persistent growth in air travel demand ensures that these emissions will continue to rise without significant interventions in aircraft fuel technologies and air traffic management [3].

Sustainable aviation fuels (SAFs) are pivotal to the future of aviation, offering a viable alternative to conventional jet fuels, which are major sources of aviation-related emissions. SAFs, derived from bio-based resources, significantly reduce the lifecycle carbon emissions associated with aviation, providing a critical solution to the pressing need for the aviation sector's decarbonization [4,5]. The adoption of SAFs not only aligns with global carbon reduction targets but also helps in mitigating the climate impact of non-CO_2 emissions, such as those from NOx and contrails, which are significant contributors to radiative forcing [6,7].

1.2. Importance of Sustainable Aviation Fuels (SAFs)

The aviation industry is undergoing a resurgence since the COVID-19 pandemic, resulting in a heightened global demand for aviation fuel. The International Air Transport Association (IATA) projects this demand to reach 6.46 million gallons per day, which is equivalent to approximately 19.8 thousand tons per day. The increased consumption of aviation fuel will impact the environment and human health by raising greenhouse gas (GHG) emissions. In 2019, air transport generated 914 million tons of CO_2, accounting for 2.1% of total human-produced CO_2 emissions and 12% of CO_2 emissions from the transport sector [8].

Achieving net zero carbon dioxide (CO_2) emissions by 2050 is a key objective for the aviation industry, governments, and regulatory bodies. High-level strategies, scenarios, and roadmaps have been proposed to support this transition. Examples include the European Green Deal with its Fit-for-55 package, IATA's "Fly Net Zero" strategy, and EUROCONTROL's aviation outlook for 2050 [9]. The transition to SAFs offers numerous benefits over traditional fuels. SAFs contribute to energy security by diversifying the fuel supply and reducing dependency on oil markets. They also support rural development and agricultural diversification through the use of non-food bio-feedstocks [10,11].

Despite these advantages, the widespread implementation of SAFs faces significant challenges, including high production costs, feedstock availability, and the need for substantial technological and infrastructural developments to support large-scale deployment [12,13]. To realize the full potential of SAFs, coordinated efforts among governments, industry stakeholders, and the scientific community are essential. This includes investments in research and development to improve fuel efficiency and reduce costs, as well as policies to support the market uptake of SAFs [14–16].

The main goal of this review paper is to thoroughly investigate the environmental challenges presented by the aviation sector, particularly its substantial contribution to global carbon emissions. The paper examines the impact of these emissions—such as carbon dioxide, nitrogen oxides, and water vapor—on global warming and climate systems, both directly and indirectly. It also explores the role of sustainable aviation fuels (SAFs) as a feasible alternative to conventional jet fuels, emphasizing their potential to lower lifecycle carbon emissions and support the decarbonization of the aviation industry.

2. Methodology

2.1. Data Sources and Selection Criteria

The selection of sources for this comprehensive review was guided by specific criteria to ensure the relevance and reliability of the information pertaining to sustainable aviation fuels and their application in aircraft engines. Initially, the primary data sources were peer-reviewed journal articles, industry reports, and authoritative publications from significant entities in the aviation and environmental sectors, such as the International Air Transport Association (IATA) and the International Civil Aviation Organization (ICAO). The relevance of each source was determined based on its publication date, preferably within the last five years, to ensure the currency of the data in the rapidly evolving field of aviation biofuels.

Reliability was assessed through the source's peer-reviewed status, citation metrics, and the recognition of the publishing journal or organization within the scientific and aviation communities. Sources were also selected based on the specificity of their focus on biofuels' lifecycle analysis, emission impacts, economic viability, and regulatory frameworks. This ensured a comprehensive collection of data that spans across the necessary dimensions of the study topic.

2.2. Approach to Literature Review and Analysis

The approach to the literature review and analysis involved a structured process to synthesize and interpret the gathered data effectively. The review started with a broad search of keywords and phrases related to "sustainable aviation fuels", "biofuel use in aviation", and "environmental impact of aviation fuels". This initial search was refined by applying the selection criteria to narrow down the most relevant and reliable sources. Each selected source was then meticulously reviewed to extract data pertinent to the environmental impacts, technical performance, economic aspects, and regulatory environments associated with biofuels in aviation.

The synthesis of information was conducted through a thematic analysis, where data were categorized into themes such as environmental benefits, technical challenges, economic feasibility, and policy implications. This thematic approach allowed for a coherent structure in which to discuss the findings comprehensively. The analysis also included a critical evaluation of conflicting data and gaps in current research, providing a balanced perspective on the potential and limitations of sustainable aviation fuels. The review culminated in integrating these insights into a narrative that reflects both the current state of knowledge and the future directions for research and policymaking in the domain of sustainable aviation fuels.

3. Overview of Aircraft Engine Types

3.1. Piston Engines

Piston engines, also known as reciprocating engines, are among the oldest types of aircraft engines and are primarily used in small general aviation aircraft. They operate similarly to automobile engines, using the expansion of burning gas to drive pistons. These engines are typically powered by avgas (aviation gasoline), which is a high-octane fuel suitable for the high compression ratios of aviation engines [17,18]. The transition to biofuels in piston engines poses challenges primarily due to the biofuels' different chemical characteristics compared to avgas, such as lower energy content per volume and variations in stoichiometric fuel-to-air ratio, which can affect engine performance and maintenance [19,20]. Research into biofuel compatibility with piston engines has indicated that modifications to the fuel delivery systems and ignition timing may be necessary to accommodate some types of biofuels [21,22]. Studies have also highlighted the potential for biofuels to reduce emissions of unburned hydrocarbons and carbon monoxide in piston aircraft engines, aligning with environmental sustainability goals [20,23].

3.2. Turboprop Engines

Turboprop engines combine aspects of both jet engines and propellers. These engines are used primarily in aircraft that require short take-offs and landings or in those flying at lower speeds for regional transport. Turboprops are known for their efficiency at speeds below 450 miles per hour and are typically more fuel-efficient than turbojets or turbofans at these lower speeds [13,15]. The use of biofuels in turboprop engines has been explored, with studies demonstrating that biofuels such as HEFAs (hydroprocessed esters and fatty acids) can be integrated without significant modifications to the engine [5,9].

Research has shown that SAFs, including those derived from jatropha, camelina, and even used cooking oils, can meet the stringent energy density and freezing point requirements necessary for turboprop applications [12,14]. These studies underline the potential for significant reductions in lifecycle greenhouse gas emissions when switching from conventional jet fuel to biofuels in turboprop engines [5,24].

3.3. Jet Engines (Turbojet and Turbofan)

Turbojet and turbofan engines are the most common engines used in commercial aviation, with turbofans being more prevalent due to their greater efficiency and reduced noise output. These engines work by drawing air into the engine, compressing it, mixing it with fuel, and igniting the mixture to produce thrust [25,26]. The compatibility of biofuels with jet engines has been extensively tested, demonstrating that biofuels can replace conventional jet fuel up to a 50% blend with minimal to no modifications required to the engines [22]. Figure 1 shows a basic illustration of a turbofan engine.

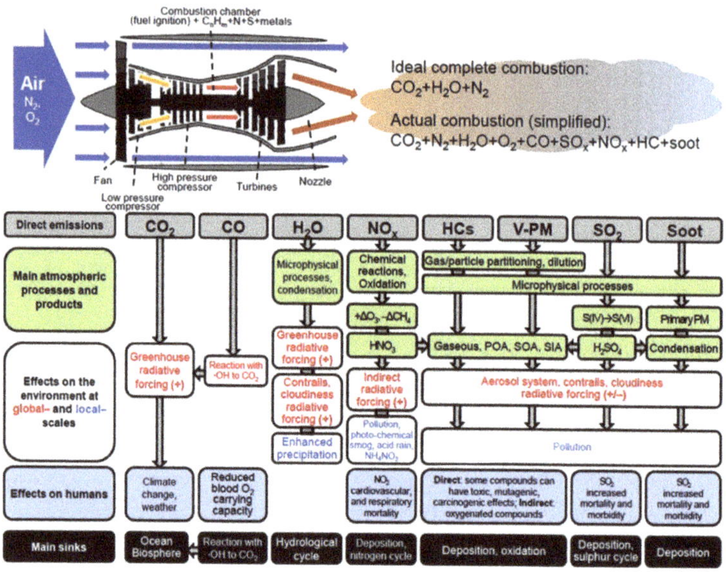

Figure 1. A simplified diagram of a turbofan engine in the upper left corner, and the byproducts of ideal and actual combustion in an aircraft engine in the upper right corner. Various atmospheric processes, products, environmental impacts, effects on human health, and sinks for emitted compounds are in the bottom section [27].

The studies have highlighted that biofuels can significantly reduce the emission of pollutants such as sulfur oxides and particulate matter, which are prevalent in emissions from conventional jet fuels [28,29]. The aviation industry sees this as a crucial step toward achieving carbon-neutral growth, with the added benefit of reducing contrail formation, which has a high climate impact due to its greenhouse gas trapping ability [20,30].

3.4. *Compatibility of Different Engines with Biofuels*

The compatibility of different engine types with biofuels is a critical area of research, particularly as the aviation industry seeks sustainable fuel alternatives to meet environmental targets. While piston and turboprop engines may require some adjustments for optimal biofuel performance, turbojet and turbofan engines have shown a higher degree of readiness for biofuel integration, particularly with SAFs [10,20]. Each engine type presents unique challenges, from material compatibility and fuel system adjustments to efficiency metrics and emission profiles.

Studies continue to focus on the long-term impacts of biofuel use on engine wear and tear, maintenance schedules, and overall lifecycle costs [19,26]. Additionally, the research emphasizes the need for industry-wide standards and certification processes to ensure that biofuels not only meet the technical requirements of various engines but also contribute positively to environmental sustainability goals without compromising safety or performance [14,15].

4. Types of Biofuels in Aviation

4.1. *Hydroprocessed Esters and Fatty Acids (HEFAs)*

Hydroprocessed esters and fatty acids (HEFAs) are among the leading types of sustainable aviation fuels (SAFs) currently being integrated into the aviation sector. HEFA fuels are produced through the hydroprocessing of vegetable oils or animal fats, which involves the removal of oxygen and other non-hydrocarbon molecules to create hydrocarbons that are structurally similar to those found in conventional jet fuels. This process results in a renewable jet fuel that can be used as a direct drop-in replacement for traditional jet fuels without requiring modifications to aircraft engines or fuel distribution infrastructure [31,32].

One of the crucial points for HEFA production is hydrogen production. This is a significant factor affecting both the economic viability and the environmental footprint of HEFA, particularly when considering green hydrogen. A recent study by Pipitone et al. (2023) addresses both the economic and environmental impacts of hydrogen production in HEFA processes, underscoring the importance of green hydrogen in reducing lifecycle emissions [33].

The appeal of HEFA fuels lies in their significant environmental benefits. They do not contain sulfur and have the potential to reduce greenhouse gas emissions by up to 80% over their lifecycle compared to conventional jet fuels. This reduction is primarily due to the sustainable nature of the feedstocks and the potential for the carbon dioxide released during combustion to be offset by the carbon dioxide absorbed during the growth of the feedstocks [34]. Technological advancements in the HEFA production process have focused on optimizing yields and improving the cold flow properties of the fuel, which are critical for aviation use. Techniques such as adjusting the isomerization process have been explored to enhance the yield while maintaining the necessary freezing point properties required for aviation fuels [31]. Additionally, ongoing research aims to expand the variety of feedstocks that can be used, thereby increasing the scalability and economic feasibility of HEFA fuels [35].

One of the critical challenges facing the adoption of HEFAs as a sustainable aviation fuel is the economic viability. The cost of production remains higher than that of conventional jet fuels, primarily due to the high cost of bio-based feedstocks and the complexity of the production process. However, economic analyses suggest that with strategic policy support and technological advancements, HEFA fuels could become more competitively priced, fostering broader adoption in the aviation industry [32]. HEFA fuels have been demonstrated to perform comparably to conventional fuels in terms of engine efficiency and emissions. Studies have shown that HEFA can significantly reduce emissions of particulate matter and other pollutants, contributing to cleaner engine operations and less environmental impact. Continued research and development are focusing on further reducing any potential toxicological impacts associated with HEFA fuel usage, ensuring safety and sustainability in its application [33].

4.2. *Synthetic Iso-Paraffin (SIP) Fuel*

Synthetic iso-paraffin (SIP) fuels, often referred to as synthetic paraffinic kerosene when used in aviation, are a type of sustainable aviation fuel (SAF) that holds significant promise for reducing the aviation sector's carbon footprint. SIP fuels are produced using the Fischer–Tropsch process or through the hydroprocessing of renewable fats and oils. These processes yield a high-quality fuel that contains no sulfur and has excellent combustion properties [36,37]. The benefits of SIP fuels include their ability to significantly reduce emissions of particulates and greenhouse gases compared to conventional jet fuels. For example, the use of SIP in aircraft engines has demonstrated a reduction in the emission of soot and other particulate matter, which are known contributors to atmospheric warming and pollution [36]. Moreover, due to their cleaner burning characteristics, SIP fuels can help improve air quality around airports and major cities.

Technologically, SIP fuels are compatible with existing aircraft engines and fuel distribution infrastructure, which allows them to be blended with traditional jet fuels without any modifications to aircraft systems. This compatibility is pivotal for the adoption of SIP fuels in the aviation industry, as it enables airlines to transition to greener fuels without incurring substantial retrofitting costs or operational changes [37]. There are challenges associated with the widespread adoption of SIP fuels. The production of SIP is currently more costly than that of conventional jet fuel, primarily due to the high cost of production facilities and the need for significant upfront investment in technology development. Moreover, while the feedstocks used in the production of SIP are renewable, the scalability of these resources and the sustainability of their supply chains are critical factors that need further exploration and optimization to meet the growing demand of the aviation industry [38].

In terms of safety and performance, studies have shown that SIP fuels perform well under various atmospheric conditions, maintaining stability and energy content comparable to traditional fuels. This aspect is crucial for ensuring that the adoption of SIP does not compromise the safety and reliability of air travel [36].

4.3. Fischer–Tropsch (FT) Fuel

Fischer–Tropsch (FT) synthesis is a versatile and well-established process used to convert syngas (a mixture of carbon monoxide and hydrogen) derived from various carbonaceous materials like coal, natural gas, or biomass into liquid hydrocarbons. This technology is critical for producing synthetic fuels, including sustainable aviation fuels (SAFs), that are virtually sulfur-free and have significantly lower greenhouse gas emissions compared to conventional jet fuels [39,40]. FT fuels are praised for their cleanliness and high performance in aircraft engines. They can be tailored to produce a wide range of hydrocarbon chains, making them highly suitable for aviation purposes. The adaptability of FT synthesis allows for the production of fuels that meet stringent aviation fuel standards regarding purity and performance. Furthermore, FT fuels have the potential to significantly reduce particulate matter emissions and other pollutants associated with jet engines, contributing to cleaner airport environments and reducing the aviation industry's overall carbon footprint [41,42].

Despite their benefits, the scale-up and economic viability of FT fuels face several challenges. The capital-intensive nature of FT synthesis plants and the variability in feedstock costs can hinder the broader deployment of this technology. However, advancements in catalysis and process engineering are continually being explored to reduce these costs and improve efficiency. For instance, innovations in catalyst formulations and reactor designs have shown potential to enhance conversion rates and selectivity towards desired fuel components, thereby optimizing production costs and environmental benefits [43,44]. Moreover, the sustainability of FT fuels largely depends on the source of the hydrogen and carbon monoxide used in the synthesis process. Utilizing biomass or waste-derived syngas aligns with circular economy principles and further enhances the environmental credentials of FT fuels. The integration of carbon capture and utilization technologies into the FT process also presents a promising avenue to reduce the overall carbon intensity of the produced fuels, aligning with global carbon reduction goals [45]. Fischer–Tropsch synthesis can be conducted at varying temperatures. In processes using iron catalysts, the synthesis temperature can range from 220 °C to 340 °C, depending on the specific technology employed [46].

4.4. Alcohol-to-Jet (ATJ)

Alcohol-to-jet (ATJ) technology represents a transformative approach to sustainable aviation fuel (SAF) production, converting alcohols such as ethanol or isobutanol into synthetic paraffinic kerosene suitable for jet engines. This pathway leverages established catalytic processes to dehydrate alcohols into alkenes, which are then oligomerized into larger hydrocarbon chains typical of jet fuels. ATJ fuels are distinguished by their ability to blend seamlessly with conventional jet fuel, meeting the stringent specifications required for commercial aviation without modifications to existing engines or fuel systems [47,48]. ATJ's appeal lies in its use of renewable feedstocks, including sugars, starches, and lignocellulosic biomass, which can be fermented into alcohols before conversion. This flexibility allows the utilization of a wide range of raw materials, including non-food crops and agricultural residues, potentially reducing competition with food sources and minimizing land use impacts [49,50]. Furthermore, the ATJ process is capable of producing jet fuel that performs comparably to conventional jet fuel in terms of energy density and cold-flow properties while significantly reducing lifecycle greenhouse gas emissions.

Figure 2 illustrates the process of converting bio-ethanol to bio-jet fuel hydrocarbons. This process includes ethanol C-C coupling, distillation of oil-phase high-carbon alcohols, and hydrodeoxygenation (HDO) of heavy alcohols. Verification of the entire process resulted in the production of bio-jet fuel hydrocarbons, predominantly composed of branched hydrocarbons. Additionally, the C4–C7 high-carbon alcohols present in the light alcohols can be further upgraded to synthesize C8–C16 high-carbon alcohols, thereby enhancing the yield of bio-jet fuel hydrocarbons [51].

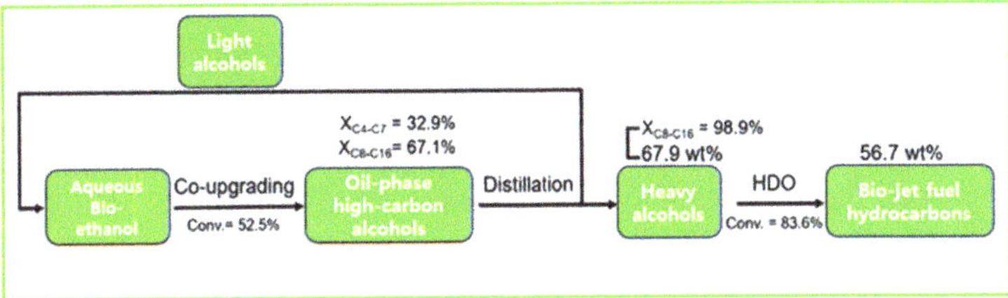

Figure 2. Comprehensive flow diagram of kilogram-scale bio-ethanol to jet fuel hydrocarbon production [51].

Despite its advantages, the adoption of ATJ technology faces challenges, primarily related to the cost-effectiveness of production at scale. The initial capital investment for ATJ facilities is significant, and the economic viability depends heavily on the cost of feedstock and the efficiency of the alcohol fermentation process [52]. Moreover, the regulatory landscape for biofuels can influence the deployment of ATJ technology, necessitating supportive policies to ensure a competitive edge against traditional jet fuels. From an environmental perspective, ATJ fuels promise substantial reductions in particulate emissions and other pollutants. Studies have shown that ATJ fuels can lower the emission of soot and sulfur oxides, contributing to cleaner combustion and less environmental impact at airports and in the upper atmosphere, where jet travel predominates [53].

4.5. *Oil from Algae*

Oil derived from algae, particularly microalgae, is a promising source of sustainable aviation fuel (SAF). Algae-based oils are regarded for their rapid growth rates and high oil yields per area, making them a highly productive source of biofuels compared to traditional bioenergy crops. Algae can be cultivated in various environments, including freshwater, marine settings, and even in non-arable lands, which minimizes competition for agricultural resources and does not contribute to deforestation [54,55]. The process of converting algae into biofuel typically involves cultivating the algae, harvesting the biomass, extracting the lipids, and then refining these lipids into biofuel through processes like hydroprocessing or transesterification. This biofuel is capable of being a "drop-in" replacement for conventional jet fuels, meaning it can be used without modifications to existing aircraft engines or fuel distribution infrastructures [56,57].

However, despite these advantages, several challenges remain in the path to commercializing algae-based aviation fuels. These include the economic costs associated with large-scale production, the energy-intensive nature of algae cultivation and harvesting, and the need for technological advancements to improve lipid extraction and conversion efficiencies. Studies also focus on optimizing the conditions under which algae grow to maximize lipid production, such as by adjusting light intensity, nutrient availability, and CO_2 supplementation [56,58].

The sustainability of algae-based aviation fuel production depends significantly on the lifecycle analysis of emissions and energy use. The integration of algae cultivation with waste management systems, such as using municipal or industrial wastewater as a nutrient source, can improve the sustainability and economic viability of this biofuel source. This integration not only helps in treating wastewater but also in reducing the operational costs associated with algae cultivation [54,59]. Algae-based fuels also present an opportunity for decarbonizing the aviation sector, which is under increasing pressure to reduce its environmental impact. As research continues, the scalability of algae as a biofuel source holds the potential to meet a significant portion of aviation fuel demand, contributing to the reduction of greenhouse gas emissions and reliance on fossil fuels [54,60].

4.6. Other Emerging Biofuels

The landscape of sustainable aviation fuels (SAFs) is rapidly evolving, with several emerging biofuel technologies showing promise for revolutionizing energy sources within the aviation industry. These include advanced biofuels such as lignin-based fuels, solar fuels, and electrofuels, each offering unique benefits and challenges to sustainable aviation [61].

Lignin-Based Biofuels: Lignin, a major component of plant biomass, provides a promising feedstock for biofuel production due to its abundance and relatively low value in the current market. Research into lignin-based biofuels primarily focuses on converting this complex polymer into simpler hydrocarbon chains that can serve as drop-in fuels for aviation. While the process is challenging due to the recalcitrant nature of lignin, advances in catalytic hydroprocessing and genetic engineering are improving the efficiency of these conversions, promising a sustainable route to high-energy-density jet fuels [61].

Solar Fuels: Solar fuels are produced through the conversion of solar energy into chemical energy, typically via processes that mimic natural photosynthesis. Although still at an experimental stage, solar fuels could potentially provide limitless carbon-neutral energy, sourced directly from sunlight and air. The integration of solar energy systems with catalytic processes to produce liquid hydrocarbons could significantly reduce the carbon footprint of aviation fuels, aligning with global decarbonization goals [62].

Electrofuels: Produced from carbon dioxide and water using electrical energy (often sourced from renewable sources), electrofuels represent a groundbreaking approach to sustainable fuel production. This technology is particularly attractive for the aviation industry due to its potential for high energy density and compatibility with existing fuel infrastructure. Electrofuels can be synthesized through various pathways, including the electrochemical reduction of CO_2, offering a versatile and potentially scalable solution to reduce aviation's reliance on fossil fuels [63].

Table 1 serves to highlight the distinct features and considerations of each biofuel type, offering a clear comparison to facilitate understanding of their potential impact on the aviation sector.

Table 1. Summary of biofuels for aviation: production processes, benefits, challenges, and technological aspects.

Type of Biofuel	Production Process	Benefits	Challenges	Technological Aspects
Hydroprocessed Esters and Fatty Acids (HEFAs)	Hydroprocessing of vegetable oils or animal fats [31,32]	Direct drop-in replacement; up to 80% reduction in lifecycle GHG emissions; no sulfur content; reduced particulate emissions [34]	Higher production costs; economic viability depends on policy support and technological advancements [32]	Improving yield and cold flow properties; exploring wider variety of feedstocks [31]
Synthetic Iso-paraffin (SIP) Fuel	Fischer–Tropsch process or hydroprocessing of renewable fats and oils [36,37]	Reduction in particulate and GHG emissions; compatible with existing engines and infrastructure [36,37]	Higher production costs; scalability and sustainability of feedstock supply chains [38]	High-quality fuel with no sulfur; excellent combustion properties [36]
Fischer–Tropsch (FT) Fuel	Conversion of syngas (from coal, natural gas, or biomass) into liquid hydrocarbons [39,40]	Sulfur-free; lower GHG emissions; high performance in engines [41,42]	Capital-intensive production plants; variability in feedstock costs [43,44]	Advances in catalysis and process engineering; integration of carbon capture and utilization technologies [45,46]
Alcohol-to-Jet (ATJ)	Conversion of alcohols (ethanol, isobutanol) into synthetic paraffinic kerosene [47,48]	Uses renewable feedstocks;sSignificant GHG emissions reduction [49,50]	High initial capital investment; economic viability depends on feedstock costs and fermentation efficiency [51]	Established catalytic processes for dehydration and oligomerization [47,48]
Oil from Algae	Cultivation of algae, lipid extraction, and refining into biofuel [54–57]	High oil yields per area; can be cultivated in diverse environments without competing with agricultural resources [54,55]	Economic costs of large-scale production; energy-intensive cultivation and harvesting processes [56–58]	Optimization of growth conditions; integration with waste management systems [59,60]

5. Production and Processing of Aviation Biofuels

5.1. Feedstock Sources and Sustainability

The production of sustainable aviation fuels (SAFs) starts with the selection of suitable feedstock, which significantly impacts both the sustainability and economic viability of the fuel produced. Feedstocks for SAFs include a wide range of biomass sources, such as agricultural residues, non-edible oil crops, algae, and municipal solid waste. Each of these sources has its own set of sustainability challenges and benefits, making the choice of feedstock crucial for the overall sustainability of the fuel production process [59]. Sustainable feedstock sourcing must consider environmental, social, and economic aspects, ensuring that the production does not adversely affect food security, biodiversity, or lead to deforestation. Technologies such as the hydrothermal liquefaction of algae and the Fischer–Tropsch synthesis using lignocellulosic biomass are being developed to increase the sustainability and efficiency of feedstock conversion. Innovations in genetic engineering and agronomic practices are also improving the yields and reducing the environmental impacts of these crops [64].

5.2. Conversion Processes

The conversion of biomass into aviation biofuel can be achieved through several pathways, each with its own technological and economic challenges. The most common processes include hydroprocessing of oils and fats, Fischer–Tropsch synthesis from syngas, and the alcohol-to-jet (ATJ) pathway. Hydroprocessing involves removing oxygen from triglycerides to form hydrocarbons that can be used directly in jet engines. Fischer–Tropsch synthesis, on the other hand, involves gasifying biomass to produce a synthesis gas, which is then converted into liquid hydrocarbons through a catalytic process. The ATJ pathway involves the conversion of alcohols, which can be produced from sugars or starches, into synthetic paraffinic kerosene [65,66].

Each of these conversion processes requires specific catalysts, temperatures, and pressures, and the efficiency of these processes is continually improving through research and development. Challenges such as catalyst deactivation, high energy requirements, and the need for large-scale infrastructure investment remain, but advances in catalytic science and process engineering are progressively overcoming these barriers [67,68].

5.3. Blending with Conventional Jet Fuels

Blending bio-based aviation fuels with conventional jet fuels is a practical approach to introducing SAF into the current fuel supply without extensive modifications to aircraft or fuel distribution infrastructure. Biofuels must meet stringent specifications for viscosity, freezing point, and energy content to be compatible with existing jet fuels. Research into the blending properties of various biofuels has shown that certain bio-derived synthetic paraffinic kerosenes can be blended at ratios of up to 50% with conventional jet fuels without adverse effects on engine performance or emissions [69,70]. Studies have demonstrated that blends of biofuels and conventional jet fuels can reduce emissions of greenhouse gases and particulate matter, contributing to cleaner combustion and reduced environmental impact. However, the stability, cold flow properties, and combustion characteristics of these blends are critical parameters that need continuous monitoring and optimization to meet the operational requirements of commercial aviation [71,72].

6. Advantages and Challenges

6.1. Economic Viability and Cost Factors

Economic viability remains a significant challenge for SAFs, with production costs typically higher than those of conventional jet fuels. Several factors contribute to these higher costs, including the expense of feedstocks, energy-intensive conversion processes, and the need for large-scale production facilities. Despite these challenges, the growing demand for sustainable fuels and supportive government policies are driving down costs through economies of scale and technological advancements [65]. To achieve economic viability,

SAF production must be scalable and cost-effective. This involves optimizing feedstock sourcing, improving conversion efficiency, and developing robust supply chains. Innovative technologies, such as hydroprocessing, Fischer–Tropsch synthesis, and alcohol-to-jet (ATJ), are becoming more efficient, helping to lower costs. Additionally, financial incentives and carbon pricing mechanisms are playing a role in making SAF more competitive with traditional fuels [61].

6.2. Scalability and Supply Chain Issues

Scalability is a critical factor in the widespread adoption of SAF. The aviation industry requires large volumes of fuel, necessitating a scalable supply chain for SAF production and distribution. Challenges in scalability often stem from feedstock availability, transportation logistics, and production capacity. A diverse range of feedstocks, such as algae, agricultural residues, and non-food crops, can improve scalability by reducing reliance on a single source [64]. Developing robust supply chains involves addressing logistical issues related to feedstock transportation, storage, and distribution. The integration of SAF production with existing infrastructure and the ability to blend with conventional jet fuels are essential for ensuring a smooth transition to sustainable aviation. Scalability also requires investment in large-scale production facilities, which can be capital-intensive but are crucial for meeting the aviation industry's demand [65].

6.3. Technical and Regulatory Challenges

Technical challenges for SAF primarily concern engine compatibility, fuel stability, and combustion characteristics. SAF must meet stringent specifications to ensure safety and performance in aircraft engines. This requires rigorous testing and certification processes to guarantee that SAF blends are safe for use in commercial aviation [67]. Research into the blending properties of SAF with conventional jet fuels is ongoing, focusing on maintaining the necessary energy content, freezing point, and viscosity for aviation applications.

Regulatory challenges also play a significant role in the adoption of SAF. The aviation industry is subject to strict regulations, and any new fuel must comply with these standards. Harmonizing regulations across different countries and regions is crucial for the global adoption of SAF. Supportive policies and financial incentives can encourage airlines to transition to SAF, while international agreements, such as those under the International Civil Aviation Organization (ICAO), are instrumental in setting industry-wide sustainability goals [65].

7. Case Studies and Current Applications

7.1. Use in Commercial Airlines

Sustainable aviation fuels (SAFs) have gained traction in commercial airlines due to their potential to reduce greenhouse gas emissions and dependence on fossil fuels. Commercial airlines have begun incorporating SAF into their operations, with many conducting test flights and some establishing long-term agreements for SAF supply [73]. For example, several major airlines, including British Airways and United Airlines, have partnered with biofuel companies to promote the use of SAFs on their flights. This shift toward biofuels is part of a broader effort to meet international sustainability goals and reduce the environmental impact of air travel [74].

Drop-in biofuels are particularly attractive to commercial airlines as they can be blended with conventional jet fuels, allowing for a smoother transition to sustainability without requiring significant infrastructure changes. Airlines have conducted successful test flights with SAF, demonstrating that these fuels do not adversely impact engine performance or maintenance schedules [75,76]. However, to increase the adoption of SAF in commercial airlines, consistent policies, reliable feedstock sources, and scalable production facilities are required to meet demand and reduce costs [77].

7.2. Military Applications

The military has also shown interest in using biofuels, primarily to increase energy security and reduce reliance on fossil fuels. The U.S. Navy, for example, has used algae-based biofuels in ships and aircraft, demonstrating that blending algal biofuels with ultra-low-sulfur diesel fuel can reduce pollutants and improve fuel economy [78]. Additionally, the military is exploring hybrid propulsion systems for aircraft and other equipment, aiming to enhance energy efficiency and reduce carbon emissions [79].

These military applications of biofuels highlight the potential for broader adoption in the aviation industry. However, challenges remain, including the high costs of biofuels, limited feedstock availability, and the need for specialized infrastructure to support these fuels. The military's efforts to use biofuels could serve as a catalyst for broader adoption across other sectors, encouraging innovation and investment in sustainable aviation fuels [78,80].

7.3. Regional Differences and Policy Impacts

Regional differences and policy impacts play a significant role in the adoption of SAF. Countries with strict environmental regulations and supportive policies are more likely to foster the growth of biofuels in aviation. For instance, Norway has implemented mandates requiring that a certain percentage of aviation fuels come from renewable sources, driving the adoption of SAF in airports like Oslo Gardermoen [16]. Similarly, Brazil has favorable conditions for SAF production due to its abundant feedstocks and policies encouraging renewable energy sources [81].

In contrast, regions with less supportive policies and limited infrastructure for biofuels face challenges in adopting SAF. The cost of production, transportation logistics, and regulatory compliance can hinder the growth of biofuels in these areas [82]. To promote broader adoption of SAF, policymakers need to address these challenges by providing incentives for biofuel production, encouraging investment in infrastructure, and harmonizing regulations across regions to facilitate the global adoption of sustainable aviation fuels [83].

8. Environmental Impact Assessment

8.1. Environmental Benefits

Sustainable aviation fuels (SAFs) offer considerable environmental benefits compared to conventional jet fuels. One of the primary advantages is the significant reduction in greenhouse gas emissions. SAF, derived from renewable resources, typically has a lower carbon footprint across its lifecycle because the carbon dioxide released during combustion can be offset by the carbon dioxide absorbed during the growth of the biofuel feedstocks [84]. This creates a more circular approach to carbon, contributing to efforts to combat climate change. SAF has the potential to reduce other pollutants such as sulfur oxides (SOx), nitrogen oxides (NOx), and particulate matter, which are common byproducts of fossil-based jet fuels. By using cleaner feedstocks and advanced processing techniques, SAF can contribute to better air quality around airports and reduce environmental impact at high altitudes where contrails and other emissions have significant climate effects [69]. These benefits make SAF an attractive option for airlines aiming to meet stricter environmental regulations and contribute to sustainability goals.

8.2. Reduction in Greenhouse Gas Emissions

Sustainable aviation fuels (SAFs) offer a significant reduction in greenhouse gas (GHG) emissions compared to conventional jet fuels. Lifecycle assessments of various SAFs reveal that they can achieve GHG reductions ranging from 40% to 80%, depending on the feedstock and production process [85]. The key to these reductions lies in the renewable nature of the feedstock, which absorbs CO_2 during growth, offsetting the CO_2 released during combustion in aircraft engines.

One critical aspect of GHG reduction is induced land use change (ILUC), which can either contribute to or detract from the net environmental benefits of SAF. Certain feedstocks, such as those cultivated on marginal or degraded lands, present lower ILUC risks, leading to significant reductions in GHG emissions [41]. Advanced biofuel pathways, like hydroprocessed esters and fatty acids (HEFAs) and Fischer–Tropsch (FT), have shown promise in reducing GHG emissions when appropriate safeguards are in place to minimize deforestation and other high-emission activities [86]. Another pathway for GHG reduction is the utilization of oilseed cover crops, which can serve as feedstocks for biofuels without requiring additional agricultural land. This approach can lead to significant land-use savings, reduce ILUC emissions, and contribute to overall GHG reduction goals [87]. The successful implementation of these pathways could contribute substantially to aviation's efforts to achieve net-zero emissions by 2050.

8.3. Lifecycle Analysis and Net Environmental Benefit

Lifecycle analysis (LCA) plays a crucial role in assessing the environmental impact of SAF. This method evaluates the entire lifecycle of aviation biofuels, from feedstock production to fuel combustion, providing a comprehensive view of the net environmental benefits. Studies focusing on LCA demonstrate that while SAF generally offers lower GHG emissions, there are significant variances depending on the production process and feedstock source [88].

A key component of LCA is understanding the broader environmental implications beyond GHG emissions. This includes evaluating the impacts on soil health, water resources, and biodiversity. For instance, biofuels derived from certain feedstocks may require large amounts of water, potentially impacting local ecosystems and water availability. Biodiversity concerns also emerge when biofuel production leads to deforestation or the conversion of natural landscapes into agricultural land [89,90]. Net environmental benefit assessments must also consider the effects of policy interventions and regulatory frameworks. Certain policies may inadvertently incentivize practices that contribute to ILUC or other environmental harms, while others encourage sustainable practices and renewable feedstock use [91]. Harmonizing these policies across regions and ensuring consistent sustainability criteria is essential to achieving a net environmental benefit from SAF.

8.4. Biodiversity and Land Use Considerations

The impact of biofuel production on biodiversity and land use is a critical consideration for the environmental sustainability of SAF. Biodiversity can be adversely affected if biofuel production leads to deforestation or disrupts existing ecosystems. Land use changes driven by biofuel production may also contribute to soil degradation, loss of habitat, and declines in regional food security [89,90].

To minimize biodiversity loss, SAF production must prioritize feedstocks that do not compete with food crops or lead to large-scale land conversion. Utilizing marginal or degraded lands for biofuel production can mitigate some of these risks, providing a sustainable source of feedstock while preserving existing ecosystems [41]. Additionally, implementing policies that promote land use efficiency and encourage the use of sustainable feedstocks can reduce the pressure on biodiversity [89]. Effective land conservation strategies, combined with sustainable agricultural practices, are essential for ensuring that SAF contributes to environmental sustainability without causing undue harm to biodiversity and land use. The integration of SAF production with existing agricultural systems and the use of cover crops can help maintain biodiversity while providing a reliable source of biofuel feedstock [86,87].

9. Future Directions and Research Needs

To further advance the field of sustainable aviation fuels (SAFs), ongoing technological innovations are crucial. Recent developments in production and processing technologies have demonstrated that emerging pathways, such as catalytic hydrothermolysis and electro-fermentation, offer new opportunities for SAF production. These innovations focus on increasing efficiency, reducing energy consumption, and maximizing yield from various feedstocks, including lignocellulosic biomass and microalgae [92,93].

Catalytic hydrothermolysis (CH) is gaining traction due to its ability to produce high-quality renewable jet fuels. It uses high temperatures and pressures to convert biomass into a mixture of hydrocarbons that can be refined into SAF. This technology shows promise for economic viability and environmental sustainability, with the potential to utilize diverse feedstocks, including agricultural waste and non-food crops [94].

Electro-fermentation, on the other hand, represents a novel approach that combines traditional fermentation with electrical stimulation to enhance the production of biofuels and biochemicals. This technology can increase the yield and efficiency of biofuel production, providing a scalable option for SAF [93]. Continued research in this area can lead to breakthroughs in SAF production, offering cost-effective and sustainable solutions for the aviation industry.

10. Conclusions

Sustainable aviation fuels (SAFs) play a pivotal role in transitioning the aviation industry towards a more environmentally responsible future. This review underscores the multifaceted benefits of SAF, from reducing greenhouse gas emissions to promoting sustainability. Through technological innovations such as catalytic hydrothermolysis and electro-fermentation, SAF production is becoming more efficient and cost-effective. Government policies and incentives are essential for driving the adoption of SAF, fostering public-private partnerships, and harmonizing regulations to ensure consistency across regions. The environmental impact assessment, including comprehensive lifecycle analyses, is crucial for evaluating the net environmental benefits of SAF and ensuring that its production does not negatively affect biodiversity, water resources, or land use.

Technological, economic, and regulatory challenges remain, but ongoing research and supportive policies can help overcome these barriers. Policies that encourage investment in SAF production and promote a robust supply chain are critical to the long-term sustainability of SAF. The aviation industry's future will depend on the collaboration among stakeholders, including airlines, fuel producers, governments, and international organizations, to create a more sustainable aviation sector. By focusing on these key areas, SAF can significantly contribute to the aviation industry's efforts to reduce its carbon footprint and achieve a greener, more environmentally conscious future.

Author Contributions: Conceptualization, R.K.; methodology, R.K.; software, R.K.; validation, R.K. and H.M.C.; formal analysis, R.K.; investigation, R.K.; resources, H.M.C.; data curation, R.K.; writing—original draft preparation, R.K.; writing—review and editing, R.K. and H.M.C.; visualization, R.K.; supervision, H.M.C.; project administration, H.M.C.; funding acquisition, H.M.C. All authors have read and agreed to the published version of the manuscript.

Funding: This research received no external funding.

Acknowledgments: This work was supported by the National Research Foundation of Korea (NRF) grant funded by the Korean government (NRF-2022H1A7A2A02000033).

Conflicts of Interest: The authors declare no conflicts of interest.

References

1. Hasan, A.; Al Mamun, A.; Rahman, S.M.; Malik, K.; Al Amran, I.U.; Khondaker, A.N.; Reshi, O.; Tiwari, S.P.; Alismail, F.S. Climate Change Mitigation Pathways for the Aviation Sector. *Sustainability* **2021**, *13*, 3656. [CrossRef]
2. Sher, F.; Raore, D.; Klemes, J.J.; Rafi-Ul-Shan, P.M.; Khzouz, M.; Marintseva, K.; Razmkhah, O. Unprecedented Impacts of Aviation Emissions on Global Environmental and Climate Change Scenario. *Curr. Pollut. Rep.* **2021**, *7*, 549–564. [CrossRef] [PubMed]
3. Yin, F.; Grewe, V.; Castino, F.; Rao, P.; Matthes, S.; Dahlmann, K.; Dietmüller, S.; Frömming, C.; Yamashita, H.; Peter, P.; et al. Predicting the climate impact of aviation for en-route emissions: The al-gorithmic climate change function submodel ACCF 1.0 of EMAC 2.53. *Geosci. Model Dev.* **2023**, *16*, 3313–3334. [CrossRef]
4. Undavalli, V.; Olatunde, O.B.G.; Boylu, R.; Wei, C.; Haeker, J.; Hamilton, J.; Khandelwal, B. Recent Advancements in Sustainable Aviation Fuels. *Prog. Aerosp. Sci.* **2023**, *136*, 100876. [CrossRef]
5. Becken, S.; Mackey, B.; Lee, D.S. Implications of preferential access to land and clean energy for Sustainable Aviation Fuels. *Sci. Total. Environ.* **2023**, *886*, 163883. [CrossRef] [PubMed]

6. Grewe, V.; Matthes, S.; Dahlmann, K. The contribution of aviation NO_x emissions to climate change: Are we ignoring methodological flaws? *Environ. Res. Lett.* **2019**, *14*, 121003. [CrossRef]
7. Frömming, C.; Grewe, V.; Brinkop, S.; Jöckel, P.; Haslerud, A.S.; Rosanka, S.; van Manen, J.; Matthes, S. Influence of weather situation on non-CO_2 aviation climate effects: The REACT4C climate change functions. *Atmos. Chem. Phys.* **2021**, *21*, 9151–9172. [CrossRef]
8. Dinkov, R.; Stratiev, D.; Andreev, I.; Georgiev, G.; Angelova, M.; Dimitrova, R.Z.; Toteva, V. Relation of the Content of Sustainable Components (HEFAs) in Blends with Hydrotreated Straight-Run Kerosene to the Properties of Aviation Fuel. *Processes* **2024**, *12*, 1045. [CrossRef]
9. Braun, M.; Grimme, W.; Oesingmann, K. Pathway to net zero: Reviewing sustainable aviation fuels, environmental impacts and pricing. *J. Air Transp. Manag.* **2024**, *117*, 102580. [CrossRef]
10. Ringbeck, J.; Koch, V. Aviation biofuels: A roadmap towards more carbon-neutral skies. *Biofuels* **2010**, *1*, 519–521. [CrossRef]
11. Burov, E.A.; Ivanova, L.V.; Koshelev, V.N.; Adam, T.A. Additives to Biodiesel Fuels. *Chem. Technol. Fuels Oils* **2023**, *59*, 998–1002. [CrossRef]
12. Panoutsou, C.; Germer, S.; Karka, P.; Papadokostantakis, S.; Kroyan, Y.; Wojcieszyk, M.; Maniatis, K.; Marchand, P.; Landalv, I. Advanced biofuels to decarbonise European transport by 2030: Markets, challenges, and policies that impact their successful market uptake. *Energy Strat. Rev.* **2021**, *34*, 100633. [CrossRef]
13. Carter, R.N.; Smith, L.L.; Karim, H.; Castaldi, M.; Etemad, S.; Muench, G.; Boorse, R.S.; Menacherry, P.; Pfefferle, W.C. Catalytic combustion technology development for gas turbine engine applications. *MRS Proc.* **1998**, *549*, 93. [CrossRef]
14. Atmowidjojo, A.H.; Rianawati, E.; Chin, B.L.; Yusup, S.; Quitain, A.T.; Assabumrungrat, S.; Yiin, C.L.; Kiatkittipong, W.; Srifa, A.; Eiad-ua, A. Supporting Clean Energy in the ASEAN: Policy Opportunities from Sustainable Aviation Fuels Initi-atives in Indonesia and Malaysia. *IOP Conf. Ser. Earth Environ. Sci.* **2021**, *940*, 012031. [CrossRef]
15. Shahriar, F.; Khanal, A. The current techno-economic, environmental, policy status and perspectives of sustainable aviation fuel (SAF). *Fuel* **2022**, *325*, 124905. [CrossRef]
16. Baxter, G. An Assessment of Sustainable energy management at a Major Scandinavian Hub Airport: The Case of Oslo Airport Gardermoen. *Int. J. Environ. Agric. Biotechnol.* **2021**, *6*, 328–351. [CrossRef]
17. Su-Ungkavatin, P.; Tiruta-Barna, L.; Hamelin, L. Framework for Life Cycle Assessment of Sustainable Aviation (Sa) Systems. *Sci. Total. Environ.* **2023**, *885*, 163881. [CrossRef] [PubMed]
18. Kumar, T.; Mohsin, R.; Majid, Z.A.; Ghafir, M.F.A.; Wash, A.M. Experimental study of the anti-knock efficiency of high-octane fuels in spark ignited aircraft engine using response surface methodology. *Appl. Energy* **2020**, *259*, 114150. [CrossRef]
19. Shao, L.; Zhou, Y.; Geng, T.; Zhao, S.; Zhu, K.; Zhong, Z.; Yan, H.; Yu, T.; Xu, Z.; Ding, S. Advanced combustion in heavy fuel aircraft piston engines: A comprehensive review and future directions. *Fuel* **2024**, *370*, 131771. [CrossRef]
20. Grim, R.G.; Ravikumar, D.; Tan, E.C.D.; Huang, Z.; Ferrell, J.R.; Resch, M.; Li, Z.; Mevawala, C.; Phillips, S.D.; Snowden-Swan, L.; et al. Electrifying the production of sustainable aviation fuel: The risks, economics, and environmental benefits of emerging pathways including CO_2. *Energy Environ. Sci.* **2022**, *15*, 4798–4812. [CrossRef]
21. Agarwal, A.K. Biofuels (alcohols and biodiesel) applications as fuels for internal combustion engines. *Prog. Energy Combust. Sci.* **2007**, *33*, 233–271. [CrossRef]
22. Richenhagen, J.; Kremer, F.; Küpper, C.; Spilker, T.; Bhardwaj, O.P.; Nijs, M. Biofuels for Combustion Engines: Aspects of Modern Engine Development in the Context of Future Biofuels. In Proceedings of the Fuels from Biomass: An Interdisciplinary Approach, Aachen, Germany, 21–22 November 2011; Springer: Berlin/Heidelberg, Germany, 2015. [CrossRef]
23. Williams, J.L., Jr. Investigation of the Performance and Emissions Characteristics of Dual Fuel Combustion in a Single Cylinder IDI Diesel Engine. 2018. Available online: https://digitalcommons.georgiasouthern.edu/cgi/viewcontent.cgi?article=2991&context=etd (accessed on 4 February 2024).
24. Khosravi, M. Developing Pyrometric and Chemiluminescence Optical Diagnostics for Investigation of Modern Alternative CI Engine Combustion Strategies. Doctoral Dissertation, University of British Columbia, Vancouver, BC, Canada, 2019. Available online: http://hdl.handle.net/2429/70880 (accessed on 4 February 2024).
25. Torenbeek, E.; Wittenberg, H. Aircraft Engines and Propulsion. In *Flight Physics: Essentials of Aeronautical Disciplines and Technology, with Historical Notes*; Springer: Berlin/Heidelberg, Germany, 2009; pp. 181–252. [CrossRef]
26. Almena, A.; Siu, R.; Chong, K.; Thornley, P.; Röder, M. Reducing the environmental impact of international aviation through sustainable aviation fuel with integrated carbon capture and storage. *Energy Convers. Manag.* **2024**, *303*, 118186. [CrossRef]
27. Masiol, M.; Harrison, R.M. Aircraft engine exhaust emissions and other airport-related contributions to ambient air pollution: A review. *Atmos. Environ.* **2014**, *95*, 409–455. [CrossRef] [PubMed]
28. Zhang, L.; Butler, T.L.; Yang, B. Recent trends, opportunities and challenges of sustainable aviation fuel. In *Green Energy to Sustainability: Strategies for Global Industries*; Wiley: Hoboken, NJ, USA, 2020; pp. 85–110. [CrossRef]
29. Ali, A.H.H.; Ibrahim, M.N. Performance and environmental impact of a turbojet engine fueled by blends of biodiesels. *Int. J. Environ. Sci. Technol.* **2017**, *14*, 1253–1266. [CrossRef]
30. Gillespie, C.W. A General Equilibrium Analysis of Climate Policy for Aviation. Doctoral Dissertation, Massachusetts Institute of Technology, Cambridge, MA, USA, 2011. Available online: https://core.ac.uk/download/pdf/4430211.pdf (accessed on 20 March 2024).

31. Starck, L.; Pidol, L.; Jeuland, N.; Chapus, T.; Bogers, P.; Bauldreay, J. Production of Hydroprocessed Esters and Fatty Acids (HEFA)—Optimisation of Process Yield. *Oil Gas Sci. Technol. Rev. d'IFP Energies Nouv.* **2014**, *71*, 10. [CrossRef]
32. Hendrawati, T.; Siswahyu, A.; Ramadhan, A.I. Analysis of Technoeconomy of Bioavtur Production with Hydroprocessed Esters And Fatty Acids (Hefa) Process In Indonesia. *Int. J. Sci. Technol. Res.* **2018**, *7*, 39–43.
33. Pipitone, G.; Zoppi, G.; Pirone, R.; Bensaid, S. Sustainable aviation fuel production using in-situ hydrogen supply via aqueous phase reforming: A techno-economic and life-cycle greenhouse gas emissions assessment. *J. Clean. Prod.* **2023**, *418*, 138141. [CrossRef]
34. Mattie, D.; Hurley, J.M.; Riccio, E.; Sterner, T.R. Acute Dermal Irritation Study and Salmonella-Escherichia coli/Microsome Plate Incorporation Assay of Hydroprocessed Esters and Fatty Acids (HEFA) Bio-Based Jet Fuels. 2013. Available online: https://apps.dtic.mil/sti/tr/pdf/ADA584649.pdf (accessed on 29 March 2024).
35. Zanata, M.; Amelia, S.T.W.; Mumtazy, M.R.; Kurniawansyah, F.; Roesyadi, A. Synthesis of Bio Jet Fuel from Crude Palm Oil by HEFA (Hydroprocessed Esters and Fatty Acids) Using Ni-Mo Catalyst Supported by Rice Husk Ash-Based SiO_2. *Mater. Sci. Forum* **2019**, *964*, 193–198. [CrossRef]
36. Parker, L.H. Investigation of the Low-Temperature Combustion Phenomena of a Fischer-Tropsch Synthetic Aerospace Fuel in a Constant Volume Combustion Chamber for Greenhouse Gas Reduction. Available at Semantic Scholar. 2022. Available online: https://digitalcommons.georgiasouthern.edu/honors-theses/687?utm_source=digitalcommons.georgiasouthern.edu/honors-theses/687&utm_medium=PDF&utm_campaign=PDFCoverPages (accessed on 29 March 2024).
37. Barbosa, J.A.; Coronado, C.J.; de Andrade, J.C.; Tuna, C.E.; Silva, M.H.; Junior, J.A.C.; Mendiburu, A.Z. Experimental Determination of Upper Flammability Limits of Synthesized Iso-Paraffins (SIP), Jet Fuel and their Mixtures with Air at Atmospheric and Sub-Atmospheric Pressures. Process Safety and Environmental Protection. *Process. Saf. Environ. Prot.* **2022**, *160*, 102–115. [CrossRef]
38. Barbosa, J.A.; Coronado, C.J.; Tuna, C.E.; Silva, M.H.; Mendiburu, A.Z.; Junior, J.A.C.; de Andrade, J.C. Experimental determination of lower flammability limits of Synthesized Iso-Paraffins (SIP), jet fuel and mixtures at atmospheric and reduced pressures with air. *Fire Saf. J.* **2021**, *121*, 103276. [CrossRef]
39. Otun, K.O.; Yao, Y.; Liu, X.; Hildebrandt, D. Synthesis, structure, and performance of carbide phases in Fischer-Tropsch synthesis: A critical review. *Fuel* **2021**, *296*, 120689. [CrossRef]
40. Liu, Z.; Tang, Z.; Yang, X. Integrating properties and conditions to predict spray performance of alternative aviation fuel by ANN model. *Biotechnol. Biofuels Bioprod.* **2023**, *16*, 171. [CrossRef] [PubMed]
41. Panoutsou, C.; Giarola, S.; Ibrahim, D.; Verzandvoort, S.; Elbersen, B.; Sandford, C.; Malins, C.; Politi, M.; Vourliotakis, G.; Zita, V.E.; et al. Opportunities for Low Indirect Land Use Biomass for Biofuels in Europe. *Appl. Sci.* **2022**, *12*, 4623. [CrossRef]
42. Capaz, R.S.; Guida, E.; Seabra, J.E.A.; Osseweijer, P.; Posada, J.A. Mitigating carbon emissions through sustainable aviation fuels: Costs and potential. *Biofuels Bioprod. Biorefining* **2020**, *15*, 502–524. [CrossRef]
43. Kranenburg, K.V.; Delft, Y.V.; Gavrilova, A.; Kler, R.D.; Schipper, C.; Smokers, R.; Verbeek, M.; Verbeek, R. E-Fuels: Towards a More Sustainable Future for Truck Transport, Shipping and Aviation. 2020. Available online: http://resolver.tudelft.nl/uuid:487a6a47-853d-4a1d-bc2e-dbe21d584cca (accessed on 10 April 2024).
44. Reksowadojo, I.K.; Duong, L.; Pham, D. Review of typical biofuel for aviation alternative fuel purposes. *ASEAN Eng. J.* **2018**, *4*, 20–30. [CrossRef]
45. Karatzos, S.; van Dyk, J.S.; McMillan, J.D.; Saddler, J. Drop-in biofuel production via conventional (lipid/fatty acid) and advanced (biomass) routes. Part I. *Biofuels Bioprod. Biorefining* **2017**, *11*, 344–362. [CrossRef]
46. de Klerk, A.; Chauhan, G.; Halmenschlager, C.; Link, F.; Sánchez, N.M.; Gartley, B.; El-Sayed, H.E.M.; Sehdev, R.; Lehoux, R. Sustainable aviation fuel: Pathways to fully formulated synthetic jet fuel via Fischer–Tropsch synthesis. *Energy Sci. Eng.* **2024**, *12*, 394–409. [CrossRef]
47. Kim, K.; Ryu, J.I.; McGann, B.; Min, K.; Temme, J.; Kweon, C.-B.; Lee, T. Correction: Data-Driven Combustion Kinetic Modeling Concept of Alternative Alcohol-to-Jet (ATJ) Fuel. In Proceedings of the AIAA Scitech 2021 Forum, virtual, 11–15 & 19–21 January 2021. [CrossRef]
48. Prak, D.J.L.; Jones, M.H.; Trulove, P.; McDaniel, A.M.; Dickerson, T.; Cowart, J.S. Physical and Chemical Analysis of Alcohol-to-Jet (ATJ) Fuel and Development of Surrogate Fuel Mixtures. *Energy Fuels* **2015**, *29*, 3760–3769. [CrossRef]
49. Ryu, J.I.; Motily, A.; Lee, T.; Scarcelli, R.; Som, S.; Kim, K.; Kweon, C.-B. Effect of Hot Probe Temperature on Ignition of Alcohol-to-Jet (ATJ) Fuel Spray under Aircraft Propulsion System Conditions. In Proceedings of the AIAA Scitech 2021 Forum, virtual, 11–15 & 19–21 January 2021. [CrossRef]
50. Dickerson, T.; McDaniel, A.; Williams, S.; Luning-Prak, D.; Hamilton, L.; Bermudez, E.; Cowart, J. Start-up and Steady-State Performance of a New Renewable Alcohol-To-Jet (ATJ) Fuel in Multiple Diesel Engines. In Proceedings of the SAE 2015 World Congress & Exhibition, Detroit, MI, USA, 21–23 April 2015. [CrossRef]
51. Liao, J.; Zhong, Q.; Gu, J.; Qiu, S.; Meng, Q.; Zhang, Q.; Wang, T. New approach for bio-jet fuels production by hydrodeoxygenation of higher alcohols derived from C-C coupling of bio-ethanol. *Appl. Energy* **2022**, *324*, 119843. [CrossRef]
52. Hutzler, S.A. Alcohol-to-Jet (ATJ) Fuel Blending Study. 2015. Available online: https://apps.dtic.mil/sti/tr/pdf/AD1001842.pdf (accessed on 12 April 2024).
53. Kurzawska, P.; Jasiński, R. Overview of Sustainable Aviation Fuels with Emission Characteristic and Particles Emission of the Turbine Engine Fueled ATJ Blends with Different Percentages of ATJ Fuel. *Energies* **2021**, *14*, 1858. [CrossRef]

54. Özdemir, I. Sustainable Aviation Fuel Investment and Feedstock Calculations by Microalgae Applications with Biorefinery Approach. In Proceedings of the Abu Dhabi International Petroleum Exhibition and Conference, Abu Dhabi, UAE, 4–7 October 2023. [CrossRef]
55. Shi, Z.; Zhao, B.; Tang, S.; Yang, X. Hydrotreating lipids for aviation biofuels derived from extraction of wet and dry algae. *J. Clean. Prod.* **2018**, *204*, 906–915. [CrossRef]
56. Bwapwa, J.K.; Akash, A.; Trois, C. Jet Fuel from Domestic Wastewater Treatment Using Microalgae: A Review. In *Green Materials for Wastewater Treatment. Environmental Chemistry for a Sustainable World*; Naushad, M., Lichtfouse, E., Eds.; Springer: Cham, Switzerland, 2020; Volume 38. [CrossRef]
57. Karnis, M.; Fang, L.; Bristow, M. Evaluating Actions Towards Sustainable Aviation: A Case Study Using Action Interdependence in Multiple Criteria Decision Making. In Proceedings of the 2015 IEEE International Conference on Systems, Man, and Cybernetics (SMC), Hong Kong, China, 9–12 October 2015; pp. 592–597. [CrossRef]
58. Gutiérrez-Antonio, C.; la Cruz, A.G.-D.; Romero-Izquierdo, A.G.; Gómez-Castro, F.I.; Hernández, S. Modeling, simulation and intensification of hydroprocessing of micro-algae oil to produce renewable aviation fuel. *Clean Technol. Environ. Policy* **2018**, *20*, 1589–1598. [CrossRef]
59. Mehariya, S.; Goswami, R.K.; Verma, P.; Lavecchia, R.; Zuorro, A. Integrated Approach for Wastewater Treatment and Biofuel Production in Microalgae Biorefineries. *Energies* **2021**, *14*, 2282. [CrossRef]
60. Darda, S.; Papalas, T.; Zabaniotou, A. Biofuels journey in Europe: Currently the way to low carbon economy sustainability is still a challenge. *J. Clean. Prod.* **2019**, *208*, 575–588. [CrossRef]
61. Chiaramonti, D.; Buffi, M.; Palmisano, P.; Redaelli, S. Lignin-based advanced biofuels: A novel route towards aviation fuels. *Chem. Eng. Trans.* **2016**, *50*, 109–114. [CrossRef]
62. Muelaner, J. *Decarbonized Fuel Options for Civil Aviation*; SAE Technical Paper EPR2023012; SAE International: Warrendale, PA, USA, 2023. [CrossRef]
63. Su-Ungkavatin, P.; Tiruta-Barna, L.; Hamelin, L. Biofuels, electrofuels, electric or hydrogen?: A review of current and emerging sustainable aviation systems. *Prog. Energy Combust. Sci.* **2023**, *96*, 101073. [CrossRef]
64. Baidoo, M.F.; Adjei, E.A.; Opoku, R.; Aidam, G.S.K. Rubber seed oil: Potential feedstock for aviation biofuel production. *Sci. Afr.* **2022**, *17*, e01393. [CrossRef]
65. Chong, J.W.; Chemmangattuvalappil, N.G.; Thangalazhy-Gopakumar, S. Aviation Biofuels: Conversion Routes and Challenges. In *Sustainable Technologies for the Oil Palm Industry*; Foo, D.C.Y., Tun Abdul Aziz, M.K., Yusup, S., Eds.; Springer: Singapore, 2023. [CrossRef]
66. Fischer, G.; Tramberend, S.; Velthuizen, H.; Bole-Rentel, T.; Reeler, J.; Sustainable Aviation Biofuel Feedstock Potential in Sub-Saharan Africa. A Systems Analysis Investigation into the Current and Future Potential for Biofuel Feedstock Production. 2019. Available online: https://pure.iiasa.ac.at/15626 (accessed on 10 February 2024).
67. Viguié, J.; Ullrich, N.; Porot, P.; Bournay, L.; Hecquet, M.; Rousseau, J. BioTfueL Project: Targeting the Development of Second-Generation Biodiesel and Biojet Fuels. *Oil Gas Sci. Technol. Rev. D'ifp Energ. Nouv.* **2013**, *68*, 935–946.
68. Sadykov, V.A.; Simonov, M.N.; Bespalko, Y.N.; Bobrova, L.N.; Eremeev, N.F.; Arapova, M.V.; Smal', E.A.; Mezentseva, N.V.; Pavlova, S.N. Design and Characterization of Nanocomposite Catalysts for Biofuel Conversion into Syngas and Hydrogen in Structured Reactors and Membranes. *Kinet. Catal.* **2019**, *60*, 582–605. [CrossRef]
69. Attia, N.K.; Abdel Kader, E.A.; ElDiwani, G.; Hussein, H.S.; El-Araby, R. Evaluation of Blending of Lowest Emission Biodiesel with Jet A for Producing Aviation Biofuels. In *Utilization and Management of Bioresources*; Ghosh, S., Ed.; Springer: Singapore, 2018. [CrossRef]
70. Duong, L.H.; Reksowardojo, I.K.; Soerawidjaja, T.H.; Pham, D.N.; Fujita, O. The sooting tendency of aviation biofuels and jet range paraffins: Effects of adding aromatics, carbon chain length of normal paraffins, and fraction of branched paraffins. *Combust. Sci. Technol.* **2018**, *190*, 1710–1721. [CrossRef]
71. Stadnichenko, V. Comparative characteristics of biocomponents based on rapeseed and camellina oils for blending with JET fuels. In Proceedings of the 21th Conference for Lithuanian Junior Researchers Science-Future of Lithuania. Transport Engineering and Management, Vilnius, Lithuania, 4–5 May 2018.
72. Donnelly, J.; Horton, R.; Gopalan, K.; Bannister, C.D.; Chuck, C.J. Branched ketone biofuels as blending agents for Jet-A1 aviation kerosene. *Energy Fuels* **2016**, *30*, 294–301. [CrossRef]
73. Noh, H.M.; Taher, M.N.M.; A Rodrigo, G.; Rahman, N.A.A.; Ismail, S.; Rani, M.M.; Salleh, I.M.; Dahdi, Y.; Wan, W.N.S.; Razak, A.; et al. Refuelling the future: Progress towards testing drop-in biofuels in replacing conventional fuel for commercial flights. *IOP Conf. Series Mater. Sci. Eng.* **2018**, *370*, 012037. [CrossRef]
74. Reals, K. Fuelling the future: Powering commercial flights with biofuels creates good PR, but to genuinely make a difference, airlines must put their money where their mouths are to ensure there is the necessary investment in production ventures. *Airl. Bus.* **2013**, *29*. Available online: https://www.library.northwestern.edu/find-borrow-request/requests-interlibrary-loan/lending-institutions.html (accessed on 17 January 2024).
75. Mohsin, R.; Kumar, T.; Majid, Z.A.; Nasri, N.S.; Sharer, Z.; Kumar, I.; Wash, A.M. Assessment of biofuels in aviation industry for environmental sustainability. *Chem. Eng. Trans.* **2017**, *56*, 1189–1194. [CrossRef]
76. Mohd Noh, H. Investigation on the Applicability of Biofuels in the Operation of Commercial Aircraft. Doctoral Dissertation, Universidad Politécnica de Madrid Escuela Técnica Superior de Ingenieros Aeronáuticos, Madrid, Spain, 2016. [CrossRef]

77. Gegg, P.K. Factors Affecting the Emergence, Development and Uptake of Aviation Biofuels. Doctoral Dissertation, Loughborough University, Loughborough, UK, 2014. Available online: https://hdl.handle.net/2134/14074 (accessed on 8 April 2024).
78. Savage, P.E. Algae under pressure and in hot water. *Science* **2012**, *338*, 1039–1040. [CrossRef] [PubMed]
79. Santarelli, M.; Cabrera, M. Hybrid solid oxide fuel cell and micro gas turbine for regional jets. *J. Aircr.* **2011**, *48*, 1216–1224. [CrossRef]
80. Miličević, Z.M.; Bojković, Z.S. Military green technology: Present and future. *Vojnoteh. Glas. Mil. Tech. Cour.* **2023**, *71*, 136–152. [CrossRef]
81. Walter, A.; Seabra, J.; Rocha, J.; Guarenghi, M.; Vieira, N.; Damame, D.; Santos, J.L. Spatially Explicit Assessment of Suitable Conditions for the Sustainable Production of Aviation Fuels in Brazil. *Land* **2021**, *10*, 705. [CrossRef]
82. Wang, Z.; Osseweijer, P.; Duque, J.P. Assessing social sustainability for biofuel supply chains: The case of aviation biofuel in Brazil. In Proceedings of the 2017 IEEE Conference on Technologies for Sustainability (SusTech), Phoenix, AZ, USA, 12–14 November 2017; pp. 1–5. [CrossRef]
83. Loran, O. The Development of Sustainability Requirements for Aviation Biofuels. Master's Thesis, Lund University, Lund, Sweden, 2013.
84. Yoo, E.; Lee, U.; Wang, M. Life-cycle greenhouse gas emissions of sustainable aviation fuel through a net-zero carbon biofuel plant design. *ACS Sustain. Chem. Eng.* **2022**, *10*, 8725–8732. [CrossRef]
85. Mawhood, R.; Gazis, E.; Hoefnagels, R.; Jong, S.D.; Slade, R. Technological and commercial maturity of aviation biofuels: Emerging options to produce jet from lignocellulosic biomass. In Proceedings of the 14th International Conference on Sustainable Engineering Technologies, Rhodes, Greece, 3–5 September 2015.
86. Fiorini, A.C.O.; Angelkorte, G.; Maia, P.L.; Bergman-Fonte, C.; Vicente, C.; Morais, T.; Carvalho, L.; Zanon-Zotin, M.; Szklo, A.; Schaeffer, R.; et al. Sustainable aviation fuels must control induced land use change: An integrated assessment modelling exercise for Brazil. *Environ. Res. Lett.* **2023**, *18*, 014036. [CrossRef]
87. Taheripour, F.; Sajedinia, E.; Karami, O. Oilseed cover crops for sustainable aviation fuels production and reduction in greenhouse gas emissions through land use savings. *Front. Energy Res.* **2022**, *9*, 790421. [CrossRef]
88. de Jong, S.; Antonissen, K.; Hoefnagels, R.; Lonza, L.; Wang, M.; Faaij, A.; Junginger, M. Life-cycle analysis of greenhouse gas emissions from renewable jet fuel production. *Biotechnol. Biofuels* **2017**, *10*, 64. [CrossRef] [PubMed]
89. Palmer, J.R. Biofuels and the politics of land-use change: Tracing the interactions of discourse and place in European policy mak-ing. *Environ. Plan. A Econ. Space* **2014**, *46*, 337–352. [CrossRef]
90. Nunez-Regueiro, M.M.; Siddiqui, S.F.; Fletcher, R.J. Effects of bioenergy on biodiversity arising from land-use change and crop type. *Conserv. Biol.* **2021**, *35*, 77–87. [CrossRef] [PubMed]
91. Ortega, A.; Gkoumas, K.; Tsakalidis, A.; Pekár, F. Low-Emission Alternative Energy for Transport in the EU: State of Play of Research and Innovation. *Energies* **2021**, *14*, 7764. [CrossRef]
92. Eswaran, S.; Subramaniam, S.; Geleynse, S.; Brandt, K.; Wolcott, M.; Zhang, X. Techno-economic analysis of catalytic hydrothermolysis pathway for jet fuel production. *Renew. Sustain. Energy Rev.* **2021**, *151*, 111516. [CrossRef]
93. Chandrasekhar, K.; Kumar, A.N.; Kumar, G.; Kim, D.-H.; Song, Y.-C.; Kim, S.-H. Electro-fermentation for biofuels and biochemicals production: Current status and future directions. *Bioresour. Technol.* **2021**, *323*, 124598. [CrossRef]
94. Stevens, J. A Stochastic Techno-Economic Analysis of Aviation Biofuels Production from Pennycress Seed Oil. Doctoral Dissertation, Purdue University, West Lafayette, IN, USA, 2019.

energies

MDPI

Article

Climate Change Mitigation in Thailand's Domestic Aviation: Mitigation Options Analysis towards 2050

Arthit Champeecharoensuk [1,*], Shobhakar Dhakal [1,*] and Nuwong Chollacoop [2]

[1] Department of Energy, Environment, and Climate Change, School of Environment, Resource and Development, Asian Institute of Technology, Pathum Thani 12120, Thailand
[2] National Energy Technology Center (ENTEC), National Science and Technology Development Agency (NSTDA), Pathum Thani 12120, Thailand
* Correspondence: st121433@ait.asia (A.C.); shobhakar@ait.asia (S.D.)

Abstract: Thailand's civil aviation industry has expanded rapidly in the past ten years resulting in increasing aviation greenhouse gas (GHG) emissions and energy consumption. The rapid growth in air transport is anticipated to continue further. Presently, domestic aviation and the economy of many countries are recovering rapidly in the post-COVID-19 period, resulting in fuel consumption and GHG emissions gradually increasing again. However, despite implementing the ICAO's CORSIA (International Civil Aviation Organization's Carbon Offsetting and Reduction Scheme for International Aviation) rule for international aviation, GHG emissions in the domestic aviation sector are largely unregulated. Moreover, the literature lacks a GHG emissions analysis that considers this sector's potential growth and mitigation policies for future GHG emissions. To close the gap, this study conducted a GHG emissions analysis from this sector under various scenarios through 2050 using historical data during 2008–2020 to forecast future trends. It evaluates the impact of the mitigation policies, such as fuel switching and aircraft technology, on improving fuel efficiency due to technological advancements in aircraft and carbon pricing. The results show that the fuel switching option would result in a significant long-term reduction in GHG emissions, whereas the carbon pricing option and aircraft technology option are desirable in reducing GHG emissions in the short term. Therefore, to meet GHG emissions reduction targets more successfully, all measures must be simultaneously executed to address short- and long-term mitigation strategies. These findings have significant implications for both present and future GHG emissions reduction measures, supporting Thailand's 2050 climate targets and energy efficiency policies as the domestic aviation industry adjusts.

Keywords: greenhouse gas emissions; aviation sector; mitigation; policy; energy consumption; scenario analysis

1. Introduction

In 2004, the global air transport sector utilized 0.19 billion metric tonnes (Mt) of fuel, producing approximately 0.59 billion tonnes of CO_2 emissions [1]. Air travel flights accounted for 2–3% of global anthropogenic CO_2 emissions in 2012 [2]. Projections indicate that global air traffic will double within 15 years from 2012, with a corresponding doubling of energy consumption and CO_2 emissions within 25 years [3]. The aviation industry is one of the largest markets worldwide, transporting approximately 2.2 billion passengers annually and employing 32 million people globally [4]. This sector has a significant economic impact, contributing approximately 3.56 trillion USD or around 7.5% of the global gross domestic product (GDP) [5].

While it has been established that transportation emissions from rail, road, and water conveyance significantly impact the environment and contribute to climate change [6–9], aviation exhaust is the second-largest contributor to emissions. This constitutes approximately 12% of the annual emissions from all modes of transportation and contributes to around 4% of the current observed global warming caused by human activities [10,11]. In addition to its role in climate change stemming from aircraft operations, the aviation industry bears responsibility for various other environmental issues associated with the supporting systems of aircraft, such as airports and fuel production, as well as all stages of its value chain, including aircraft manufacturing and disposal. Examples of these impacts encompass noise pollution, the limited use of rare metals (like cobalt or chromium in specialized alloys), and potential harmful effects from the release of chemicals, all of which have the potential to harm human health, ecosystems, and deplete natural resources [12,13]. The aviation industry has experienced steady growth in recent years until it was abruptly halted by the onset of the COVID-19 pandemic [14]. In 2018, global CO_2 emissions from aircraft operations, encompassing passenger and cargo transport, amounted to 0.92 billion metric tonnes [15]. This accounted for 2.4% of the estimated 37.9 gigatonnes of CO_2 emitted globally that year due to fossil fuel usage [16]. Commercial aviation's CO_2 emissions increased by 32%, from 0.69 billion tonnes in 2013 to 0.92 billion tonnes in 2018 [17]. Consequently, CO_2 emissions from international aviation are projected to triple by 2050, surpassing the rate forecasted by ICAO, corresponding to a compound annual growth rate of 5.7% [18]. Furthermore, there was a resurgence in worldwide international travel starting around June/July 2020, with notable improvements particularly observed by November 2020. Substantial expansion has persisted since June 2021, and this growth has been consistently maintained in numerous aviation markets. Domestic travel has displayed greater resilience in passenger volume when compared to the international tourism sector [19].

In Thailand, commercial aviation is one of the transportation sectors that contributes significantly to CO_2 emissions. The emissions from international commercial aviation experienced a 29% growth, rising from 10.2 million tonnes of CO_2 in 2014 to 13.2 million tonnes of CO_2 in 2018 (an overall increase of 29%). Another notable source of CO_2 emissions in Thailand is commercial domestic aviation. Emissions from passenger and freight flights grew by 42%, increasing from 1.9 million tonnes of CO_2 in 2014 to 2.7 million tonnes of CO_2 in 2018 [20]. The rapid increase in fuel consumption for commercial, domestic aviation generates substantial CO_2 emissions. The issue of reducing CO_2 emissions in Thailand's domestic aviation industry is gaining more attention, particularly following Thailand's government commitment to reducing GHG emissions during the UNFCC's COP-20 [21].

The Thai government aims for its Nationally Determined Contribution (NDC) in line with global climate efforts. As a result, Thailand aims to reduce its GHG emissions by 20–25% (555 Mt-CO_2-eq) in 2030, compared to the business-as-usual (BAU) scenario [22]. Thailand aims to decrease GHG emissions by approximately 111 to 139 Mt-CO_2-eq. Climate policies are necessary across all industries and countries to achieve the objectives outlined in the 2015 Paris Agreement and limit global warming to 2 °C. Furthermore, a comprehensive study of greenhouse gas emissions in the aviation sector is required [23].

In 2016, following a comprehensive research-driven examination of the aviation industry, the ICAO Assembly reached a consensus to implement a global, market-driven initiative aimed at mitigating greenhouse gas emissions, arising from international aviation. This initiative is known as the Carbon Offsetting and Reduction Scheme for International Aviation (CORSIA) [24]. Furthermore, CORSIA mandates that airlines must offset CO_2-eq emissions exceeding the levels recorded in 2019. The CORSIA framework is designed to facilitate offsetting through the use of credits or CORSIA Eligible Fuels (CEFs), with these choices determined by impact assessments and current scientific understanding. The overarching goal of this approach is to ensure that international aviation achieves carbon-neutral growth, commencing in the year 2020 [25].

The current focus on reducing GHG emissions in the aviation sector, particularly seen in ICAO's CORSIA initiative, does not extend to domestic aviation [26,27]. However, the domestic aviation sector plays a role in national GHG mitigation efforts and contributes to the Nationally Determined Contributions (NDC) under the UNFCCC. Unfortunately, policymakers do not actively monitor the emissions from commercial domestic aviation. Thus, this study emphasizes the importance of forecasting energy use and GHG emissions in the commercial, domestic aviation industry. This study highlights the need for appropriate policies to effectively reduce future GHG emissions from the aviation sector, aligning with environmental and climate-related solutions.

Our literature review reveals that numerous studies have examined the issue of CO_2 emissions from the aviation industry, both on a global and national scale. Some of these studies have focused on medium- to long-term scenario analysis [28–30]. Additionally, past research was concentrated on policies aimed at reducing GHG emissions in the medium to long term [29,31–35] and the different factors that influence the growth of GHG emissions [36–41].

Pathways for the mitigation of aviation sector have adopted a comprehensive mitigation strategy that relies on the successful implementation of individual policies or measures and other strategies. The International Civil Aviation Organization (ICAO) proposes five significant measures for reducing emissions in commercial aviation: Operational efficiency improvement; Use of alternative fuels; Demand shift engineering; Technological efficiency improvement; and Carbon pricing (market-based incentives) [33]. These efforts include advancing aircraft-related technologies and standards to encourage their adoption, enhancing air traffic management and aircraft operations (with a primary focus on reducing fuel consumption per flight through measures such as more direct cruise paths and more efficient altitude profiles in air traffic management), promoting the development and use of Sustainable Aviation Fuels (SAFs), and implementing market-based measures (MBMs) at both global and regional scales [42]. The initial two aspects, aircraft technology and operational enhancements, primarily aim to reduce fuel consumption as their main objective, ultimately resulting in a reduction in CO_2 emissions [42]. However, all measures proposed by ICAO can be applied in the domestic aviation sector to reduce greenhouse gas emissions, aligning with national strategies for emission reduction.

This study is timely considering the current COVID-19 situation and the ongoing market restructuring in the aviation industry. This restructuring will impact energy efficiency and technology, bringing changes in the domestic aviation market. The aviation industry was significantly affected by the COVID-19 pandemic between 2020 and 2021, with the government implementing lockdown measures during this period, resulting in a direct impact on the decline in aviation activity. Domestic air travel within Thailand experienced a downturn, leading to a significant decline in both passenger numbers and flight frequencies, affecting both the domestic and international segments of the aviation industry [43]. The domestic aviation sector is currently experiencing a recovery phase. This recovery is expected to lead to structural adjustments and changes in the domestic aviation market. Previous studies have shown that the increase in the tourism industry and domestic air travel demand will correspond to a rise in energy consumption and emissions [44]. Therefore, this period of market adjustment in the domestic aviation sector presents a crucial opportunity to assess future trends in GHG emissions following the market restructuring and address the impact of the COVID-19 situation and the pandemic measures, factors that contribute to making the forecasting of energy consumption in the aviation sector more accurate and precise. Furthermore, it serves as a starting point for implementing policies to reduce greenhouse gas (GHG) emissions in the domestic aviation sector, aligning with the country's targets for GHG reduction under the UNFCCC. Considering the current context, this study holds significant importance in understanding and addressing the future implications of market changes and the recovery of GHG emissions in the domestic aviation sector. It also highlights the need for appropriate policy interventions.

Over the past decade, Thailand has witnessed a significant increase in the number of travelers. The compound annual growth rate (CAGR) for all travelers from 2010 to 2019 was 11.38%, with a CAGR of 10.77% for international travelers and 12.13% for domestic travelers [45]. Before the COVID-19 pandemic, Thailand experienced a notable surge in domestic aviation passengers, rising from 3.482 million in 2008 to 76.256 million in 2019 [46]. These figures demonstrate domestic aviation's substantial growth and importance in Thailand, driven by international and domestic travel demand.

In parallel, jet fuel consumption in Thailand's domestic aviation sector witnessed a notable increase, rising from 246 kilotonnes of oil equivalent (ktoe) in 2008 to 856 ktoe in 2018 [47]. It is noteworthy that Thailand's Nationally Determined Contribution (NDC) sets a greenhouse gas reduction target of 20–25% lower than the Business-As-Usual prediction for 2030 across all sectors [48]. Furthermore, during the UNFCCC's COP-26 in Glasgow, Thailand committed to achieving carbon neutrality by 2050 and net-zero emissions by 2065. It indicates that Thailand's domestic aviation sector can align with its climate commitments even during growth. However, medium-term to long-term GHG emissions forecasting and policies to reduce future emissions in the domestic aviation sector have yet to be developed and implemented. These observations highlight the need for comprehensive GHG emissions forecasting and the implementation of policies to reduce emissions in the domestic aviation sector to align with Thailand's climate targets.

This study aims to contribute to understanding reference future scenarios, quantification of GHG emissions, and the mitigation potential of policies in Thailand's commercial and domestic aviation sector for the medium-term (2030) and long-term (2050). It seeks to fill the research gap by being the first study to assess medium- and long-term mitigation policies tailored to Thailand's commercial domestic aviation sector. Furthermore, the study aims to provide novel insights into the benefits of these policies in reducing GHG emissions. For example, Sustainable Aviation Fuel (SAF) can help mitigate the aviation sector's impact on global warming [49]. Additionally, improving the auxiliary power unit (APU) can contribute to reducing greenhouse gas emissions within the aviation sector [50], aligning with Thailand's greenhouse gas emission reduction targets under the UNFCCC. By conducting this analysis, the study intends to inform decision-makers and stakeholders about the effectiveness and implications of various mitigation policies in the domestic aviation sector. The findings can guide the development and implementation of targeted strategies to reduce GHG emissions in line with Thailand's climate goals.

2. Methodology and Data

This study employs a comprehensive framework, as depicted in Figure 1, to forecast medium-term (2030) and long-term (2050) GHG emissions in Thailand's commercial domestic aviation sector. The forecasting of GHG emissions is based on key factors such as the number of passengers, freight, gross domestic product (GDP), and jet fuel price.

Considering the above variables, this study uses a multiple linear regression method to forecast fuel consumption. By applying this approach, the study aims to provide an estimation of future fuel consumption in the commercial domestic aviation sector. This study also evaluated the impacts of three key policies in shaping reference future scenarios for GHG emissions reduction in Thailand's commercial domestic aviation sector. These policies are fuel switching, carbon pricing, and advancements in aircraft technology. By analyzing these policies, the study aims to identify their potential impacts on reducing GHG emissions.

2.1. Key Data Sources

The data used in this study encompass all 39 airports in Thailand, as reported by [45]. Several key variables were gathered from official data sources from 2008 to 2020 to forecast fuel consumption in Thailand's commercial domestic aviation sector. These variables include the number of passengers, the amount of freight, gross domestic product (GDP), jet fuel prices, and emission factors. The data sources are summarized and presented in Table 1. The data sources provide the necessary information to estimate fuel consumption and assess the factors influencing GHG emissions in Thailand's commercial domestic aviation sector. This comprehensive data set allows for robust analysis and forecasting of reference future scenarios.

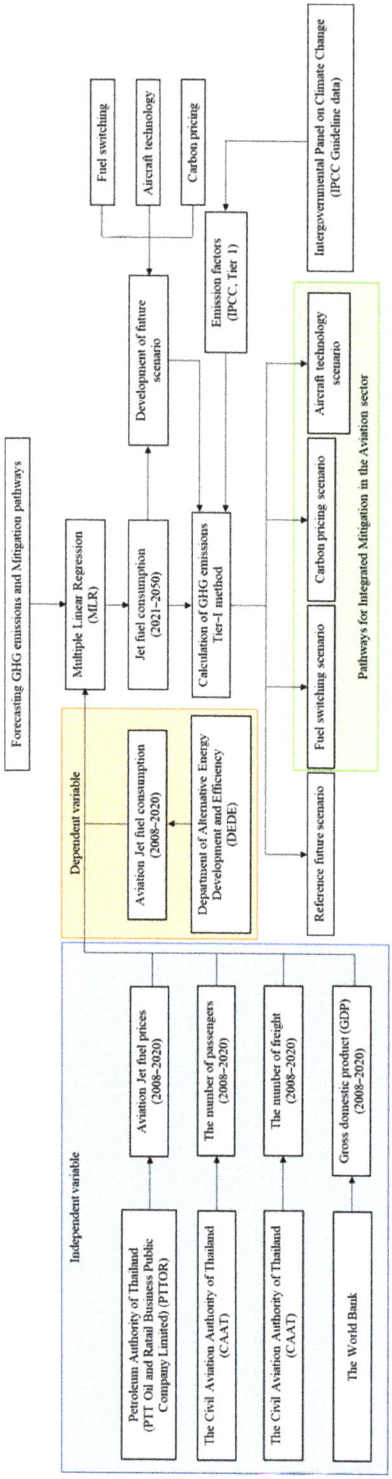

Figure 1. The overall framework.

Table 1. Variables, description, period, sources factors of GHG emissions forecasting, and sources of the estimation of GHG emissions.

Variables	Period	Description	Sources
3Jet fuel prices (USD/Litre)	2008–2020	Aviation jet fuel prices	Petroleum Authority of Thailand (PTT Oil and Retail Business Public Company Limited) [51]
Number of passengers in domestic flight (person)	2008–2020	The number of passengers in domestic aviation sector	The Civil Aviation Authority of Thailand [46]
Emissions factors (kg/TJ), (kg/LTO)	2006	Emission factors for calculating GHG emissions from fuel consumption	Intergovernmental Panel on Climate Change [52]
Aviation Jet fuel use (ktoe)	2008–2020	Jet fuel use in domestic aviation sector	Department of Alternative Energy Development and Efficiency [47]
Number of freights in domestic flight (kg)	2008–2020	The number of freights in domestic aviation sector	The Civil Aviation Authority of Thailand [46]
Gross domestic product (GDP) (Billion Baht)	2008–2020	Thailand's gross domestic product (Base year is 2008)	Bank of Thailand [53] The World bank [54]

2.2. Forecasting Model of Jet Fuel Consumption

2.2.1. Multiple Linear Regression (MLR)

To forecast fuel consumption in the commercial domestic aviation sector, the Multiple Linear Regression (MLR) method was employed in this study. The future estimates were based on 2008, considering planning horizons up to 2030 and 2050. The analysis results were presented at five-year intervals, corresponding to 2025, 2030, 2035, 2040, and 2050. The MLR method has been widely utilized in previous research to predict energy consumption [40,55–59]. In this study, the dependent variable is the aviation jet fuel use in Thailand's commercial domestic aviation sector, while the independent variables include aviation jet fuel prices, the amount of freight, the number of passengers, and GDP. The formula for Multiple Linear Regression is shown in Equation (1).

$$B = c + m_1a_1 + m_2a_2 + m_3a_3 + m_4a_4 \quad (1)$$

where

B = Dependent variable of aviation jet fuel consumption (Liter);
c = Intercept (constant term);
m_1, m_2, m_3, m_4 = Regression coefficient;
a_1, a_2, a_3, a_4 = Independent variable of aviation jet fuel prices (USD/Liter), the amount of freight (kg), the number of passengers (person), and GDP (Billion USD).

The MLR method enables the study to establish a quantitative relationship between fuel consumption and the independent variables, allowing for estimating future fuel consumption based on changes in the identified factors.

Linear regression is a statistical technique to model the relationship between an outcome variable and one or more explanatory factors. In this study, linear regression was employed to forecast the amount of freight, the number of passengers, aviation jet fuel prices, and GDP. The relationship is expressed mathematically through Equation (2):

$$B_i = c + m_ia_i \quad (2)$$

where

B_i = Dependent variable;
c = Intercept (constant value);
m_i = Coefficient values;
a_i = Independent variable.

By estimating the coefficients through the linear regression analysis, the study could quantify the relationship between the explanatory factors and fuel consumption, enabling the prediction of future fuel consumption based on changes in the identified factors.

The correlation coefficient (r) is a statistical measure that evaluates the strength and direction of the relationship between two quantitative variables. It quantifies the degree to which the variables are linearly related. The correlation coefficient (r) was calculated using Equation (3):

$$r = \frac{[n(\sum ab)] - [(\sum a)(\sum b)]}{\left[\sqrt{\left[n\sum a^2 - (\sum a)^2\right] \times \left[n\sum b^2 - (\sum b)^2\right]}\right]} \quad (3)$$

where

r = Correlation coefficient;
n = The amount of data;
$\sum a$ = The first variable's total value;
$\sum b$ = The second variable's total value;
$\sum ab$ = The second value and the sum of the products of;
$\sum a^2$ = Sum of the squares of the first value;
$\sum b^2$ = Sum of the squares of the second value.

The coefficient of multiple determination (R^2) is a commonly used statistical metric in multiple regression analysis. It assesses the proportion of the variance in the dependent variable that the independent variables in the model can explain. R^2 compares the model's accuracy to a basic benchmark model, where the forecast is the data's average. It was calculated using Equation (4):

$$R^2 = 1 - \left(\frac{\sum_i (y_i - \hat{y}_i)^2}{\sum_i (y_i - \bar{y}_i)^2}\right) \quad (4)$$

where

R^2 = The coefficient of multiple determination;
y_i = The actual value;
$\hat{y}_i$ = The predicted value of y;
$\bar{y}_i$ = The mean of y values.

In linear models, the Adjusted R^2 is a measure commonly used to assess the proportion of variation in the target variable that can be explained by the input or inputs in the model while considering the number of predictors and the sample size. The Adjusted R^2 is calculated using Equation (5):

$$\text{Adjusted } R^2 = 1 - \left(\frac{\left(1 - R^2\right)(T - 1)}{T - v - 1}\right) \quad (5)$$

where

R^2 = R-squared sample;
T = Total sample size;
v = The number of independent variables.

2.2.2. Formula for Future Projection

Table 2 presents the Multiple Linear Regression forecasting models for aviation jet fuel consumption in the commercial domestic aviation sector. It includes the sample R-squared (r), which measures the strength of the correlation, and the coefficient of multiple determination (R^2), which indicates the proportion of variation in fuel consumption explained by the independent variables in the model.

Table 2. Forecasting model.

Type of Fuel Used	R^2	R-Value	Adjusted-R^2	Equation
Aviation Jet fuel	0.987	0.994	0.982	$Y = (14.170 \times X_1) - (0.924 \times X_2) + (190{,}013.386 \times X_3) - (132{,}057{,}994.567 \times X_4)$

2.3. *Assumptions and Scenarios for Analysis of Policies or Measure*

It is essential to consider the potential reduction in energy-related greenhouse gas (GHG) emissions for each option to evaluate the effectiveness of various mitigation options and inform the development of appropriate regulations. Considering the reference future scenario, this study forecasted the fuel consumption and GHG emissions of Thailand's commercial domestic aviation sector from 2021 to 2050. Three mitigation scenarios were assessed for their potential to reduce GHG emissions in Thailand's domestic commercial aviation sector. The assumptions for each scenario are described below.

2.3.1. Reference Future Scenario

In the reference future scenario, after the recovery of the domestic economy and domestic transportation from the COVID-19 crisis, this study assumed that no changes would happen in the medium-term and long-term trends in demand for commercial domestic aviation fuel consumption. Additionally, no mitigation options are considered from 2021 to 2050. The study analyzed four-factor variables related to GHG emissions growth to create a realistic time series and project future fuel consumption. The study aims to capture each factor's real behavior and realism by reproducing the historical time series of each factor's growth. This approach creates a time series that reflects the continuous behavior cycle observed in the historical data. The analysis of these factor variables helps in understanding their impact on fuel consumption and enables the projection of future fuel consumption based on their historical trends. Figure 2 shows the historical data of four-factor variables related to GHG emissions growth for the analysis and time series.

2.3.2. Fuel Switching Scenario

Using alternative fuels in the aviation industry is of utmost importance for two key reasons: reducing dependency on fossil fuels and mitigating GHG emissions. Aviation specialists and researchers have been actively exploring alternatives to conventional jet fuel [60,61]. Numerous studies have demonstrated the potential of biofuels and synthetic fuels as substitutes for traditional petroleum-derived jet fuels to reduce emissions of pollutants [62,63]. Adopting alternative fuels in aviation presents an opportunity to address environmental concerns and contribute to reducing pollution emissions. Biofuels and synthetic fuels are being investigated as viable options for the industry.

This study focuses on reducing greenhouse gas (GHG) emissions in commercial domestic aviation by using hydro-processed esters and fatty acids (HEFA) as an alternative to conventional jet A-1 fuel. The choice of HEFA was based on the suitability of oil-to-jet technology, which utilizes biomass as a raw material. This approach takes advantage of the underutilized biomass resources available in the country and benefits from indirect biomass utilization, contributing to the reduction of GHG emissions [64]. Palm has a high agricultural yield, with a production rate of 15.7 tons/ha and an oil content ranging from 21% to 37%. Globally, palm production amounts to 282.2 million tons, with major producers including Indonesia (45%), Malaysia (37%), and Thailand (5%) [65]. The European Commission guidelines utilize Sustainable Aviation Fuel (SAF) volume shares. These volume shares indicate the proportion of SAF used in the aviation sector from 2021 to 2050. The European Commission guidelines serve as a reference for reducing conventional jet fuel use and transitioning to SAF to mitigate GHG emissions. The volume shares of SAF are presented in Table 3. This represents the recommended SAF ratios according to European Commission guidelines, illustrating the growing proportion of SAF usage in the domestic aviation sector over time.

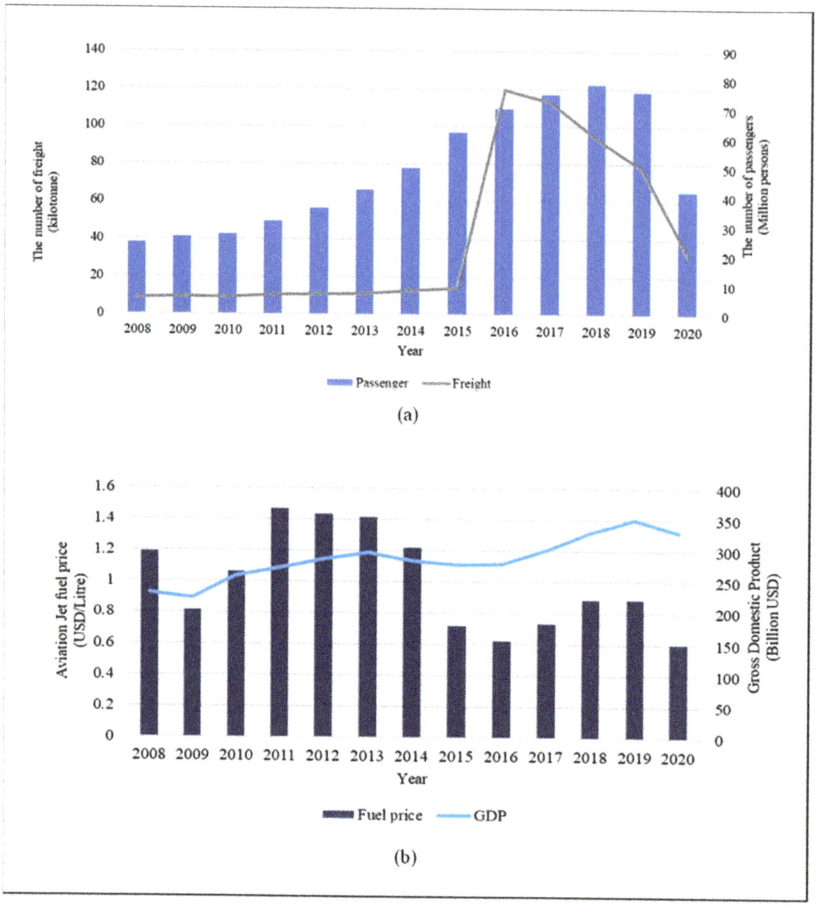

Figure 2. The historical data of four-factor variables related to the growth in GHG emissions. (**a**) The number of passengers and the amount of freight 2008–2020. (**b**) Aviation Jet fuel price and GDP 2008–2020.

Table 3. Volume shares of SAF fuel from 2021 to 2050.

Year	SAF Percentage (%)
2021	0
2025	2
2030	5
2035	20
2040	32
2045	38
2050	63

Sources: [66].

2.3.3. Aircraft Technology Scenario

The primary factor contributing to emissions reduction in the aviation sector is the improvement in fuel efficiency. Effective strategies for enhancing efficiency include using fuel-efficient next-generation aircraft, advancements in air traffic management (ATM), re-engineering processes, and implementing technologically enhanced flight patterns [67]. The introduction of advanced aircraft designs has increased fuel economy in the aviation industry. ICAO's predictions for international aviation activities indicate a future decrease in fuel demand. With a projected reduction of 10 MCMOED (million cubic meters of oil equivalent per day) by 2050, the scenario that achieves the 2% yearly fuel efficiency target yields the most significant savings. The ICAO has set a target of 2% improved efficiency for the global aviation fleet from 2021 to 2050 through non-engine-based efficiency enhancements and fuel efficiency improvements. Non-engine-based efficiency enhancements, such as lighter aircraft components and improved fuel efficiency, are considered crucial in significantly reducing aviation sector GHG emissions [33].

Improving fuel efficiency is crucial for reducing CO_2 emissions in the aviation industry. Fuel costs account for approximately 20% of total operating costs for modern airplanes [68]. Over the years, significant advancements in aircraft fuel efficiency have been achieved. Since 1960, aircraft fuel efficiency has improved by approximately 70–80% [69]. Projections indicate that 40–50% improvements are possible by 2050 [68]. To incorporate the potential fuel efficiency improvements in this study, it is assumed that aircraft technology will continue to advance. Specific engine performance enhancements have been achieved through upgrading programs in recent years, reducing fuel usage by up to 2% [70].

Additionally, new engines and auxiliary power units (APUs) are expected to consume at least 15% less fuel than the aircraft they replace. These improvements are made possible through innovations in engine technologies, including materials, coatings, combustion methods, sensors, and cooling methods [70]. Based on the ICAO target, this study assumes that aircraft technology will continue to improve fuel efficiency by 2% annually from 2021 to 2050. This assumption reflects the ongoing efforts and advancements in aircraft design and technology to achieve greater fuel efficiency and reduce CO_2 emissions.

2.3.4. Carbon Pricing Scenario

Carbon pricing is widely recognized as the primary policy tool for countries to reduce GHG emissions globally [71]. Imposing a price on carbon through taxes, emissions trading, or regulation is considered the foremost objective of mitigation policies [72]. Carbon pricing creates incentives for investments in new technologies that can mitigate global warming and holds emitters accountable for the environmental impact of their emissions [73,74]. By increasing the cost of carbon-intensive production, carbon pricing can reduce demand for fuel-based products, encouraging the substitution of technologies. Therefore, in theory, an effective carbon price should lower aircraft emissions by raising prices and subsequently reducing demand. Thus, market-based measures (MBM) are recognized as one of the international guidelines for reducing GHG emissions. This measure is in line with the existing international policy framework provided by the International Civil Aviation Organization (ICAO). MBMs encompass mechanisms such as carbon offset programs or emissions trading schemes, which create economic incentives and market mechanisms to encourage emissions reductions in the aviation sector. Carbon pricing and market-based measures play significant roles in global efforts to reduce GHG emissions and are particularly relevant in the aviation industry.

Carbon pricing mechanisms, such as cap and trade or direct taxation systems, can impact fuel use and CO_2 emissions. As the price of carbon increases, the effective fuel price also rises, leading to a reduction in consumption due to the demand–price relationship. Hasan et al. [33] anticipated that increasing the high carbon price could result in a 12% reduction in overall emissions, equivalent to 180 $MtCO_2$ of emissions in the mitigation pathway. To estimate the price elasticity of fuel demand, income elasticities were used to adjust the price elasticity. Hasan et al. [33] estimated that price elasticity of fuel demand is -0.48, indicating that a 10% increase in price will result in a 4.8% decrease in demand. This relationship demonstrates the responsiveness of fuel consumption to changes in price. In Thailand's domestic aviation sector, this study examined the change in demand for fuel consumption and aviation jet fuel prices during 2009–2010 and 2016–2017. Figure 3 illustrates the relationship between jet fuel price increases and the corresponding change in demand for fuel consumption. The jet fuel prices increased and had an impact on decreased demand twice: first, between 2009 and 2010, when jet fuel prices increased by around 12%, resulting in an average 6% decrease in demand; and second, between 2016 and 2017, when jet fuel prices increased by around 7%, resulting in an average 4% decrease in demand. Therefore, the results show that the average increase in jet fuel prices in both periods was 10%, resulting in an average 4.9% decrease in demand. Based on these observations, the study assumes an annual increase in jet fuel prices by 10% from 2021 to 2050 will result in a 4.9% decrease in fuel demand. This assumption reflects the anticipated impact of carbon pricing on fuel consumption in the domestic aviation sector.

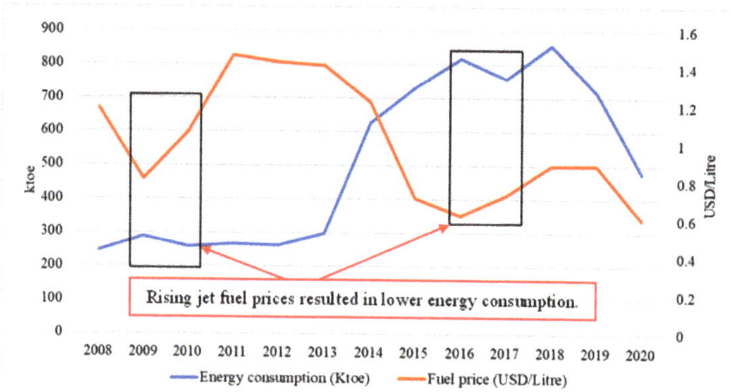

Figure 3. The relationship between fuel consumption and jet fuel prices.

2.4. Calculation Method for GHG Emissions

The Intergovernmental Panel on Climate Change (IPCC) used the tier-I method to calculate GHG emissions from the domestic aviation sector's jet fuel consumption activities [75]. This method is based on the quantity of energy used (aviation fuel consumption) multiplied by an average emission factor. Equation (6) shows the formula to estimate GHG emissions. Firstly, the activity data for aviation mode was established. After that, the amounts of aviation fuel consumed in physical units or other energy units (such as l, ktoe, kg, etc.) were converted to the terajoule (TJ), the common international energy unit; and the fuel consumption was multiplied by the factor for carbon emission to determine the carbon emission. Finally, only the fraction of oxidized carbon was utilized to determine the actual CO_2 emission, as not all the carbon in the fuel is oxidized to generate carbon dioxide.

$$CO_2 = E_f \times N_f \times CE_c \quad (6)$$

where

CO_2 = An emission of carbon dioxide (t CO_2);
E = A specific energy usage (ktoe);
N = A net calorific value specific to fuel (TJ/ktoe);
CE_c = A factor for carbon emissions (kg/TJ);
f = Type of fuel.

The global warming potential (GWP) conversion factors calculated GHG emissions to CO_2 equivalent units. The impact of individual greenhouse gas emissions on global warming cannot be easily compared on a mass basis since the gases' physical and chemical properties vary. Therefore, IPCC advises converting all computed greenhouse gas emissions to CO_2 equivalent units using the GWP conversion factor to precisely compare the global warming effects between individual GHGs. The Global Warming Potential (GWP) value from IPCC, AR5 [76] and IPCC, AR6 [77] is used in this study. Table 4 shows the global warming potential (GWP). Moreover, the default parameters adopted in this study are those recommended by the IPCC [52], as shown in Table 5 for CO_2 emission and non-CO_2 emission factors and Table 6 for net calorific values.

Table 4. The global warming potential (GWP) in IPCC 5th and 6th Assessment Reports.

Industrial Designation or Standard Name	The Formula for Chemical	The GWP Value for the 100-Year Time Horizon	
		Fifth Assessment (AR5)	Sixth Assessment (AR6)
Nitrous oxide	N_2O	265	273
Methane	CH_4	28	29.8
Carbon dioxide	CO_2	1	1

Table 5. CO_2 emission factors.

Fuel	CO_2 Default (kg/TJ)	CH_4 Default (kg/TJ)	N_2O Default (kg/TJ)
Jet fuel (Jet Kerosene)	71,500	0.5	2

Table 6. Net calorific values.

Fuel	Factors ($TJ/10^3$ tonnes)
Jet Kerosene	44.59

2.5. Empirical Data

In this study, the forecasting of GHG emissions for 2021–2050 was calculated by forecasting fuel consumption and calculating GHG emissions using IPCC, Tier-I [52]. The data on fuel consumption for the domestic aviation sector was derived from the Department of Alternative Energy Development and Efficiency's statistical data reports for 2008–2020 [47]. Gross domestic product (GDP) figures are available from the World Bank—Thailand Bureau through their statistical data report on gross domestic product [54]. However, GDP was converted into terms representing 'real prices'. The conversion was performed using a deflator derived from the World Bank's national accounts and the Organization for Economic Cooperation and Development's (OECD) national accounts data [71,78]. The number of passengers and freights in the domestic aviation sector was obtained from the Civil Aviation Authority of Thailand (CAAT) [46]. Information on jet fuel prices is sourced from the Petroleum Authority of Thailand's statistics data reports for 2015–2020 (Petroleum Authority of Thailand Oil and Retail Business Public Company Limited) [51]. However, jet fuel prices from 2008 to 2014 were calculated from the estimated rate of change of Brent crude oil futures. Moreover, all jet fuel prices include interior tax, a 7% VAT, and excise tax. For consistency in calculations, prices were converted to USD per liter.

3. Results

3.1. Historical Trends

The results highlight the rapid growth of domestic commercial aviation activities in Thailand, which can be attributed to various strategies to stimulate the economy, particularly targeting air travel and tourism. The implementation of Thailand's Transport Infrastructure Development Strategy 2015–2022 has played a significant role in supporting the growth of the aviation sector. This strategy focuses on constructing domestic airports that meet international standards and cater to the increasing travel needs of the population. Additionally, the strategy emphasizes the utilization of regional airports to enhance their contribution to the country's aviation industry. Through these efforts, the number of passengers in the domestic commercial aviation industry has witnessed a substantial increase of approximately 214% from 2008 to 2019.

Similarly, freight volume experienced a remarkable growth of about 793% during the same period. These statistics demonstrate the growing demand and importance of the aviation industry in Thailand. Consequently, the increased activity in the commercial domestic aviation sector has led to a significant rise in aviation fuel consumption. From 2008 to 2019, aviation fuel consumption in this sector increased from 246 ktoe (kilo-tonnes of oil equivalent) to 716 ktoe, representing a surge of approximately 190%.

In 2020, aviation fuel consumption decreased from 716 ktoe in 2019 to 477 ktoe in 2020 due to the COVID-19 pandemic, and Thailand proclaimed a state of emergency in all areas in March 2020 [79]. The pandemic affected the domestic and international aviation sectors [43]. The number of passengers and freight, GDP, fuel price, and aviation fuel consumption from sources mentioned in Table 1 are presented in Table 7.

Table 7. Historical trends of the number of passengers and amount of freight, GDP, fuel price, and aviation fuel consumption in Thailand's commercial domestic aviation industry from 2008 to 2020.

Year	Passengers (Person)	Freight (kg)	GDP Billion Baht (at 2008 Price)	Fuel Prices (USD/Litre)	Fuel Consumptions (ktoe)
2008	24,310,188	8,706,271	7722	1.19	246
2009	26,219,477	9,273,089	7668	0.81	288
2010	27,208,643	9,109,330	8243	1.06	258
2011	31,623,503	10,238,865	8302	1.47	265
2012	36,192,158	10,777,970	8903	1.43	261
2013	42,427,923	11,210,853	9143	1.41	295
2014	50,059,872	12,800,068	9233	1.22	625
2015	62,216,533	14,292,021	9523	0.72	732
2016	70,327,980	119,490,892	9866	0.62	818
2017	75,342,243	112,653,693	10,260	0.73	757
2018	78,625,622	93,682,980	10,692	0.89	856
2019	76,253,599	77,828,059	10,887	0.89	716
2020	41,996,665	32,214,457	10,348	0.60	477

3.2. Forecasting of Fuel Use and GHG Emissions

The analysis of GHG emissions in the reference future scenario is presented in Figure 4, reflecting forecasted trends based on factors such as aviation jet fuel price, the number of passengers and amount of freight, and GDP. Using Equation (6), the study calculated the predicted GHG emissions in the domestic aviation sector. The results indicate that GHG emissions and aviation fuel consumption trends will increase from 2021 to 2050. GHG emissions are expected to rise from 2340 thousand tonnes of CO_2-eq in 2021 to 7200 thousand tonnes of CO_2-eq in 2050, representing an average annual growth rate of 3.99%. This is consistent with other research on future increases in GHG emissions [31,80]. This growth rate signifies a substantial GHG emissions increase over the predicted period. The emissions are projected to increase by more than 9.2 times, reflecting the substantial impact of the sector on overall emissions in the long term. The forecasted increase in GHG emissions emphasizes the urgency of implementing measures to reduce emissions, aligning with the country's climate targets.

3.3. Pathways for Mitigation in the Domestic Aviation Industry

The following sections describe mitigation measures or policies that were analyzed and considered to reduce the domestic aviation sector's GHG emissions that could pave the path for sustainable reduction of aviation industry GHG emissions and support Thailand's goal of reducing GHG emissions by 2050. Figure 5 shows the forecasting of the usage of aviation fuel and the mitigation pathways for greenhouse gas (GHG) emissions for each scenario.

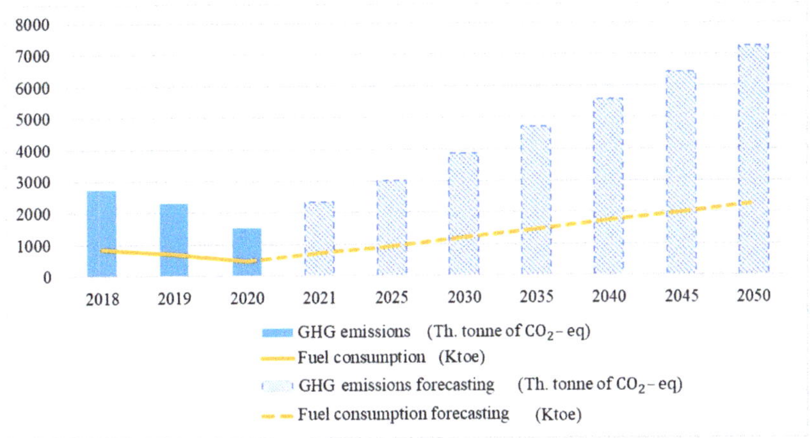

Figure 4. Forecasting fuel consumption and GHG emissions for the reference future scenario.

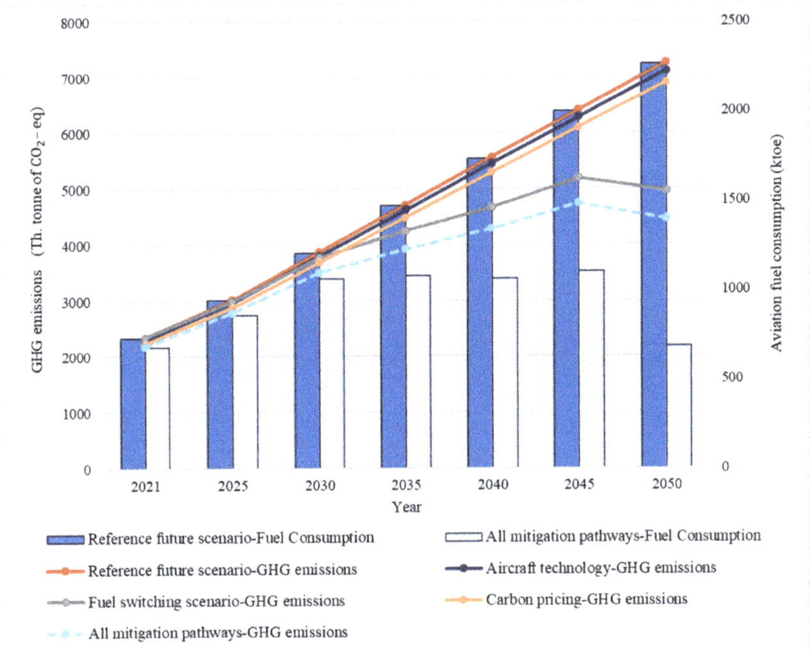

Figure 5. Forecasting aviation fuel consumption and GHG emissions mitigation pathways for each scenario.

3.3.1. Fuel Switching

According to the European Commission regarding volume shares of SAF fuel in Section 2.3.2, the use of alternative fuels in 2025 will be 2% and increase every five years. The results in Figure 5 indicate that GHG emissions in this scenario show a downward trend from 2025 to 2050. The GHG emissions reduction depends on the proportion of alternative fuels. The most considerable GHG emissions reduction is in 2050, when the proportion of alternative fuels is around 63%. It decreased by around 2300 thousand tonnes of CO_2-eq compared to the reference scenario 2050. In addition, with the reduction of GHG emissions from alternative fuels compared to other scenarios, alternative fuels can reduce GHG emissions the most, according to the results of calculations and analysis.

3.3.2. Aircraft Technology

One method of reducing GHG emissions in the aviation industry is through increased fuel efficiency and non-engine-based efficiency improvements. This is the main factor and an important factor in reducing GHG emissions. Figure 5 shows that the goal is to enhance fuel efficiency by 2% annually regardless of costs or obstacles. It begins from 2021 to 2050, according to the ICAO target. The trends in GHG emissions are projected to decrease from 2021 to 2050. It decreased from 2340 thousand tonnes of CO_2-eq to 2293 thousand tonnes of CO_2-eq compared to the reference future scenario 2021, decreasing yearly. The short-term reductions in GHG emissions will be very effective because the first period (2021–2025) is the beginning of the aviation sector's growth after the COVID-19 situation. As a result, the fuel consumption during that period was not intense compared to the improving fuel efficiency. However, improving fuel efficiency will be less efficient in the long term due to increased fuel consumption, because historical data on the growth of the aviation sector shows that Thailand's aviation sector is growing rapidly, but the efficiency improvement goals will remain stagnant.

3.3.3. Carbon Pricing

Through the demand-price relationship, rising carbon prices lead to rising effective fuel prices, reducing consumption [41]. As mentioned in Section 2.3.4, a 10% increase in jet fuel price will lead to a 4.9% reduction in fuel consumed on demand. The results in Figure 5 show that GHG emissions will decrease from 2021 to 2050. GHG emissions began to decrease in 2021, from 2340 thousand tonnes of CO_2-eq to 2225 thousand tonnes of CO_2-eq, decreasing every year. The reduction in GHG emissions from this scenario depends on carbon pricing and the demand–price relationship. However, carbon pricing will be less efficient in the long term due to the forecasted increase in fuel consumption from the reference future scenario (the rapid growth of the aviation sector in the past). It will increase fuel consumption and GHG emissions; however, carbon pricing remains the same, causing a decrease in the efficiency of this scenario.

3.3.4. Creating Multi-Policy Scenarios by Combining Different Policies

The policies outlined in Section 2.3, aimed at reducing GHG emissions in the domestic aviation sector (e.g., fuel switching, aircraft technology, and carbon pricing), have been examined individually in specific areas of improvement. However, in this section, we analyze all these policies concurrently, as this represents the primary strategy for reducing GHG emissions from 2021 to 2050.

The three policies considered in this study have the potential to significantly reduce domestic aviation emissions by 2050 compared to employing only one policy to reduce greenhouse gas emissions. Combining all three scenarios leads to a greater reduction in GHG emissions than any single scenario alone. Specifically, it results in a decrease of approximately 2796 thousand tonnes of CO_2-eq, equivalent to a 39% reduction compared to the reference future scenario projected for 2050.

4. Discussion

Figure 5 illustrates the mitigation pathways, each with its advantages and challenges. For instance, alternative fuels offer substantial potential for reducing emissions. However, there are limitations, including constraints on arable lands for biofuel production, the high cost of alternative fuels, and the need for policy support to facilitate large-scale production and utilization. Another effective approach to curbing GHG emissions is the implementation of carbon pricing [81]. Carbon pricing, often enacted as a carbon tax, has proven effective in reducing greenhouse gas (GHG) emissions by discouraging carbon emissions through increased fossil fuel prices, thereby reducing their use [35]. Additionally, carbon pricing mechanisms have been shown to promote biofuels [82].

Despite efforts to enhance fuel efficiency in the aviation industry, GHG emissions from this sector are projected to increase. Thailand's aviation industry has undergone rapid growth, driven by the country's economic development and the implementation of the Transport Infrastructure Development Strategy 2015–2022. Addressing rising GHG emissions requires accurate estimation of future emissions and fuel consumption in the commercial domestic aviation sector. This estimation forms the basis for implementing mitigation strategies over the next three decades, which include fuel switching, aircraft technology improvement, and carbon pricing. The objective of these mitigation measures is to reduce GHG emissions in Thailand's aviation sector, aligning with the country's commitment to decreasing GHG emissions under the UNFCCC. Accurate estimation of GHG emissions and fuel usage provides essential guidance for policymakers and stakeholders, aiding in the development of effective strategies to mitigate emissions and attain climate targets.

5. Conclusions

The forecasted growth in aviation fuel consumption in Thailand's domestic commercial aviation sector indicates an average annual increase of approximately 4% over the past decade. In the reference future scenario, GHG emissions from the domestic commercial aviation industry are anticipated to more than triple, which is in line with other research that predicts a significant increase in the aviation sector's GHG emissions [31,80]. This signifies a substantial increase in emissions over the three-decade period, highlighting the urgent need for mitigation strategies to address the environmental impact of the aviation sector. Adopting fuel switching, particularly alternative fuels, can significantly reduce greenhouse gas (GHG) emissions from Thailand's domestic aviation sector. By 2050, the potential reduction in GHG emissions is projected to be as high as 2279.4 thousand tonnes of CO_2 equivalent. This signifies a substantial decrease in emissions, highlighting the effectiveness of fuel switching as a mitigation strategy for reducing the environmental impact of the domestic aviation sector. The achievement of these reductions depends on the implementation of fuel switching measures and the gradual increase in the utilization of alternative fuels.

The aircraft technology and carbon pricing scenarios play important roles in reducing greenhouse gas (GHG) emissions from the domestic aviation sector in the short term. In the case of aircraft technology, the scenario will reduce 145.2 thousand tonnes of CO_2 equivalent in GHG emissions by 2050. Similarly, the carbon pricing scenario aims to impose a price on carbon, incentivizing fuel consumption and GHG emissions reduction. This scenario will reduce 355.8 thousand tonnes of CO_2 equivalent in GHG emissions by 2050. However, it is important to note that the effectiveness of these scenarios diminishes in the long term due to the constant goals of aircraft technology and carbon pricing.

All of the individual scenarios for reducing GHG emissions have differences in efficiency in short- and long-term reductions. However, each scenario has limitations and relevance to reducing GHG emissions in the future. For example, alternative fuels require arable lands that are limited for biofuel production, alternative fuels are currently expensive, and policy support is required for large-scale use and production. In addition, carbon pricing has been effective in promoting biofuels [82]. Therefore, combining all scenarios can reduce GHG emissions by about 2796 thousand tonnes of CO_2-eq or 39% compared with the reference future scenario in 2050.

Indeed, reducing GHG emissions in Thailand's aviation sector requires implementing significant ICAO measures and collaborating with various stakeholders. While the identified policies can potentially reduce emissions, it is important to acknowledge the obstacles and limitations associated with their implementation. Factors such as the availability of personnel and expertise, budgetary support, technological advancements, societal acceptance, and policy frameworks play significant roles in determining the success of emission reduction measures. Governmental policymakers and industry executives should engage in ongoing discussions, surveys, and assessments to identify and address barriers to implementation, prioritize effective policies, and promote best practices. By actively involving relevant stakeholders, including government bodies, industry players, research institutions, and civil society, developing comprehensive strategies that consider the unique context of Thailand's aviation sector will be possible. Continuous dialogue and collaboration are essential to foster a supportive environment for adopting effective policies and measures to drive significant GHG emission reductions while aligning with national goals and international commitments. Through these efforts, Thailand can progress in achieving its GHG emission reduction targets, contributing to global climate change mitigation efforts and creating a more sustainable aviation sector.

Author Contributions: Conceptualization, A.C. and S.D.; methodology, A.C.; software, A.C.; validation, A.C., S.D. and N.C.; formal analysis, A.C.; investigation, S.D.; resources, A.C. and N.C.; data curation, A.C.; writing—original draft preparation, A.C.; writing—review and editing, S.D.; visualization, A.C.; supervision, S.D. and N.C.; project administration, A.C.; funding acquisition, A.C. All authors have read and agreed to the published version of the manuscript.

Funding: This research received no external funding.

Data Availability Statement: Data are available in a publicly accessible repository.

Conflicts of Interest: The authors declare no conflict of interest.

References

1. Kim, B.; Fleming, G.; Balasubramanian, S.; Malwitz, A.; Lee, J.; Waitz, I.; Klima, K.; Locke, M.; Holsclaw, C.; Morales, A.; et al. *SAGE System for Assessing Aviation's Global Emissions, Version 1.5*; Federal Aviation Administration Office of Environment and Energy: Washington, DC, USA, 2005.
2. Edwards, H.A.; Dixon-Hardy, D.; Wadud, Z. Aircraft cost index and the future of carbon emissions from air travel. *Appl. Energy* **2016**, *164*, 553–562. [CrossRef]
3. Chiaramonti, D.; Prussi, M.; Buffi, M.; Tacconi, D. Sustainable bio kerosene: Process routes and industrial demonstration activities in aviation biofuels. *Appl. Energy* **2014**, *136*, 767–774. [CrossRef]
4. Álvarez-Gil, M.J.; Yan, W. Is Environmental Innovation Worth It? The Case of the Civil Aviation Industry of Emerging Markets. *IFIP Adv. Inf. Commun. Technol.* **2013**, *415*, 294–301. [CrossRef]
5. Norton, T.M. Aircraft Greenhouse Gas Emissions during the Landing and Takeoff Cycle at Bay Area Airports. 2014. Available online: https://repository.usfca.edu/capstone/15 (accessed on 24 May 2022).
6. Uherek, E.; Halenka, T.; Borken-Kleefeld, J.; Balkanski, Y.; Berntsen, T.; Borrego, C.; Gauss, M.; Hoor, P.; Juda-Rezler, K.; Lelieveld, J.; et al. Transport impacts on atmosphere and climate: Land transport. *Atmos. Environ.* **2010**, *44*, 4772–4816. [CrossRef]
7. Shon, Z.H.; Kim, K.H.; Song, S.K. Long-term trend in NO_2 and NO_x levels and their emission ratio in relation to road traffic activities in East Asia. *Atmos. Environ.* **2011**, *45*, 3120–3131. [CrossRef]
8. Lee, D.S.; Pitari, G.; Grewe, V.; Gierens, K.; Penner, J.E.; Petzold, A.; Prather, M.J.; Schumann, U.; Bais, A.; Berntsen, T.; et al. Transport impacts on atmosphere and climate: Aviation. *Atmos. Environ.* **2010**, *44*, 4678–4734. [CrossRef]
9. Lee, D.S.; Fahey, D.W.; Skowron, A.; Allen, M.R.; Burkhardt, U.; Chen, Q.; Doherty, S.J.; Freeman, S.; Forster, P.M.; Fuglestvedt, J.; et al. The contribution of global aviation to anthropogenic climate forcing for 2000 to 2018. *Atmos. Environ.* **2021**, *244*, 117834. [CrossRef]
10. Klöwer, M.; Allen, M.R.; Lee, D.S.; Proud, S.R.; Gallagher, L.; Skowron, A. Quantifying aviation's contribution to global warming. *Environ. Res. Lett.* **2021**, *16*, 104027. [CrossRef]
11. Planès, T.; Delbecq, S.; Pommier-Budinger, V.; Bénard, E. Simulation and evaluation of sustainable climate trajectories for aviation. *J. Environ. Manag.* **2021**, *295*, 113079. [CrossRef]
12. Rupcic, L.; Pierrat, E.; Fricke, K.; Moll, T.; Hauschild, M.Z.; Laurent, A. Improving environmental performances of integrated bladed rotors for aircraft. *CIRP Ann.* **2022**, *71*, 13–16. [CrossRef]
13. Héroux, M.-E.; Babisch, W.; Belojevic, G.; Brink, M.; Janssen, S.; Lercher, P.; Paviotti, M.; Pershagen, G.; Waye, K.P.; Preis, A.; et al. WHO Environmental noise guidelines for the European Region. In Proceedings of the Euronoise, 10th European Congress and Exposition on Noise Control Engineering, Maastricht, The Netherlands, 31 May–3 June 2015; pp. 2589–2593.
14. Air Transport Bureau. *Effects of Novel Coronavirus (COVID-19) on Civil Aviation: Economic Impact Analysis*; ICAO–International Civil Aviation Organization: Montréal, QC, Canada, 2020.
15. Brandon, G.; Kevin, Z.; Dan, R. *CO_2 Emissions from Commercial Aviation*; International Council on Clean Transportation: Washington, DC, USA, 2018. Available online: https://theicct.org/publication/co2-emissions-from-commercial-aviation-2018/ (accessed on 24 May 2022).

16. Crippa, M.; Oreggioni, G.; Guizzardi, D.; Muntean, M.; Schaaf, E.; Lo Vullo, E.; Solazzo, E.; Monforti-Ferrario, F.; Olivier, J.G.J.; Vignati, E. Fossil CO2 and GHG Emissions of all World Countries. 2019. Available online: https://op.europa.eu/en/publication-detail/-/publication/9d09ccd1-e0dd-11e9-9c4e-01aa75ed71a1/language-en (accessed on 24 May 2022).
17. IATA—International Air Transport Association. Economic Performance of the Airline Industry: 2015 Mid-Year Report. 2015. Available online: https://www.iata.org/en/publications/economics/ (accessed on 24 May 2022).
18. ICAO—International Civil Aviation Organization. *Assembly-40th Session Executive Committee Agenda Item 15: Environmental Protection-General Provisions, Aircraft Noise and Local Air Quality-Policy and Standardization Icao Global Environmental Trends-Present and Future Aircraft Noise and Emissions*; ICAO: Montreal, QC, Canada, 2019.
19. Dube, K. Emerging from the COVID-19 Pandemic: Aviation Recovery, Challenges and Opportunities, 2022. *Aerospace* **2023**, *10*, 19. [CrossRef]
20. OECD—The Organisation for Economic Co-Operation and Development. Air Transport CO_2 Emissions. 2022. Available online: https://stats.oecd.org/Index.aspx?DataSetCode=AIRTRANS_CO2# (accessed on 18 July 2022).
21. C2ES—Center for Climate and Energy Solutions. Outcomes of the U.N. *Climate Change Conference in Lima.* 2016. Available online: https://www.c2es.org/wp-content/uploads/2017/10/outcomes-of-the-u-n-climate-change-conference-in-lima.pdf (accessed on 24 July 2022).
22. UNFCCC—United Nations Framework Convention on Climate Change. Thailand's Intended Nationally Determined Contribution (INDC). 2016. Available online: https://www4.unfccc.int/sites/SubmissionsStaging/NationalReports/Documents/347251_Thailand-BUR2-1-SBUR%20THAILAND.pdf (accessed on 26 July 2022).
23. IATA—International Air Transport Association. Fact Sheet Climate Change & CORSIA a CO2 Standard for Aircraft. 2017. Available online: www.iata.org/policy/environment (accessed on 22 July 2022).
24. ICAO—International Civil Aviation Organization. *Introduction to the ICAO Basket of Measures to Mitigate Climate Change*; ICAO: Montreal, QC, Canada, 2007.
25. ICAO—International Civil Aviation Organization. CORSIA Eligible Fuels. Available online: https://www.icao.int/environmental-protection/CORSIA/Pages/CORSIA-Eligible-Fuels.aspx (accessed on 9 October 2023).
26. Maertens, S.; Grimme, W.; Scheelhaase, J.; Jung, M. Options to Continue the EU ETS for Aviation in a CORSIA-World. *Sustainability* **2019**, *11*, 5703. [CrossRef]
27. Tawatchai, S. *Global Commitment of ICAO to the Development of the Low-Carbon Aviation Industry (Global Aspirational Goal)*; Thailand Greenhouse Gas Management Organization: Bangkok, Thailand, 2019.
28. Gudmundsson, S.V.; Anger, A. Global carbon dioxide emissions scenarios for aviation derived from IPCC storylines: A meta-analysis. *Transp. Res. Part D Transp. Environ.* **2012**, *17*, 61–65. [CrossRef]
29. Owen, B.; Lee, D.S.; Lim, L. Flying into the future: Aviation emissions scenarios to 2050. *Environ. Sci. Technol.* **2010**, *44*, 2255–2260. [CrossRef]
30. Macintosh, A.; Wallace, L. International aviation emissions to 2025: Can emissions be stabilised without restricting demand? *Energy Policy* **2009**, *37*, 264–273. [CrossRef]
31. Zhou, W.; Wang, T.; Yu, Y.; Chen, D.; Zhu, B. Scenario analysis of CO_2 emissions from China's civil aviation industry through 2030. *Appl. Energy* **2016**, *175*, 100–108. [CrossRef]
32. Kousoulidou, M.; Lonza, L. Biofuels in aviation: Fuel demand and CO_2 emissions evolution in Europe toward 2030. *Transp. Res. Part D Transp. Environ.* **2016**, *46*, 166–181. [CrossRef]
33. Hasan, M.A.; Mamun, A.A.; Rahman, S.M.; Malik, K.; Al Amran, M.I.U.; Khondaker, A.N.; Reshi, O.; Tiwari, S.P.; Alismail, F.S. Climate Change Mitigation Pathways for the Aviation Sector. *Sustainability* **2021**, *13*, 3656. [CrossRef]
34. Bows-Larkin, A. All adrift: Aviation, shipping, and climate change policy. *Clim. Policy* **2015**, *15*, 681–702. [CrossRef]
35. Fageda, X.; Teixidó, J.J. Pricing carbon in the aviation sector: Evidence from the European emissions trading system. *J. Environ. Econ. Manag.* **2022**, *111*, 102591. [CrossRef]
36. Agencia Portuguesa do Ambiente, Portuguese National Inventory Report on Greenhouse Gases, 1990–2016. Submitted Under the United Nations Framework Convention on Climate Change and the Kyoto Protocol. 2019. Available online: https://unfccc.int/documents/194464 (accessed on 30 August 2022).
37. Andrés, L.; Padilla, E. Driving factors of GHG emissions in the EU transport activity. *Transp. Policy* **2019**, *61*, 60–74. [CrossRef]
38. Environment Agency Austria. Austria's National Inventory Report 2019, Submission to the UNFCCC Secretariat. 2019. Available online: https://unfccc.int/documents/194891 (accessed on 30 August 2022).
39. Environmental Protection Agency. Ireland's National Inventory Report 2019. *Greenhouse Gas Emissions 1990–2017. Submission to the UNFCCC Secretariat.* 2019. Available online: https://unfccc.int/documents/194638 (accessed on 30 August 2022).
40. Tavakoli, A. A journey among top ten emitter country, decomposition of 'Kaya Identity. *Sustain. Cities Soc.* **2018**, *38*, 254–264. [CrossRef]
41. Sgouridis, S.; Bonnefoy, P.A.; Hansman, R.J. Air transportation in a carbon constrained world: Long-term dynamics of policies and strategies for mitigating the carbon footprint of commercial aviation. *Transp. Res. Part A Policy Pract.* **2011**, *45*, 1077–1091. [CrossRef]
42. Zaporozhets, O.; Isaienko, V.; Synylo, K. Trends on current and forecasted aircraft hybrid electric architectures and their impact on environment. *Energy* **2020**, *211*, 118814. [CrossRef]

43. CAAT—The Civil Aviation Authority of Thailand, State of Thai Aviation Industry, 2019 report. 2019. Available online: https://www.caat.or.th/wp-content/uploads/2020/06/STATE-OF-THAI-AVIATION-INDUSTRY-2019.pdf (accessed on 24 May 2022).
44. Ozturk, I.; Al-Mulali, U.; Saboori, B. Investigating the environmental Kuznets curve hypothesis: The role of tourism and ecological footprint. *Environ. Sci. Pollut. Res.* **2015**, *23*, 1916–1928. [CrossRef]
45. CAAT—The Civil Aviation Authority of Thailand. Announcement of the Civil Aviation Authority of Thailand: Guidelines for Airport Operators and Air Operators on Domestic Routes during the Coronavirus Disease 2019 Epidemic Situation. 2020. Available online: https://www.caat.or.th/th/archives/59631 (accessed on 24 May 2022).
46. CAAT—The Civil Aviation Authority of Thailand. *State of Thai Aviation Industry, 2020 Report*; CAAT: Bangkok, Thailand, 2020.
47. DEDE—Department of Alternative Energy Development and Efficiency. Energy Statistics, Energy Balance of Thailand. 2019. Available online: https://webkc.dede.go.th/testmax/sites/default/files/Energy_Balance_of_Thailand_2019.pdf (accessed on 24 May 2022).
48. TGO—Thailand Greenhouse Gas Management Organization. Greenhouse Gas Management Organization Strategic Plan 2018–2022. 2019. Available online: http://www.tgo.or.th/2020/index.php/en/page/tgo-strategy-plan (accessed on 24 May 2022).
49. Narciso, M.; Melo de Sousa, J.M. Influence of Sustainable Aviation Fuels on the Formation of Contrails and Their Properties. *Energies* **2021**, *14*, 5557. [CrossRef]
50. Baxter, G. Mitigating Aircraft Auxiliary Power Unit Carbon Dioxide (CO_2) Emissions During the Aircraft Turnaround Process from the Use of Solar Power at the Airport Gate: The Case of Moi International Airport, Kenya. *Int. J. Environ. Agric. Biotechnol.* **2022**, *7*, 14–22. [CrossRef]
51. PTTOR—Petroleum Authority of Thailand Oil and Retail Business Public Company Limited. Aviation Jet Fuel Prices Report. 2020. Available online: https://www.pttplc.com/en/Products/Ourbusinessbyaffiliates/Oilandretailbusiness.aspx (accessed on 5 February 2023).
52. IPCC—Intergovernmental Panel on Climate Change. Guidelines for National Greenhouse Gas Inventories. Available online: https://www.ipcc.ch/report/2006-ipcc-guidelines-for-national-greenhouse-gas-inventories/ (accessed on 26 July 2022).
53. BOT—Bank of Thailand. Rates of Exchange of Commercial Banks. 2021. Available online: https://www.bot.or.th/english/_layouts/application/exchangerate/exchangerate.aspx (accessed on 24 May 2022).
54. The World Bank. GDP (Current US$)—Thailand | Data. 2021. Available online: https://data.worldbank.org/indicator/NY.GDP.MKTP.CD?locations=TH (accessed on 24 May 2022).
55. Yin, L.; Yao, T.; Zhou, J.; Liu, G.; Liao, Y.; Ma, X. Prediction of CO_2 Emissions Based on Multiple Linear Regression Analysis. *Energy Procedia* **2017**, *105*, 4222–4228. [CrossRef]
56. Zhou, Y.; Zhang, J.; Hu, S. Regression analysis and driving force model building of CO_2 emissions in China. *Sci. Rep.* **2021**, *11*, 6715. [CrossRef]
57. Maaouane, M.; Zouggar, S.; Krajačić, G.; Zahboune, H. Modelling industry energy demand using multiple linear regression analysis based on consumed quantity of goods. *Energy* **2021**, *225*, 120270. [CrossRef]
58. Geem, Z.W.; Roper, W.E. Energy demand estimation of South Korea using artificial neural network. *Energy Policy* **2009**, *37*, 4049–4054. [CrossRef]
59. Parikh, J.; Purohit, P.; Maitra, P. Demand projections of petroleum products and natural gas in India. *Energy* **2007**, *32*, 1825–1837. [CrossRef]
60. Blakey, S.; Rye, L.; Wilson, C.W. Aviation gas turbine alternative fuels: A review. *Proc. Combust. Inst.* **2011**, *33*, 2863–2885. [CrossRef]
61. Williams, P.I.; Allan, J.D.; Lobo, P.; Coe, H.; Christie, S.; Wilson, C.; Hagen, D.; Whitefield, P.; Raper, D.; Rye, L. Impact of alternative fuels on emissions characteristics of a gas turbine engine—Part 2: Volatile and semivolatile particulate matter emissions. *Environ. Sci. Technol.* **2012**, *46*, 10812–10819. [CrossRef]
62. Corporan, E.; Edwards, T.; Shafer, L.; Dewitt, M.J.; Klingshirn, C.; Zabarnick, S.; West, Z.; Striebich, R.; Graham, J.; Klein, J. Chemical, Thermal Stability, Seal Swell, and Emissions Studies of Alternative Jet Fuels. *Energy Fuels* **2011**, *25*, 955–966. [CrossRef]
63. Murgan, M.; Hamid, A.; Mustapha, M. Rethinking the potency of ICAO SARPS on global reduction of aviation emission and protection of global environment. *Bratisl. Law Rev.* **2017**, *1*, 28–37. [CrossRef]
64. DEDE—Department of Alternative Energy Development and Efficiency. The Study of Sustainable Biojet Promotion Plan for Thailand. 2020. Available online: https://www.statista.com/statistics/655057/fuel-consumption-of-airlines-worldwide/ (accessed on 15 August 2022).
65. ICAO—International Civil Aviation Organization. Sustainable Aviation Fuels Guide Version 2. 2018. Available online: https://www.icao.int/environmental-protection/Documents/Sustainable%20Aviation%20Fuels%20Guide_100519.pdf (accessed on 1 September 2022).
66. European Commission. Regulation of the European Parliament and of the Council on Ensuring a Level Playing Field for Sustainable Air Transport. 2021. Available online: https://ec.europa.eu/transport/themes/mobilitystrategy_en (accessed on 1 September 2022).
67. Keramidas, K.; Diaz-Vazquez, A.R.; Vandyck, T.; Rey Los Santos, L.; Schade, B.; Soria-Ramirez, A. Global Energy and Climate Outlook 2018: Sectoral Mitigation Options towards a Low-Emissions Economy. 2018. Available online: https://op.europa.eu/en/publication-detail/-/publication/edff2046-f2c1-11e8-9982-01aa75ed71a1/language-en (accessed on 16 August 2022).

68. Kahn Ribeiro, S.; Kobayashi, S.; Beuthe, M.; Gasca, J.; Greene, D.; Lee, D.S.; Muromachi, Y.; Newton, P.J.; Plotkin, S.; Sperling, D.; et al. Transport and Its Infrastructure. In *Climate Change 2007: Mitigation, Contribution of Working Group III to the Fourth Assessment Report of the Intergovernmental Panel on Climate Change*; Cambridge University Press: Cambridge, UK, 2007. Available online: https://archive.ipcc.ch/publications_and_data/ar4/wg3/en/ch5.html (accessed on 4 September 2022).
69. OECD—The Organisation for Economic Co-Operation and Development. Transport Outlook 2012: Seamless Transport for Greener Growth. 2012. Available online: https://www.itf-oecd.org/transport-outlook-2012-seamless-transport-greener-growth (accessed on 16 January 2023).
70. ICAO—International Civil Aviation Organization. Environmental Report Aviation and Climate Change. 2010. Available online: https://www.icao.int/environmental-protection/Documents/Publications/ENV_Report_2010.pdf (accessed on 4 September 2022).
71. World Bank. *State and Trends of Carbon Pricing 2014*; The World Bank: Washington, DC, USA, 2014. [CrossRef]
72. Stern, N. *The Stern Review: The Economics of Climate Change*; Cambridge University Press: Cambridge, UK, 2007. Available online: https://www.cambridge.org/core/books/economics-of-climate-change/A1E0BBF2F0ED8E2E4142A9C878052204 (accessed on 17 August 2022).
73. Baranzini, A.; Van den Bergh, J.C.J.M.; Carattini, S.; Howarth, R.B.; Padilla, E.; Roca, J. Carbon pricing in climate policy: Seven reasons, complementary instruments, and political economy considerations. *Wiley Interdiscip. Rev. Clim. Chang.* **2017**, *8*, e462. [CrossRef]
74. Shen, X.J.; Liu, B.H.; Zhou, D.W. Spatiotemporal changes in the length and heating degree days of the heating period in Northeast China. *Meteorol. Appl.* **2017**, *24*, 135–141. [CrossRef]
75. IPCC—Intergovernmental Panel on Climate Change. *Guidelines for National Greenhouse Gas Inventories, Prepared by the National Greenhouse Gas Inventories Programme*; IGES: Hayama, Japan, 2006. Available online: https://www.ipcc-nggip.iges.or.jp/support/Primer_2006GLs.pdf (accessed on 24 May 2022).
76. IPCC—Intergovernmental Panel on Climate Change. Anthropogenic and Natural Radiative Forcing. 2018. Chapter 8. Available online: https://www.ipcc.ch/site/assets/uploads/2018/02/WG1AR5_Chapter08_FINAL.pdf (accessed on 24 May 2022).
77. IPCC—Intergovernmental Panel on Climate Change. The Earth's Energy Budget, Climate Feedbacks and Climate Sensitivity. In *Climate Change 2021–The Physical Science Basis: Working Group I Contribution to the Sixth Assessment Report of the Intergovernmental Panel on Climate Change*; Cambridge University Press: Cambridge, UK, 2021. [CrossRef]
78. OECD—The Organisation for Economic Co-Operation and Development. Economic Surveys Economic Assessment. 2020. Available online: www.oecd.org/economy/thailand-economic-snapshot/ (accessed on 24 May 2022).
79. CAAT—The Civil Aviation Authority of Thailand. The Report Predicts the Demand for Air Travel of the Country. 2020. Available online: https://www.caat.or.th/th/archives/53358 (accessed on 24 May 2022).
80. Guo, X.; Ning, C.; Shen, Y.; Yao, C.; Chen, D.; Cheng, S. Projection of the Co-Reduced Emissions of CO_2 and Air Pollutants from Civil Aviation in China. *Sustainability* **2023**, *15*, 7082. [CrossRef]
81. Hasan, M.A.; Frame, D.J.; Chapman, R.; Archie, K.M. Curbing the car: The mitigation potential of a higher carbon price in the New Zealand transport sector. *Clim. Policy* **2020**, *20*, 563–576. [CrossRef]
82. Rahman, S.M.; Khondaker, A.N.; Hasan, M.A.; Reza, I. Greenhouse gas emissions from road transportation in Saudi Arabia—A challenging frontier. *Renew. Sustain. Energy Rev.* **2017**, *69*, 812–821. [CrossRef]

energies

Article

Study on the Dynamic Combustion Characteristics of a Staged High-Temperature Rise Combustor

Meng Li [1,2,3], Jinhu Yang [4,*], Cunxi Liu [2,3], Fuqiang Liu [2,3,4], Kaixing Wang [2,3], Changlong Ruan [2,3,4], Yong Mu [2,3,4] and Gang Xu [2,3]

1 Fluid Machinery Engineering Technology Research Center, Jiangsu University, Zhenjiang 212013, China; limeng_ujs@163.com
2 Institute of Engineering Thermophysics, Chinese Academy of Sciences, Beijing 100190, China; liucunxi@iet.cn (C.L.); liufuqiang@iet.cn (F.L.); wangkaixing@iet.cn (K.W.); ruanchanglong@iet.cn (C.R.); muyong@iet.cn (Y.M.); xugang@iet.cn (G.X.)
3 National Key Laboratory of Science and Technology on Advanced Light-Duty Gas-Turbine, Beijing 100190, China
4 Qingdao Aeronautical Technology Research Institute, Qingdao 266000, China
* Correspondence: yangjinhu@iet.cn

Abstract: Currently, steady-state analysis predominates in combustion chamber design, while dynamic combustion characteristics remain underexplored, and there is a lack of a comprehensive index system to assess dynamic combustion behavior. This study conducts a numerical simulation of the dynamic characteristics of the combustion chamber, employing a method combining large eddy simulation (LES) and Flamelet Generated Manifold (FGM). The inlet air temperature, air flow rate, and fuel flow rate were varied by 1%, 2%, and 3%, respectively, with a pulsation period of 0.008 s. The effects of nine different inlet parameter pulsations on both time-averaged and instantaneous combustion performance were analyzed and compared to benchmark conditions. The results indicate that small pulsations in the inlet parameters have minimal impact on the steady-state time-averaged performance. In the region near the cyclone outlet, which corresponds to the flame root area, pronounced unsteady flame characteristics were observed. Fluctuations in inlet parameters led to an increase in temperature fluctuations near the flame root. Analysis of the outlet temperature results for each operating condition reveals that inlet parameter fluctuations can mitigate the inherent combustion instability of the combustion chamber and reduce temperature fluctuations at the outlet hot spot.

Keywords: large eddy simulation; staged combustion chamber; dynamic combustion characteristics

1. Introduction

The combustion process in the main combustion chamber of an aero-engine is inherently dynamic. Under steady operating conditions, the physical quantities and their spatial distribution within the combustion chamber fluctuate over time. Under transient operating conditions, the physical fields in the combustion chamber exhibit dynamic variations. As performance requirements for both military and civil aircraft continue to evolve, the design of hot-end components is increasingly focused on precision, with the performance indices now extending beyond steady-state parameters to include both steady-state and dynamic performance metrics. However, research on the dynamic combustion characteristics of the combustion chamber remains insufficient, and a comprehensive index system for evaluating these dynamic characteristics is lacking. To address this gap, this study

https://doi.org/10.3390/en18030662

presents a numerical simulation of the dynamic combustion characteristics of the combustion chamber, providing both data and methodological support for future fine-tuned design of combustion chambers based on dynamic combustion behavior [1].

With advancements in laser and combustion diagnostic technologies, detailed studies of dynamic combustion within combustion chambers have become increasingly feasible [2–5]. Fu et al. [6] utilized a high-speed camera and laser to measure OH* Chemiluminescence (CL), CH_2O, and SO_2-PLIF at a frequency of 20 kHz to capture the two-dimensional structure and heat release distribution of the flame. They observed that fluctuations in fuel mass flow rate fluctuation could extend blowoff limits. Stohr et al. [7,8] investigated the Lean Blowout (LBO) characteristics of partially premixed combustors using synchronized stereo Particle Image Velocimetry (PIV) and OH Planar Laser-Induced Fluorescence (PLIF) techniques. Their findings revealed that the flame root exhibits inherent instability near the LBO, characterized by frequent extinction and reignition events. These results underscore the critical role of the flame root and suggest that targeted modifications to the flow field or mixture fraction in this region could potentially extend the LBO limit. Zhao et al. [9] employed CH* chemiluminescence high-speed imaging technology to visualize spray flames in a multi-swirl staged combustor and applied Fast Fourier Transform (FFT) and Proper Orthogonal Decomposition (POD) methods to analyze high-speed imaging data collected at various time points prior to flame extinction. Their results identified key precursor phenomena leading to flame extinction. Broda et al. [10] utilized CH* chemiluminescence imaging to capture unstable flame dynamics in their experiments and investigated the impact of inlet temperature variations on the flame morphology of a premixed swirl burner. The experimental results demonstrated that, as the inlet temperature gradually increased beyond a critical threshold, the flame became unstable, with a shortened flame length that eventually entered the corner vortex recirculation zone. Guiberti and Boyette [11,12] conducted a series of experimental studies on high-pressure turbulent non-premixed jet flames, examining the variations in flame lift-off height and flame length with pressure at different inlet Reynolds numbers. Their findings indicated that an increase in pressure enhanced the stability of jet flames at high Reynolds numbers. Additionally, they observed that, as pressure increased, the flame reaction zone became thinner, and the flame front exhibited significantly enhanced wrinkling.

In numerical simulations, although the Reynolds-Averaged Navier–Stokes (RANS) method is widely used, it is unable to capture the detailed characteristics of complex transient flows. In contrast, LES can directly resolve the large-scale structures within the flow, accurately model the combustion field in the combustion chamber, and capture the transient behavior of the flow. LES has become a critical tool for dynamic combustion research [13–17]. Xu Jianguo et al. [18] conducted a large eddy simulation study on methane-air turbulent combustion in a dual-swirl combustor. They found that the recirculation zone velocity increased with pressure, the flame structure became more compact at higher pressures, and both the Precessing Vortex Core (PVC) frequency and the outlet gas temperature significantly increased with rising pressure. Nan Meng et al. [19] investigated the effects of flame tube structure on flame propagation and pressure fluctuations in the combustion field using LES and Power Spectral Density (PSD) coupling. Their results demonstrated that the presence of primary combustion and mixing holes in the combustion chamber increased the amplitude of pressure oscillations and heat release rate, with peak values occurring in the shear layer and vortex structure regions. Lv et al. [20] employed the FGM combustion model in combination with a Weighted-Model Large Eddy Simulation (WMLES) to examine the dynamic combustion characteristics of a coaxial staged combustor, varying the equivalence ratio of the pilot stage by adjusting the pilot fuel flow rate. Chen et al. [21] used LES to study the dynamic flame structure during stable and unstable combustion in a dual-swirl combustor, accurately capturing various flow dynamics induced by PVC and thermoacoustic oscillations. They identified the stagnation point caused by PVC as a critical factor in flame stabilization.

At present, most of the research on dynamic combustion explores the influence of combustion chamber structure and large-scale changes in inlet parameters on dynamic combustion. However, the influence of engine inlet parameter pulsations or speed fluctuations on the pulsation characteristics of the combustion chamber under stable operating conditions remains underexplored. Therefore, this paper presents a large eddy simulation study of a staged combustion chamber, coupled with the Flamelet Generated Manifold combustion model, to investigate the effects of inlet air flow, temperature, and fuel flow pulsations on dynamic combustion behavior. The findings of this study provide both data and methodological support for the future design optimization of combustion chambers based on dynamic combustion characteristics.

2. Computational Models and Methods

2.1. Physical Model and Meshing

This study investigates a single-head fan-shaped model of a three-swirl high-temperature rise combustor, as illustrated in Figure 1. The model consists primarily of a casing, head swirler, nozzle, and flame tube. Air enters the flame tube through the head swirler, the primary holes, the dilution holes, and the cooling holes, creating the required flow field structure and outlet temperature distribution for combustion. Specifically, 45.05% of the air enters through the head of the flame tube, forming a recirculation zone that organizes the combustion process. A total of 17% of the air enters through the primary holes, with part of it participating in the combustion within the main combustion zone, while the remainder contributes to the initial mixing of the combustion chamber's temperature field. Another 17% of the air enters through the dilution holes, where it mixes with the upstream-burned high-temperature gas to achieve a uniform temperature distribution suitable for turbine acceptance. The remaining 21.05% of the air enters through the cooling holes, where it provides protection to the flame tube via adherent film cooling. There are 4 primary holes, 4 dilution holes, and 1655 cooling holes. The design parameters of the swirler are provided in Table 1. In the swirler design, the two-stage fuel nozzles consist of centrifugal nozzles and multi-point injection air atomizing nozzles, arranged from the innermost to the outermost stages. The pre-combustion fuel is discharged through the centrifugal nozzle and atomized by the air shear generated by the first and second-stage cyclones. The main combustion fuel is released through small holes in the wall of the third-stage cyclone. The fuel distribution between the pre-combustion and main combustion stages is 30% and 70%, respectively. The swirl number of the three-stage axial cyclones increases from the innermost to the outermost stages. The rotation directions are clockwise for the first stage, counterclockwise for the second, and counterclockwise for the third.

Due to the dense, multi-inclined hole design of the flame tube cooling holes, unstructured grids are employed to mesh the cyclone, combustion zone, and cooling holes, thereby reducing the overall grid count. This approach facilitates the meshing of the cyclone and flame tube, enabling accurate capture of the dynamic combustion process. Structured grids are used for the remaining areas.

Table 1. Design characteristics of the triple swirler.

	Swirler Vane Angle	Swirl Number	Air Split Ratio
1st swirler	35°	0.59	13.1%
2nd swirler	−45°	0.86	29.8%
3rd swirler	−50°	1.06	57.1%

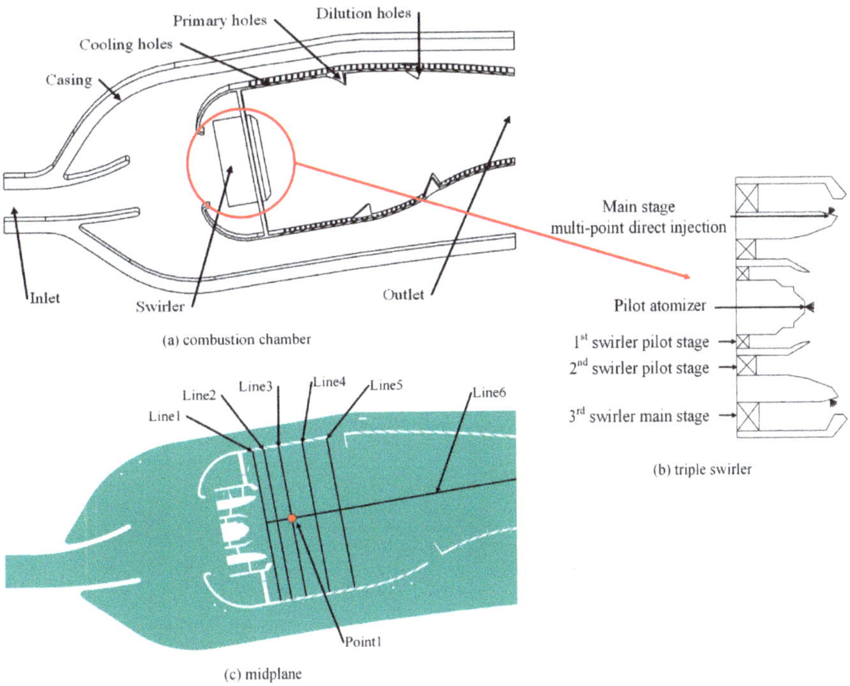

Figure 1. Schematic of the combustion chamber: (**a**) combustion chamber; (**b**) triple swirler; (**c**) midplane.

Three sets of grids with varying numbers of elements were generated using the same method, and the axial velocity profile at Line 2 was selected to assess grid independence. The results are presented in Figure 2. After comparison, it was determined that the grid with 17.89 million elements was the most suitable. The grid distribution for the cross-section and the swirler within the combustion chamber is shown in Figure 3. The grid size for the flame tube and the head cooling hole is set to 0.2 mm, while the maximum grid size overall is 1.6 mm. The final grid contains 17.89 million elements.

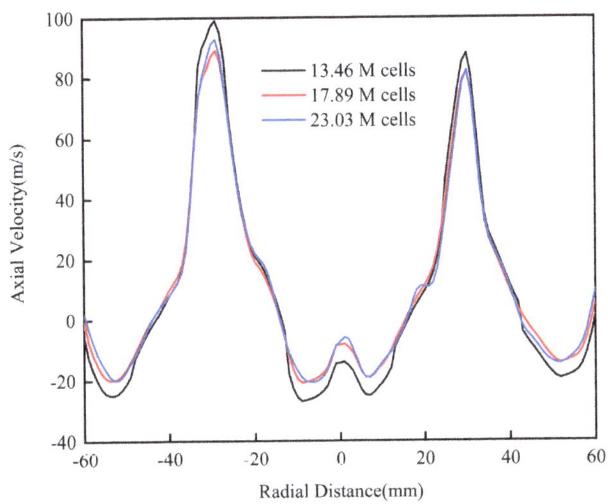

Figure 2. Grid independence verification.

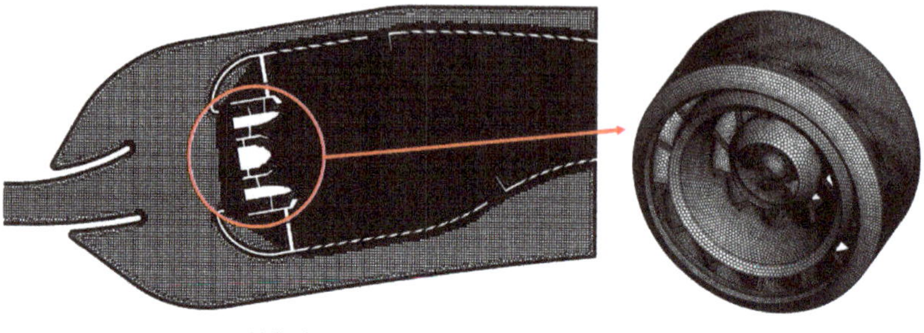

Figure 3. Mesh distributions: (**a**) combustor; (**b**) swirler.

2.2. Calculation Methods and Operating Conditions

In this paper, large eddy simulation is used for turbulence calculation. The filtered control equation is as follows:

$$\frac{\partial \overline{\rho}}{\partial t} + \frac{\partial \overline{\rho}\widetilde{u}_i}{\partial x_i} = 0 \quad (1)$$

$$\frac{\partial \overline{\rho}\widetilde{u}_i}{\partial t} + \frac{\partial (\overline{\rho}\widetilde{u}_i\widetilde{u}_j)}{\partial x_j} = \frac{\partial \sigma_{ij}}{\partial x_j} - \frac{\partial \overline{p}}{\partial x_j} - \frac{\partial \tau_{ij}^{sgs}}{\partial x_j} \quad (2)$$

$$\frac{\partial \overline{\rho}\widetilde{h}}{\partial t} + \frac{\partial (\overline{\rho}\widetilde{u}_j\widetilde{h})}{\partial x_j} = \frac{D\overline{p}}{Dt} - \frac{\partial}{\partial x_j}(\overline{\rho}\widetilde{u_j h} - \overline{\rho}\widetilde{u}_j\widetilde{h}) + \frac{\partial}{\partial x_j}\left(\overline{\rho}\alpha\frac{\partial \widetilde{h}}{\partial x_j}\right) + \overline{\tau_{ij}\frac{\partial u_i}{\partial x_j}} \quad (3)$$

$$\frac{\partial \overline{\rho}\widetilde{Y}_k}{\partial t} + \frac{\partial (\overline{\rho}\widetilde{Y}_k\widetilde{u}_j)}{\partial x_j} = \frac{\partial}{\partial x_j}\left(\overline{\rho}\widetilde{D}_k\frac{\partial \widetilde{Y}_k}{\partial x_j}\right) - \frac{\partial}{\partial x_j}(\overline{\rho}\widetilde{u_j Y_k} - \overline{\rho}\widetilde{u}_j\widetilde{Y}_k) + \overline{\omega}_k \quad (4)$$

The Wale model is used to close the sub-grid stress, the combustion chamber is partially premixed combustion, and the FGM model is used to close the chemical reaction source term. The fuel is represented by n-decane, with a skeletal mechanism containing 77 components and 359 reactions [22]. The SIMPLE algorithm is employed for the numerical solution. The near-wall region is handled using the standard wall function. Fuel droplets are simulated using the Discrete Phase Model (DPM). The main combustion stage nozzle is a hollow cone, while the pre-combustion stage nozzle is a solid cone. The side walls of the fluid domain in the single-head annular combustor are modeled with a rotating periodic boundary condition, with a rotation angle of 18°.

Due to the detailed computational grid required by LES, the time step must be kept relatively small. The maximum time step for LES calculations must satisfy the following condition:

$$\Delta t \leq \frac{\Delta x}{U} \quad (5)$$

Δx represents the grid size, and U denotes the average flow velocity within the grid. Taking into account the difference between the instantaneous velocity and the average velocity, it is assumed that the Courant number is less than 0.5, which satisfies the requirements for LES calculations. The calculation formula for the Courant number is as follows:

$$C_o = \frac{U\Delta t}{\Delta x} \quad (6)$$

To improve computational efficiency, LES calculations are performed based on the results of the RANS calculations. A time step of 10^{-6} s is selected, and the Courant number distribution across the combustion chamber cross-section is obtained. As shown in Figure 4, the global Courant number is less than 0.5, which satisfies the requirements for LES calculations.

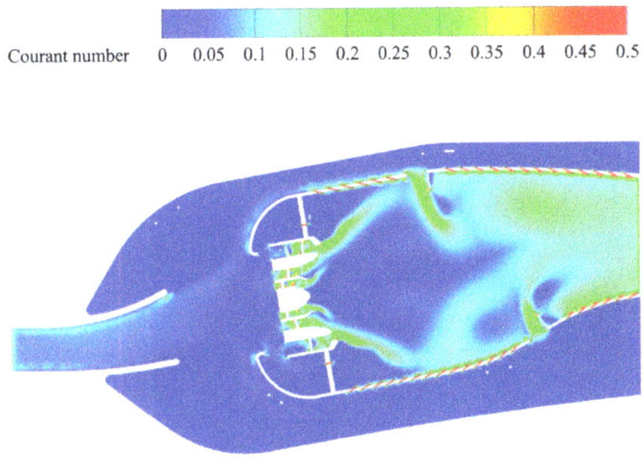

Figure 4. Courant number distribution.

The simulation conditions are summarized in Table 2. The inlet pressure of the combustion chamber is 500 kpa and the pressure drop is 6%. To investigate the impact of small pulsations in inlet parameters on the dynamic combustion characteristics of the combustion chamber, the inlet conditions are varied based on the benchmark conditions, with pulsations introduced as sine waves at amplitudes of 1%, 2%, and 3%. The pulsation frequency of the air and fuel boundary conditions, representing typical system-level pulsations, is simulated. The typical dynamic pulsation frequency of the combustion chamber is around 100 Hz, and the pulsation level does not exceed 3% of the mean value. Therefore, the pulsation period is selected as 0.008 s.

Table 2. Operating conditions.

Case	P/(kpa)	T/(K)	ma/(kg/s)	mf/(kg/s)
0	500	600	0.739	0.02733
1	500	600 ± 6	0.739	0.02733
2	500	600 ± 12	0.739	0.02733
3	500	600 ± 18	0.739	0.02733
4	500	600	0.739 ± 0.0074	0.02733
5	500	600	0.739 ± 0.0148	0.02733
6	500	600	0.739 ± 0.0222	0.02733
7	500	600	0.739	0.02733 ± 0.000273
8	500	600	0.739	0.02733 ± 0.000546
9	500	600	0.739	0.02733 ± 0.000819

2.3. Model Evaluation

To verify the accuracy of the large eddy simulation and obtain high-resolution reactive flow simulation results, grid quality is crucial. Two LES turbulence criteria proposed by Professor Pope [23] are employed to assess the quality of the simulation: the grid resolution criterion and the power spectral density of the velocity. The turbulent kinetic energy of the large-scale structures is resolved by the LES, while the turbulent kinetic energy of the small-scale structures is resolved by the sub-grid scale model. The ratio of the turbulent kinetic energy of the large-scale structures to the total turbulent kinetic energy in LES is referred to as the turbulent kinetic energy resolution percentage:

$$percentage = \frac{k_{res}}{k_{res} + k_{sgs}} \quad (7)$$

where k_{sgs} is the sub-grid turbulent kinetic energy, and k_{res} is the resolved turbulent kinetic energy. k_{res} can be obtained using the following formula:

$$k_{res} = \frac{1}{2}\left(\overline{u'u'} + \overline{v'v'} + \overline{w'w'}\right) \quad (8)$$

where u' is the root mean square error of axial velocity, v' is the root mean square error of radial velocity, and w' is the root mean square error of tangential velocity.

In this study, LES calculations are performed for the reference condition. The turbulent kinetic energy resolution ratio at the cross-section of the combustion chamber and the velocity energy spectrum at a monitoring point 20 mm downstream of the pre-combustion stage nozzle are obtained, as shown in Figure 5. According to Professor Pope, a well-converged LES should resolve at least 80% of the turbulent kinetic energy. As shown in Figure 4, the turbulent kinetic energy resolution ratio in most regions of the combustion chamber exceeds 80%, indicating that the current grid quality and solver settings meet the requirements for LES calculations. Additionally, a region exhibiting a slope close to $-5/3$ is observed in the velocity energy spectrum in Figure 6, which aligns with the Kolmogorov turbulence theory [24]. This suggests that the current grid quality and solver settings are capable of capturing the multi-scale characteristics of the flow field.

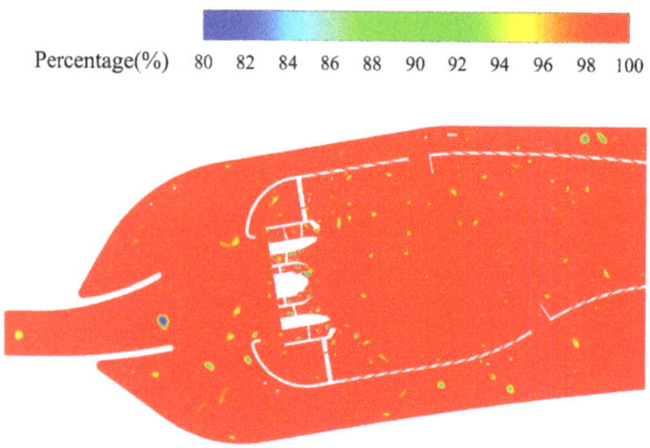

Figure 5. Percentage of resolved turbulent kinetic energy.

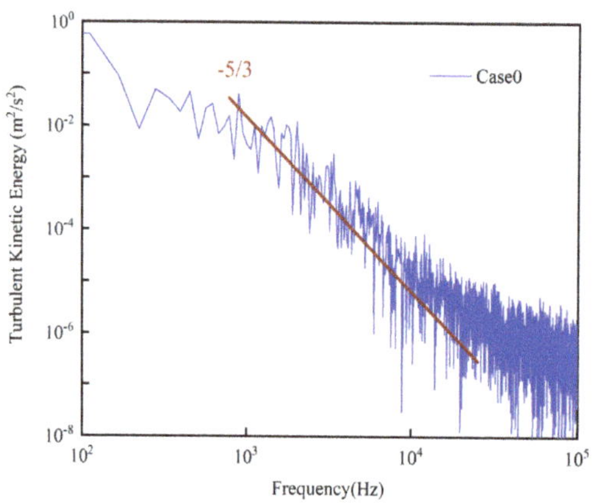

Figure 6. Kinetic energy spectra.

To further validate the accuracy of the large eddy simulation method, both experiments and simulations were conducted for the multi-swirl staged combustion chamber. The numerical approach and model used are consistent with those presented in this paper. The experimental methodology is detailed in Ref. [25]. The multi-swirl staged combustion chamber adopts the centrally staged method, and the injector consists of a pilot stage and a main stage. The pilot stage consists of two counter-rotating axial swirlers and an air atomizing nozzle. The main stage contains one axial swirler and one radial swirler. Plain-orifice atomizers are equidistantly arranged on the sidewall of the 4th swirler. Figure 7 shows the distribution of the average axial velocity. The left side is the experimental result, and the right side is the simulation result. The red solid line marks the location where the average axial velocity is zero. From the comparison, it is evident that typical swirl flow field features, such as the recirculation zone and high-speed shear layers, are present in the combustion chamber. The size and position of the recirculation zone in the simulation results are in good agreement with the experimental observations, and the variation of the average axial velocity within the combustion chamber is accurately captured. These verification results demonstrate that the meshing and numerical simulation methods for the staged combustion chamber are reliable.

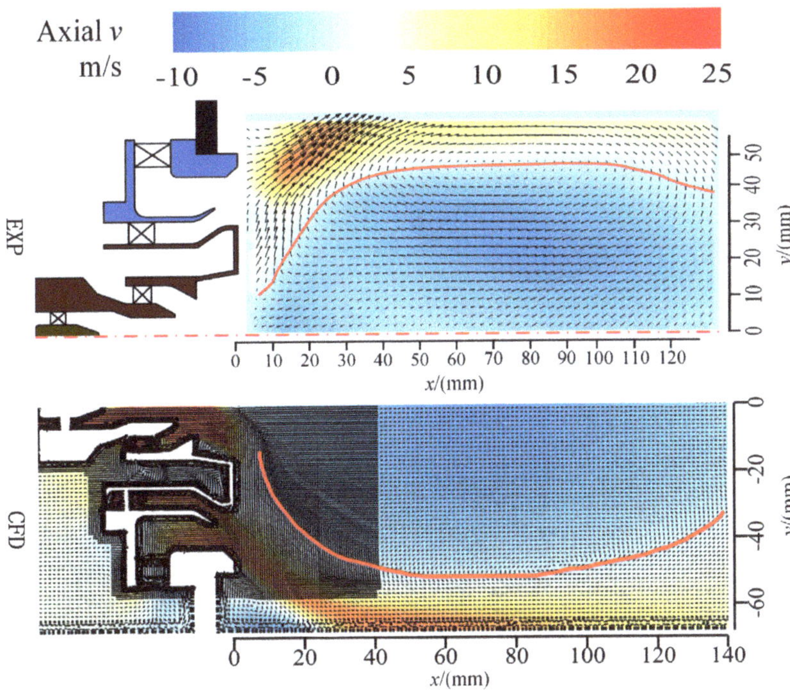

Figure 7. Comparison of experimental results and simulation results of average velocity field.

3. Results

Based on the reference condition (case 0), the inlet temperature, air flow rate, and fuel flow rate pulsate in the form of a sine wave by 1%, 2%, and 3%, respectively, and the pulsation period is 0.008 s. The results are compared to analyze the influence of small pulsations of inlet parameters on flow field characteristics, local high-temperature zone evolution, and outlet temperature distribution.

3.1. Time-Averaged Analysis

The characteristics of the recirculation zone are critical indicators of combustion performance within the combustion chamber. Figure 8 presents the time-averaged velocity field of the combustion chamber under reference conditions. The left panel shows the axial velocity field, while the right panel depicts the radial velocity field. Due to the combined effects of the three-stage swirler, a distinct confined swirl flow is generated downstream of the swirler head, exhibiting typical swirl combustion chamber flow features, including pilot swirl jets (Pilot SWJs), main swirl jets (Main SWJs), the central toroidal recirculation zone (CTRZ), the corner recirculation zone (CRZ), and the shear layer (SL). The red line in the axial velocity field indicates the location where the axial velocity is zero. A large-scale, low-velocity zone is formed downstream of the cyclone outlet, which promotes flame stability.

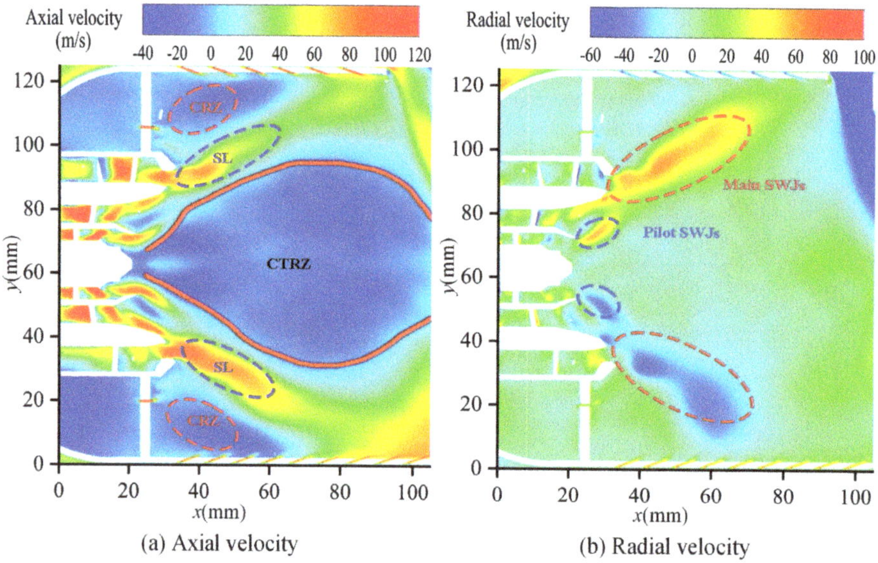

Figure 8. Time-averaged velocity field (Case0): (**a**) axial velocity; (**b**) radial velocity.

Figure 9 illustrates the axial distribution of the time-averaged axial velocity within the combustor under inlet temperature fluctuations, as well as the radial distribution curves at various axial positions along the flame tube. The dashed line indicates the point at which the axial velocity is zero. The position distribution of Line1–Line5 can be seen in Figure 1. From the figure, it is evident that the time-averaged axial velocity distributions of the combustion chamber under different temperature pulsations exhibit similar trends. The axial velocity distribution along the axial direction shows that the axial length of the recirculation zone remains relatively constant under different operating conditions, and the axial velocity profile within the main recirculation zone exhibits minimal variation.

The radial distribution curve of axial velocity at different axial positions reveals multiple velocity peaks at Line 1, which are attributed to the outlet velocities from the first, second, and third-stage cyclones, as well as the head cooling hole. Among these, the outlet velocity from the third-stage cyclone is the largest. As the axial distance increases, the velocity differences gradually diminish, and the radial distribution of axial velocity adopts an 'M' shape. Furthermore, as the axial distance increases, the asymmetry of the velocity distribution becomes more pronounced, with the axial velocity peak on the lower side of the flame tube significantly higher than that on the upper side. This asymmetry is primarily attributed to the structural design of the flame tube, where the number and area of cooling holes on the upper and lower sides are uneven. Additionally, the cross-sectional area of the inner ring flow is smaller than that of the outer ring, leading to an uneven distribution of air intake across the upper and lower sides of the flame tube. This results in a higher axial velocity on the lower side. At Line 4, the radial extent of the main recirculation zone reaches its maximum. The influence of intake flow and fuel flow pulsation on the steady-state flow field structure is similar, and is not analyzed further.

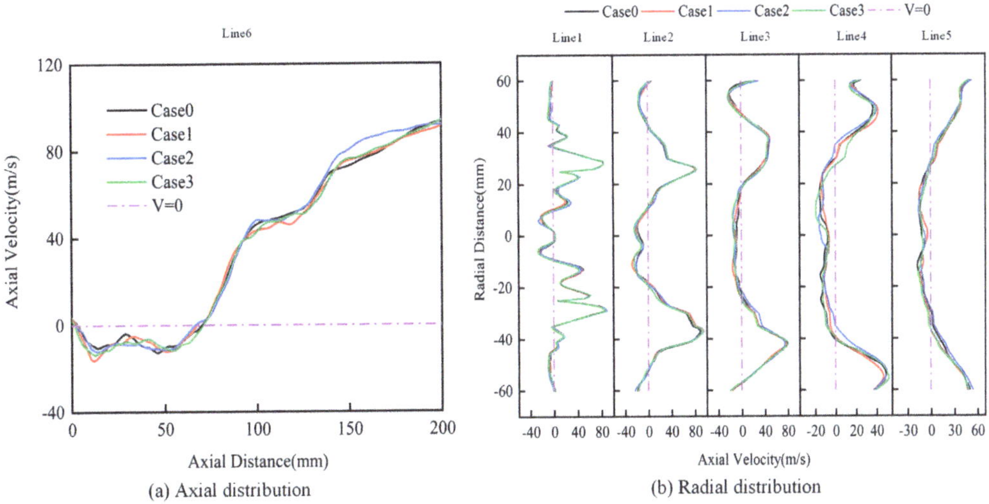

Figure 9. Axial velocity distribution under temperature fluctuation: (**a**) axial distribution; (**b**) radial distribution.

To quantitatively analyze the influence of different inlet parameters on the extent of the main recirculation zone, the length and width of the recirculation zone under various operating conditions were calculated, as presented in Table 3. From the table, it is evident that the width of the main recirculation zone under pulsating inlet conditions is smaller than that under the baseline operating conditions. Furthermore, as the degree of pulsation increases, the width of the main recirculation zone decreases. When the fuel flow pulsation reaches 3%, the width of the main recirculation zone attains its minimum value. By observing the length and width of the main recirculation zone in each operating condition, it is found that the difference is very small, and the change is not regular, which has little effect on the overall structure of the recirculation zone.

Table 3. The range of the main recirculation zone under different operating conditions.

Case	0	1	2	3	4	5	6	7	8	9
Length/mm	66	64	64	68	67	62	67	63	64	66
Width/mm	65	64	62	62	64	63	62	61	65	56

Figure 10 presents the time-averaged temperature distribution in the middle section of the combustion chamber under the reference condition (Case0). As shown in the figure, a 'W'-shaped low-temperature zone is formed at the outlet of the head swirler due to the two-stage fuel injection and the swirling airflow. This swirling airflow creates a cooling film near the cooling hole of the flame tube, which serves to protect the flame tube and enhance its service life. The jet from the main combustion hole significantly reduces the high-temperature zone in the primary combustion region. Following the mixing jet, the high-temperature zone at the outlet of the combustion chamber is substantially diminished, resulting in a more uniform temperature distribution at the outlet. The fluctuation of inlet parameters has a minimal impact on the time-averaged temperature distribution within the combustion chamber and is therefore not analyzed further.

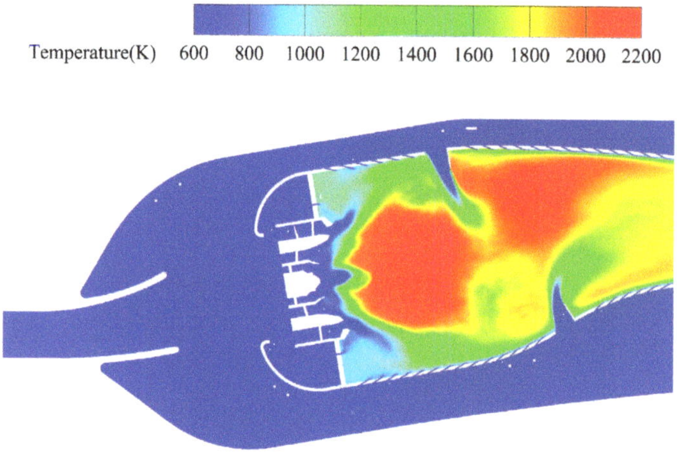

Figure 10. Time-averaged temperature distribution in the midplane of the combustor (Case0).

The uniformity of the outlet temperature distribution in the combustion chamber directly influences the service life of the turbine. The outlet temperature distribution OTDF and RTDF of each pulsating condition of the research object in this paper are analyzed. The temperature distributions for each operating condition are presented in Table 4. It is evident that the outlet temperature of the combustion chamber exhibits only slight variations across different conditions, suggesting that small pulsations in the inlet parameters have minimal impact on the time-averaged outlet temperature distribution. Compared with the reference condition, the fluctuation of inlet temperature and flow rate increases the steady-state OTDF, resulting in the increase of outlet hot spot temperature. Fuel flow pulsation has little effect on steady-state OTDF.

Table 4. Time-averaged outlet temperature distribution.

Case	0	1	2	3	4	5	6	7	8	9
RTDF	0.10	0.10	0.10	0.10	0.11	0.11	0.11	0.10	0.12	0.09
OTDF	0.17	0.20	0.19	0.18	0.20	0.20	0.21	0.16	0.18	0.16

Figure 11 presents a comparison of the average radial temperature distribution at the combustion chamber outlet under different inlet parameter pulsations. The radial temperature distribution curves across the various operating conditions exhibit a similar 'bimodal' pattern, with high-temperature regions located at the upper and lower sections of the flame tube outlet. Notably, when the inlet fuel flow pulsation is 3%, the average radial temperature distribution at the combustion chamber outlet is more uniform compared to the other operating conditions.

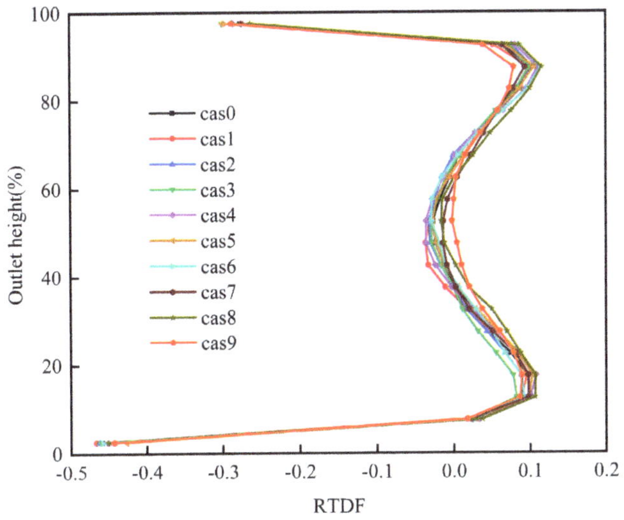

Figure 11. Comparison of time-averaged RTDF of combustion chamber outlet section.

3.2. Instantaneous Temperature Field Analysis

Figure 12 illustrates the instantaneous temperature evolution in the middle section of the combustion chamber over a cycle under inlet temperature fluctuations. The diagram reveals that the temperature distribution in the combustion chamber is similar across different operating conditions. In the main recirculation zone, the high-temperature region is more concentrated. Further downstream, the temperature fluctuations become more pronounced, primarily due to the mixing effects of the main combustion hole jet and the mixing hole jet. The mixing holes located in the rear section of the flame tube are arranged in a cross-pattern, allowing air to enter the flame tube through these holes and mix with the high-temperature gases. This mixing process helps to achieve a more uniform outlet temperature distribution. The interaction between the jet air around the primary holes and the dilution holes and the small eddy current in the combustion field inevitably causes a certain degree of temperature fluctuation.

To more intuitively illustrate the effect of inlet parameter pulsations on the temperature within the combustion chamber, the root mean square (RMS) value of the axial temperature over a period of 0.008 s is calculated. The formula is as follows:

$$T_{RMS} = \sqrt{\frac{I}{N}\sum_{i=1}^{N}(T-\overline{T})^2} \tag{9}$$

where T is the instantaneous temperature and $\overline{T}$ is the time-averaged temperature.

Figure 13 presents the axial distribution curve of temperature fluctuations over a cycle under different operating conditions. From the diagram, the temperature fluctuations are more clearly observed, with the root mean square value of the temperature being largest near the outlet of the cyclone. This is attributed to the stability of the flame in this region, which corresponds to the flame root, where unsteady characteristics are more pronounced. Fluctuations in the inlet parameters lead to enhanced temperature variations near the flame root. In the vicinity of the primary combustion hole, the root mean square value of temperature decreases overall; however, the axial temperature profile still exhibits significant fluctuations. This suggests that the fuel is influenced by the jet mixing near the primary combustion hole, which intensifies the temperature fluctuations. Conversely, near the mixing hole and towards the outlet of the flame tube, the temperature profile becomes smoother, indicating a reduction in localized high-temperature areas due to the mixing of cold and hot gases. As a result, the temperature distribution at the outlet becomes more uniform.

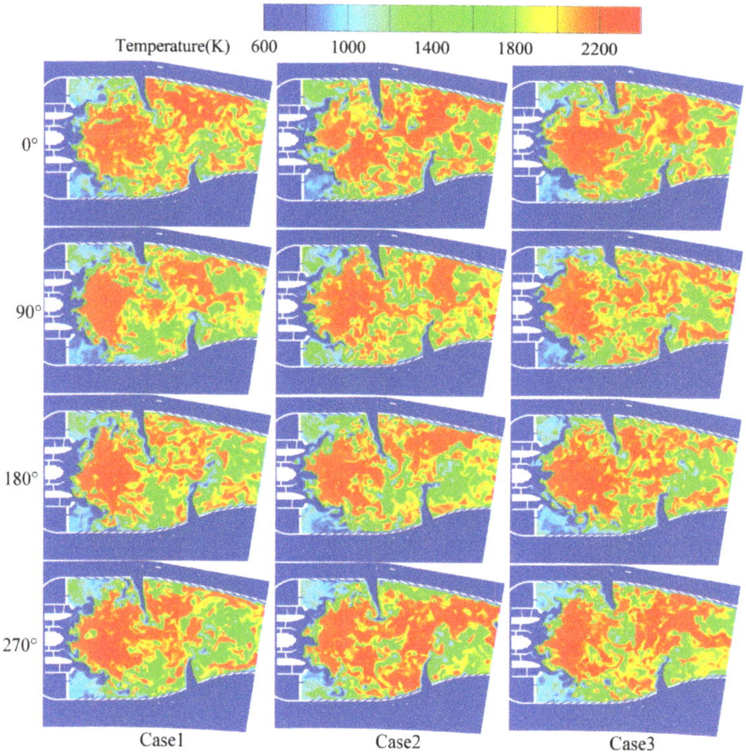

Figure 12. Instantaneous temperature evolution in the middle section of the combustor under inlet temperature fluctuation.

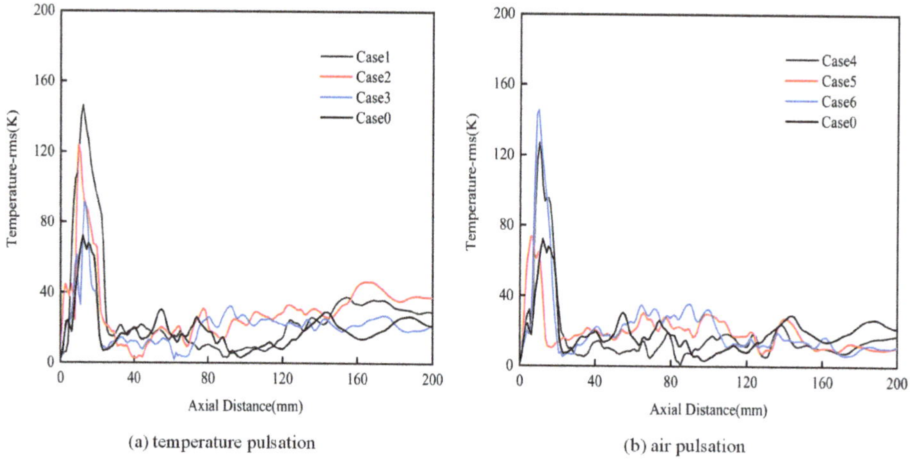

Figure 13. *Cont.*

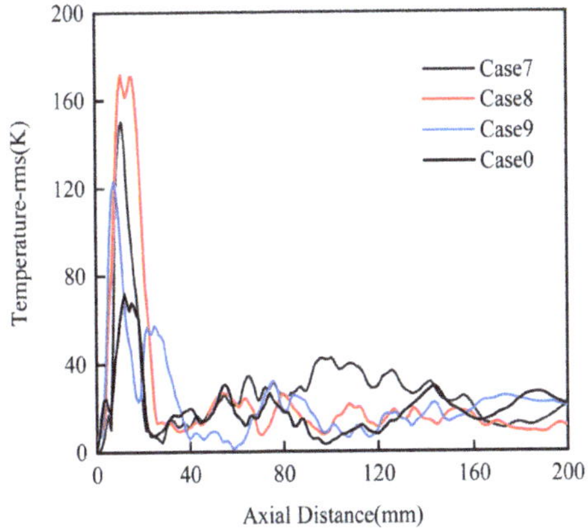

(c) fuel pulsation

Figure 13. Comparison of axial distribution of temperature fluctuation under different working conditions: (**a**) temperature pulsation; (**b**) air pulsation; (**c**) fuel pulsation.

Figure 14 shows the frequency domain diagram of heat release rate pulsations under different operating conditions. The monitoring point is point 1, within the central recirculation zone. As seen in the figure, there are slight differences in the frequency domain characteristics of the heat release rate across the various operating conditions; however, the overall trends are similar. The primary difference is observed in the peak values. When the inlet fuel flow pulsation is 3%, the peak amplitude of the heat release rate reaches a maximum value of 5.6×10^9 W·m^{-3} in each working condition. This shows that the fuel pulsation increases the unsteady pulsation characteristics of the flame, resulting in severe pulsation of the heat release rate. The primary peak frequency for each operating condition ranges from 100 to 200 Hz, which is consistent with the pulsation frequency of the inlet parameters. This suggests that the pulsations in this event are primarily governed by the inlet fluctuations.

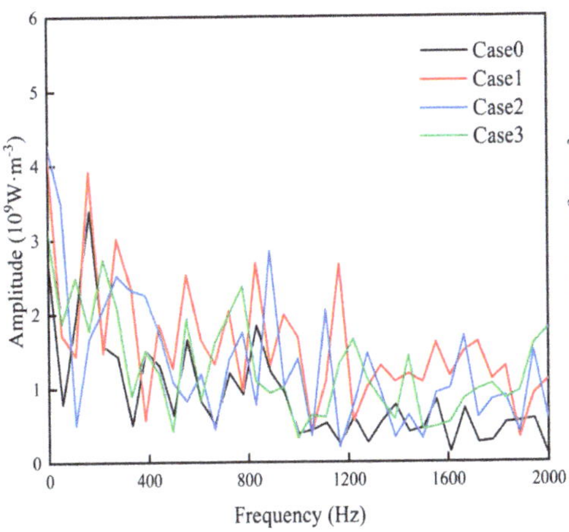

(a) temperature pulsation

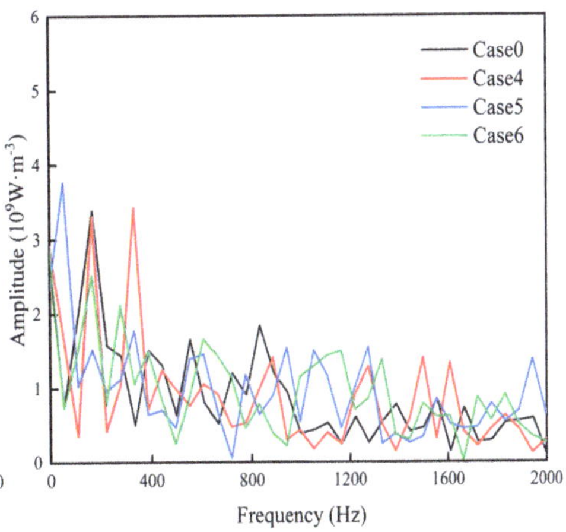

(b) air pulsation

Figure 14. *Cont.*

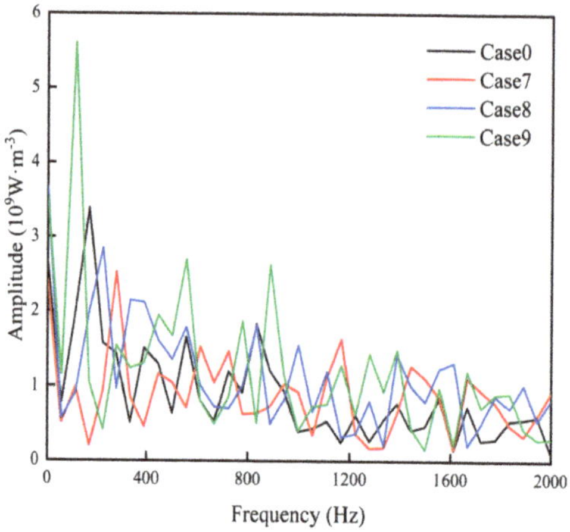

(c) fuel pulsation

Figure 14. Frequency domain diagram of heat release rate pulsation under different working conditions: (**a**) temperature pulsation; (**b**) air pulsation; (**c**) fuel pulsation.

3.3. Analysis of Outlet Temperature Fluctuation

Figure 15 illustrates the temporal variations of the inlet parameters and the average outlet temperature of the combustion chamber under different operating conditions. As shown in the figure, over a large time scale, the average outlet temperature for all conditions remains close to 1800 K, exhibiting varying degrees of pulsation over time. On a smaller time scale, the time distribution of the average outlet temperature for different inlet parameters follows a pattern similar to that of the reference condition. The key difference is that there is a phase shift, either "lead" or "lag", relative to the reference condition, which is attributed to the average temperature variation induced by fluctuations in the inlet parameters. Positive fluctuations in inlet airflow, along with negative fluctuations in inlet temperature and fuel flow, result in a decrease in the average outlet temperature, and vice versa. Additionally, the average outlet temperature displays a sine wave-like fluctuation with a "decrease-increase-decrease" trend over time, which correlates with the sine wave oscillations of the inlet parameters, particularly in the case of inlet airflow and fuel flow rate.

To quantitatively characterize the response of the average outlet temperature fluctuation to the inlet parameter fluctuations, the transfer function (*TF*) is employed to describe the system behavior. The calculation formula is given as follows:

$$TF = \frac{\hat{T}/\overline{T}}{\hat{X}/\overline{X}} = Ge^{i\Delta\varphi} \tag{10}$$

In the formula, $\hat{T}$ and $\overline{T}$ represent the fluctuation and average values of the average outlet temperature of the combustion chamber, respectively; $\hat{X}$ and $\overline{X}$ represent the fluctuation and average values of the inlet parameters (inlet temperature, air flow rate, and fuel flow rate), respectively; G denotes the gain; i is the imaginary unit; and $\Delta\varphi$ is the phase difference. The transfer functions for different operating conditions are calculated, and the corresponding gain and phase values are presented in Table 5. In general, under conditions of inlet temperature fluctuation, the amplitude of the outlet temperature fluctuation is larger, and the transfer function gain G is correspondingly higher. However, as the pulsation degree of the inlet parameters increases, the transfer function gain G exhibits a decreasing trend. This suggests that, when the pulsation magnitude is small, the inherent combustion instability of the combustion chamber amplifies the inlet pulsations, leading to a larger gain. Conversely, as the amplitude of the inlet pulsations increases, the gain does not increase proportionally. This shows that the pulsation of the inlet parameters has an inhibitory effect on the inherent combustion instability of the combustion chamber. The flow pulsation exhibits a phase difference of approximately 0.5 π, while the fuel pulsation phase difference is close to π. This is consistent with the phase difference observed between the outlet temperature fitting curve and the inlet pulsation curve in Figure 14. As the amplitude of the inlet parameter pulsations increases, the phase difference gradually decreases.

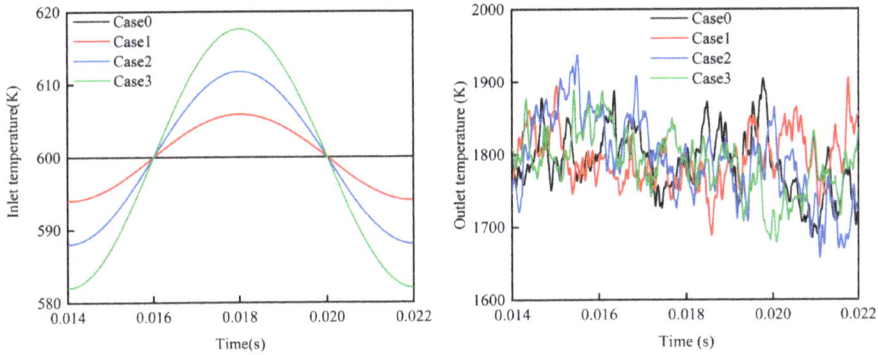

(a) temperature pulsation

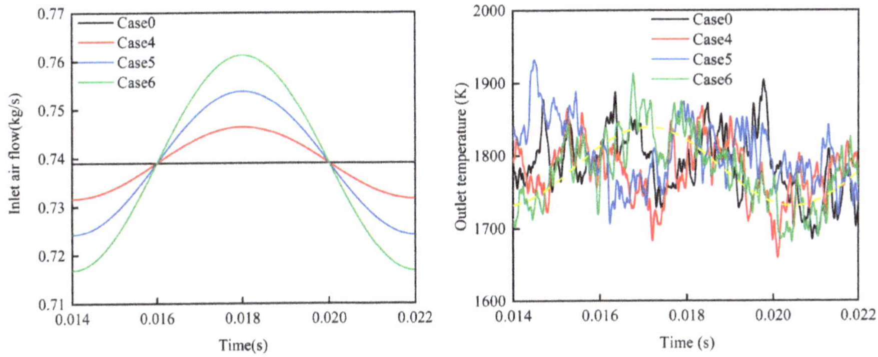

(b) air pulsation

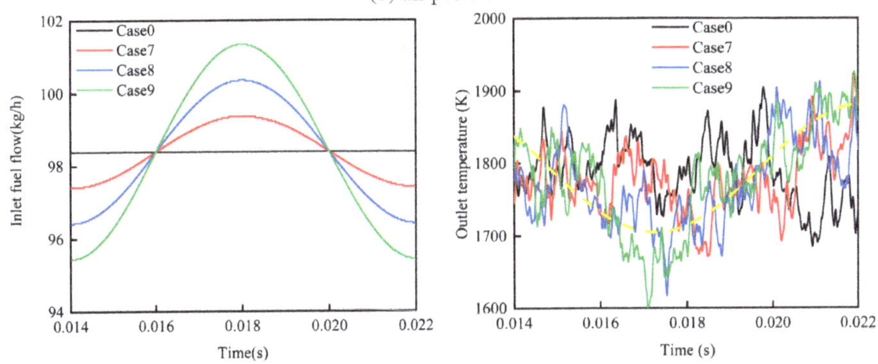

(c) fuel pulsation

Figure 15. Comparison of inlet parameters and outlet average temperature of combustion chamber with time under different operating conditions: (**a**) temperature pulsation; (**b**) air pulsation; (**c**) fuel pulsation.

Table 5. Gain and phase of outlet temperature response under different operating conditions.

Case	G	$\Delta\varphi$
1	1.51	0.92 π
2	1.37	0.45 π
3	0.76	0.48 π
4	1.68	0.44 π
5	1.12	0.62 π
6	0.99	0.25 π
7	1.77	0.99 π
8	1.63	0.80 π
9	1.56	0.82 π

Figure 16 shows the distribution of the outlet time-averaged temperature under different operating conditions, with the blue point representing the location of the maximum outlet time-averaged temperature. As observed, the outlet temperature distribution of the combustion chamber exhibits similar patterns across various operating conditions. To quantitatively analyze the impact of inlet parameter fluctuations on outlet temperature variations, the dynamic outlet temperature distribution coefficient (DOTDF) is defined. The calculation formula is as follows:

$$DOTDF = \frac{T_{4max,rms}}{T_{4ave} - T_{3ave}} \tag{11}$$

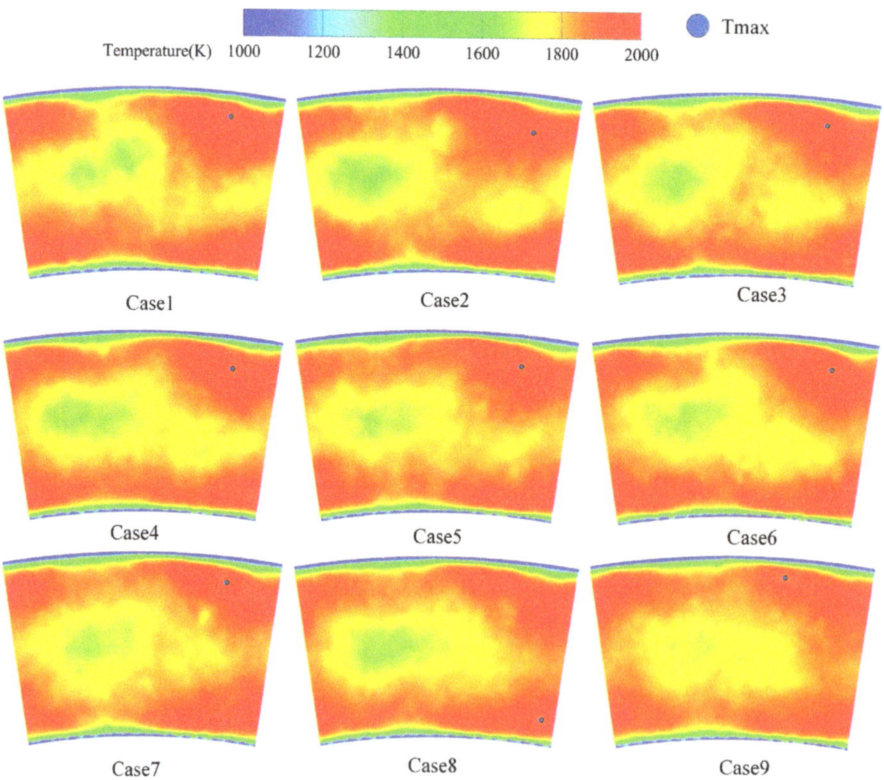

Figure 16. The outlet time-averaged temperature distribution under different operating conditions.

Here, $T_{4max,rms}$ represents the root mean square value at the location of the maximum time-averaged outlet temperature of the combustion chamber, T_{3ave} is the average inlet temperature, and T_{4ave} is the average outlet temperature of the combustion chamber. The dynamic outlet temperature distribution coefficient is calculated for each operating condition, with the results shown in Figure 17. According to the definition of the formula, a higher root mean square value for the maximum outlet temperature corresponds to a higher DOTDF, indicating a greater degree of outlet temperature fluctuation. As seen in the figure, the DOTDF for inlet parameter pulsations is lower than that for the reference condition. Additionally, under conditions involving inlet temperature and fuel flow pulsations, the DOTDF decreases with increasing pulsation amplitude. However, under airflow pulsation conditions, the DOTDF increases as the pulsation amplitude increases. This suggests that fluctuations in inlet parameters help suppress outlet hot spot temperature fluctuations, although excessive fluctuations in airflow should be avoided.

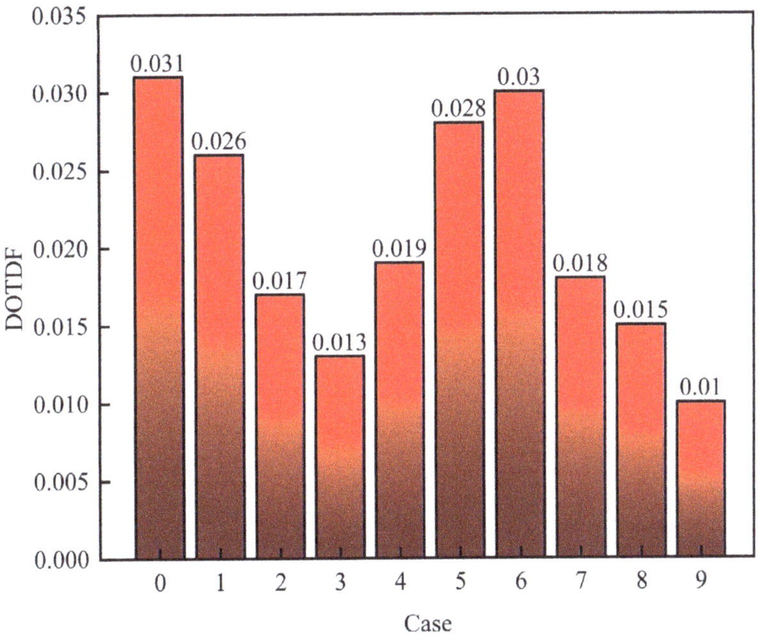

Figure 17. Distribution of DOTDF.

4. Conclusions

In this study, a method combining large eddy simulation and Flamelet Generated Manifold is employed to numerically simulate a three-swirl staged combustion chamber. The effects of nine different inlet parameter pulsations on the dynamic combustion behavior of the chamber are analyzed and compared with the reference conditions. The results are as follows:

(1) The steady-state time-averaged results of each working condition were analyzed. The three-swirl staged combustion chamber forms a large, low-speed recirculation zone that enhances flame stability. The flow field and temperature distribution for all operating conditions show minimal differences, indicating that, on a large time scale, small pulsations of the inlet parameters have no significant impact on the steady-state, time-averaged performance. Furthermore, as the pulsation intensity of the inlet parameters increases, the recirculation zone tends to expand horizontally.

(2) The transient temperature field results for each operating condition were analyzed, revealing a similar temperature distribution across all conditions. Near the outlet of the cyclone, the flame root region exhibits pronounced unsteady flame characteristics. Fluctuations in the inlet parameters lead to an increase in the temperature variations in this flame root area.
(3) The outlet temperature results for each operating condition were analyzed. It was observed that, when the inlet pulsation amplitude is small, the inherent combustion instability within the combustion chamber amplifies the inlet pulsations. As the inlet pulsation amplitude increases, the influence of inlet parameter fluctuations on the outlet pulsations becomes more pronounced, suggesting that inlet pulsations have a suppressive effect on the inherent combustion instability of the chamber. Analysis of the outlet dynamic OTDF reveals that fluctuations in the inlet parameters can mitigate the fluctuations in the outlet hot spot temperature.

Author Contributions: Conceptualization, J.Y.; methodology, M.L. and J.Y.; validation, C.L., Y.M., and G.X.; formal analysis, M.L.; investigation, M.L.; data curation M.L.; writing—original draft preparation, M.L.; writing—review and editing, C.L., Y.M., G.X., F.L., C.R., and K.W.; supervision, J.Y. and C.L.; project administration, J.Y.; funding acquisition, J.Y. All authors have read and agreed to the published version of the manuscript.

Funding: This research was supported by the National Science and Technology Major Project (J2019-III-0002-0045), National Natural Science Foundation of China (No. 52276141) and the Taishan Scholars Program.

Data Availability Statement: The data presented in this study are available on request from the corresponding author due to data confidentiality restrictions.

Conflicts of Interest: The authors declare no conflicts of interest.

References

1. Wang, B.; Hui, X.; Li, F.; Cheng, M.; Zhang, M.; Yang, J. Research status and key technology analysis of dynamic combustion in aero-engine main combustor. *J. Aerosp. Power* **2022**, *37*, 2479–2487.
2. Kapucu, M.; Kok, J.B.W.; Pozarlik, A.K. Experimental investigation of thermoacoustics and high-frequency combustion dynamics with band stop characteristics in a pressurized combustor. *Energies* **2024**, *17*, 1680. [CrossRef]
3. Tang, A.; Cai, T.; Li, C.; Zhou, C.; Gao, L. Flame visualization and spectral analysis of combustion instability in a premixed methane/air-fueled micro-combustor. *Energy* **2024**, *294*, 130793. [CrossRef]
4. Hwang, D.; Ahn, K. Experimental Study on Dynamic Combustion Characteristics in Swirl-Stabilized Combustors. *Energies* **2021**, *14*, 1609. [CrossRef]
5. Jia, D.; Pan, Y.; Wang, N.; Liu, C.; Yang, K. Combustion Modes and Unsteady Characteristics during the Condition Transition of a Scramjet Combustor. *Energies* **2021**, *14*, 2522. [CrossRef]
6. Fu, C.; Wang, X.; Wu, Y.; Gao, Y. 20 kHz CH_2O- and SO_2-PLIF/OH*-chemiluminescence measurements on blowoff in a non-premixed swirling flame under fuel mass flow rate fluctuations. *Appl. Sci.* **2024**, *14*, 9419. [CrossRef]
7. Sthr, M.; Boxx, I.; Carter, C.; Meier, W. Dynamics of lean blowout of a swirl-stabilized flame in a gas turbine model combustor. *Proc. Combust. Inst.* **2011**, *33*, 2953–2960. [CrossRef]
8. Slabaugh, C.; Boxx, I.; Werner, S.; Meier, W.; Lucht, R. High-speed measurements in partially-premixed swirl flames at elevated temperature and pressure. In Proceedings of the AIAA Aerospace Sciences Meeting 2015, Kissimmee, FL, USA, 5–9 January 2015.
9. Zhao, Q.; Yang, J.; Liu, C.; Liu, F.; Wang, S.; Mu, Y.; Xu, G.; Zhu, J. Experimental investigation on lean blowout dynamics of spray flame in a multi-swirl staged combustor. *Therm. Sci. Eng. Prog.* **2023**, *37*, 101551. [CrossRef]
10. Broda, J.; Seo, S.; Santoro, R.; Shirhattikar, G.; Yang, V. An experimental study of combustion dynamics of a premixed swirl injector. *Symp. Int. Combust.* **1998**, *27*, 1849–1856. [CrossRef]
11. Guiberti, T.; Boyette, W.; Masri, A.; Roberts, W. An experimental study of turbulent lifted flames at elevated pressures. *Combust. Flame* **2019**, *203*, 301–312. [CrossRef]
12. Guiberti, T.; Boyette, W.; Roberts, W. Height of turbulent non-premixed jet flames at elevated pressure. *Combust. Flame* **2020**, *220*, 407–409. [CrossRef]

13. Boudier, G.; Gicquel, L.; Poinsot, T.; Bissières, D.; Institute, C. Comparison of LES, RANS and experiments in an aeronautical gas turbine combustion chamber. *Proc. Combust. Inst.* **2007**, *31*, 3075–3082. [CrossRef]
14. Han, X.; Krajnović, S. An efficient very large eddy simulation model for simulation of turbulent flow. *Int. J. Numer. Methods Fluids* **2013**, *71*, 1341–1360. [CrossRef]
15. James, S.; Zhu, J.; Anand, M. Large-eddy simulations as a design tool for gas turbine combustion systems. *AIAA J.* **2006**, *44*, 674–686. [CrossRef]
16. Kawai, S.; Lele, S. Large-eddy simulation of jet mixing in supersonic crossflows. *AIAA J.* **2010**, *48*, 2063–2083. [CrossRef]
17. Wang, S.; Yang, V.; Hsiao, G.; Hsieh, S.; Mongia, H. Large-eddy simulations of gas-turbine swirl injector flow dynamics. *J. Fluid Mech.* **2007**, *583*, 99–122. [CrossRef]
18. Xu, J.; Ma, T.; Huang, D.; Wang, D.; Meng, H. Large eddy simulations of dual swirl injection and combustion of methane-air under different pressures. *J. Propuls. Technol.* **2024**, 1–24. [CrossRef]
19. Meng, N.; Li, F. Large eddy simulations of unsteady non-reaction flow characteristics using different geometrical combustor models. *Aerosp. Sci. Technol.* **2022**, *126*, 107638. [CrossRef]
20. Lv, G.; Liv, X.; Zhang, Z.; Li, S.; Gao, Y. Influences of pilot-stage equivalence ratio on dynamic combustion characteristics in a co-staged combustor. *J. Eng. Thermophys.* **2022**, *43*, 2163–2169.
21. Chen, Z.; Langella, I.; Swaminathan, N.; Stöhr, M.; Meier, W.; Kolla, H. Large eddy simulation of a dual swirl gas turbine combustor: Flame/flow structures and stabilisation under thermoacoustically stable and unstable conditions. *Combust. Flame* **2019**, *203*, 279–300. [CrossRef]
22. Yin, R.; Xue, J.; Wang, J.; Li, X. Reduced mechanism for n-Decane combustion and its application in numerical simulation of aeroengine combustor. *J. Propuls. Technol.* **2021**, *42*, 1876–1882.
23. Pope, S. Turbulent flows. *Technology* **2001**, *12*, 2020–2021.
24. Batchelor, G. *The Theory of Homogeneous Turbulence*; Cambridge University Press: Cambridge, UK, 1953.
25. Zhao, Q.; Yang, J.; Liu, C.; Liu, F.; Wang, S.; Yong, M.; Gang, X.; Zhu, J. Lean blowout characteristics of spray flame in a multi-swirl staged combustor under different fuel decreasing rates. *Chin. J. Aeronaut.* **2022**, *35*, 130–143. [CrossRef]

energies

MDPI

Review

Predictive Models for Biodiesel Performance and Emission Characteristics in Diesel Engines: A Review

Wenbo Ai and Haeng Muk Cho *

Department of Mechanical Engineering, Kongju National University, Cheonan 31080, Republic of Korea; aiwb0712@163.com
* Correspondence: hmcho@kongju.ac.kr; Tel.: +82-(10)-87113252

Abstract: With the increasing global demand for renewable energy, biodiesel has become a promising alternative to fossil fuels with significant environmental benefits. This article systematically reviews the latest advances in predictive modeling techniques for estimating the characteristics of biodiesel and its impact on diesel engine performance. Various methods for predicting the key performance of biodiesel and the performance and emissions of diesel engines have been summarized. According to the categories of parameters, research cases in recent years have been listed and discussed separately. This review provides a comprehensive overview and serves as a reference for future research and development of biodiesel.

Keywords: biodiesel; predictive modeling; diesel engine performance; emissions; statistical methods

1. Introduction

With the increasing global demand for renewable energy, biodiesel as a clean energy source is being applied in an increasing number of fields. It can directly replace or be mixed with petroleum diesel and is widely used in transportation, agricultural machinery, power generation equipment, and other fields. In recent years, many countries and regions have promoted the production and use of biodiesel through policy incentives and regulatory requirements to reduce dependence on fossil fuels and reduce greenhouse gas emissions [1–3].

Among various alternative energy sources, biodiesel has become a highly promising alternative fuel due to its renewability, environmental friendliness, and compatibility with existing diesel engines [4]. Biodiesel is a fuel obtained through the transesterification reaction of vegetable oil, animal fat, or waste oil with alcohols (such as methanol or ethanol), and its main component is fatty acid methyl esters (FAMEs). Compared with traditional petroleum diesel, biodiesel can not only reduce greenhouse gas emissions but also effectively reduce the emissions of carbon monoxide (CO), hydrocarbons (HC), and particulate matter (PM), thereby reducing air pollution and improving environmental quality [5].

On a global scale, the application of biodiesel not only promotes the development of agriculture and industry but also effectively reduces dependence on petrochemical fuels [6]. However, despite the advantages demonstrated by biodiesel in the fields of environment and energy, there are still many problems in its practical application. For example, the physicochemical properties of biodiesel vary significantly due to differences in raw materials and production processes, which directly affect its performance in diesel engines [7]. In addition, the combustion characteristics, emission characteristics, and engine performance impact of biodiesel in engines also vary depending on the fuel composition. Thus, accurately predicting biodiesel properties and its performance in diesel engines is a key focus of current research [8].

The properties of biodiesel directly affect its performance in diesel engines. The physicochemical properties of biodiesel, such as viscosity, density, heating value, and oxidation stability, not only determine the combustion efficiency of the fuel in the engine but also affect the engine's starting performance, fuel consumption, emission characteristics, and long-term reliability. In diesel engine systems, the combustion process of biodiesel directly affects the thermal efficiency, power output, and emission generation of the engine. For example, the high oxygen content of biodiesel can promote the completeness of the combustion process and reduce the emissions of HC and CO [9]. However, this characteristic may also lead to an increase in nitrogen oxide (NO_X) emissions. Therefore, during engine optimization design, it is crucial to consider both the physicochemical properties and combustion characteristics of the fuel. Accurate prediction models must be established to evaluate the overall performance of biodiesel.

In traditional research methods, the properties of biodiesel are mainly obtained through experimental measurements, but this method is not only time-consuming and costly but also has certain experimental errors [10]. As computing technology and data science advance, predictive methods based on mathematical models and machine learning (ML) have emerged as prominent areas of research. These methods can rapidly predict biodiesel properties by leveraging existing experimental data, providing a robust scientific basis for fuel development and application.

The establishment of predictive models can not only save a lot of experimental time and costs but also explore the impact of different raw materials and production processes on fuel performance, thereby optimizing the production process and improving fuel quality [11]. For example, prediction models based on linear regression, support vector machines (SVMs), and artificial neural networks (ANNs) can accurately predict key attributes such as heating value, viscosity, and oxidation stability of biodiesel [12]. The application of these models enables researchers to better understand the characteristics of different biodiesel fuels, providing effective tools for fuel optimization design. At the same time, an increasing number of researchers are adopting data-driven methods to predict the performance of biodiesel in engines. By analyzing a large amount of experimental data and training models, ML methods can capture the complex nonlinear relationships between fuel properties and engine performance, thereby achieving high-precision predictions. These methods provide new perspectives for studying biodiesel–engine interactions. They also offer robust support for optimizing fuel blends and engine design.

Due to its renewability, lower carbon dioxide emissions, and good biodegradability, biodiesel, as an important renewable energy source, offers significant environmental benefits and application prospects. Research related to biodiesel has rapidly developed, with researchers continually expanding the raw material sources, fuel blend formulations, and application scenarios for biodiesel [13,14]. The performance of biodiesel varies depending on the type of raw materials used [15]. Additionally, biodiesel is typically blended with diesel using additives for practical use. Therefore, the properties of biodiesel blends and their performance in diesel engines are more complex compared to other alternative fuels.

This article aims to systematically organize and review the research progress in predicting the properties of biodiesel and its performance in diesel engines in recent years. Starting from the key attributes and performance parameters of biodiesel, specific research cases were cited, covering application examples of different prediction models and experimental methods. This review highlights the advantages and limitations of these methods in practical research. These cases help identify best practices and potential improvements in current methods, provide guidance for future research, and promote further development of biodiesel applications.

2. Biodiesel

2.1. Production and Main Components of Biodiesel

Biodiesel is a renewable fuel obtained through the transesterification process from vegetable oils or waste edible oils. Compared with traditional petroleum diesel, biodiesel is a more environmentally friendly alternative fuel, mainly composed of FAMEs. The production of biodiesel is primarily achieved through transesterification reactions using raw materials such as edible oils, inedible oils, waste oils, and algae [16–18]. This process requires mixing the raw oil and fat with methanol or ethanol under the action of a catalyst, heating, and reacting to produce FAMEs and glycerol. Common catalysts include NaOH, KOH, and acidic catalysts such as H_2SO_4. The efficiency of the transesterification reaction is influenced by various factors, including reaction temperature, reaction time, molar ratio of alcohol to oil, and catalyst amount.

The composition of FAMEs varies depending on different raw oils and fats and usually includes the following main components [19–21]:

Saturated FAMEs, such as palmitic acid methyl ester (C16:0) and stearic acid methyl ester (C18:0), typically have high melting points and viscosities. Unsaturated FAMEs, such as oleic acid methyl ester (C18:1) and linoleic acid methyl ester (C18:2), have lower melting points and better fluidity. Linolenic acid methyl ester (C18:3) has a lower melting point and poorer oxidation stability.

The physical and chemical properties of biodiesel, such as density, viscosity, combustion characteristics, etc., are mainly determined by its fatty acid composition [22]. The structure of fatty acid components, including carbon chain length, saturation, number, and position of branched or double bonds, plays a decisive role in the final performance of fuels [23]. The composition of biodiesel results in distinct physical and chemical properties compared to petroleum-based diesel [24]. The advantages of biodiesel include high oxygen content, high cetane number, high flash point, and excellent lubricity [25]. Biodiesel molecules contain higher oxygen content than petroleum diesel, allowing for more complete combustion and reduced emissions of HC and CO [26]. The sulfur content of biodiesel is extremely low, which greatly reduces the emission of SO_2 and helps alleviate the formation of acid rain [27].

In addition to the aforementioned advantages, the drawbacks of biodiesel are also evident. The long fatty acid chains in biodiesel contribute to its high viscosity and low fluidity, particularly at low temperatures [28]. This characteristic is particularly evident in low-temperature environments and may cause clogging in fuel injection systems, which may be mitigated by using additives or blending with petroleum diesel. Meanwhile, unsaturated FAMEs are prone to oxidation, leading to the deterioration of biodiesel during storage [29]. Depending on the storage environment and the composition of biodiesel itself, this oxidative deterioration may accelerate [30]. To enhance its stability, it is typically necessary to add antioxidants and implement an appropriate storage strategy [31].

2.2. *Main Properties Affecting the Performance of Biodiesel in Diesel Engines*

2.2.1. Density and Kinematic Viscosity

The density of biodiesel is higher than that of diesel. The density increases as the fatty acid chain length decreases and the degree of unsaturation increases.

The density of biodiesel has a significant impact on its combustion characteristics and performance in diesel engines [32]. High-density biodiesel can lead to increased injection volume, poor fuel atomization quality, and thus affect the mixing efficiency and combustion completeness of air and fuel [33]. This phenomenon usually manifests as prolonged ignition delay and combustion duration. Ultimately, it will lead to a decrease in combustion efficiency and an increase in the generation of incomplete combustion products, affecting the overall performance and emission characteristics of the engine.

Kinematic viscosity significantly affects the spray characteristics and combustion quality of the fuel [34]. If the viscosity is too high, it may result in poor flow of the fluid between certain components, causing energy loss and reduced engine efficiency. At the same time, high viscosity often leads to the formation of larger droplets during injection, resulting in poor atomization [35].

2.2.2. Cetane Number

The cetane number (CN) is the key indicator for measuring the spontaneous combustion quality of fuel. The CN of biodiesel is affected by the chain length and degree of unsaturation of the FAMEs. Biodiesel usually has a higher CN than diesel fuel. A high CN helps reduce the formation of white smoke and improves cold start performance. A low CN will increase ignition delay, thereby increasing the likelihood of knocking in diesel engines [36].

2.2.3. Thermophysical Properties

The thermophysical properties of biodiesel, including heating value, thermal conductivity, specific heat capacity, and coefficient of thermal expansion (CTE), are the key factors affecting its combustion characteristics, heat transfer efficiency, and overall engine performance. These thermal properties are not only important for the energy output of fuel but also affect the temperature distribution and combustion efficiency in the combustion process.

Heating value is the energy released during fuel combustion, and it is an important index to measure the energy density of biodiesel. Biodiesel with a high heating value usually has a higher energy density and can provide greater power output [37]. Thermal conductivity determines the heat transfer efficiency of biodiesel in the combustion process and has a direct impact on the temperature distribution and combustion stability of the combustion chamber [38]. Specific heat capacity is an index describing the ability of fuel to absorb or release heat during heating, which directly affects the heating rate and combustion efficiency of fuel. The CTE, which measures the volume change of fuel when the temperature changes, has an important impact on the design and safety of fuel systems [39].

3. Classification of Prediction and Optimization Techniques for Biodiesel Properties and Engine Performance

3.1. Statistical Modeling and Regression Methods

Statistical modeling and regression methods are traditional analytical tools used to reveal and quantify relationships between variables. These methods describe patterns in data by establishing mathematical models that provide a basis for prediction. The following will introduce several commonly used statistical modeling and regression methods and their specific applications in biodiesel research.

Response surface methodology (RSM) is a method that combines experimental design, regression modeling, and optimization techniques, particularly suitable for handling multivariate problems in complex systems. In biodiesel research, RSM is commonly used to optimize fuel formulations, evaluate the effects of different production processes on biodiesel properties, and predict the performance and emission characteristics of diesel engines [40]. The construction of RSM models typically involves conducting quadratic regression analysis on experimental data to generate a mathematical equation that describes the relationship between input variables and output response. By analyzing these models, researchers can identify the factors that have the greatest impact on the properties of biodiesel and determine the optimal production conditions [41]. RSM has also demonstrated strong application potential in predicting diesel engine performance. The performance and emission characteristics of diesel engines are influenced by multiple factors, including fuel composition and engine operating conditions (such as speed and load), injection timing, and injection pressure [42,43]. RSM can systematically study the interaction of these factors and predict their impact on engine performance by designing a reasonable experimental plan. Table 1 shows the comparison between the predicted results and experimental validation results after optimizing the three parameters: injection pressure, injection timing, and exhaust gas recirculation using the RSM model by Saravanan, S et al. [44].

Table 1. Comparison of predicted results from the RSM model and experimental validation results [44].

	NO_x (g/kWh)	Smoke Opacity	CO_2 (kg/kWh)	BTE (%)	BSFC (kg/kWh)
Predicted	16.918	10.228	0.624	35.874	0.253
Actual	17.178	10.55	0.646	35.5	0.2572
% Error	1.51	3.05	3.41	1.05	1.63

Regression analysis methods are used to quantify the linear relationship between one or more independent variables and the dependent variable. The most basic linear regression can be used to estimate the production of biodiesel [45]. However, the limitation of linear regression is that it assumes that the relationship between variables is linear, while actual combustion processes often have complex nonlinear characteristics. Nonlinear regression is suitable for modeling complex relationships between variables using nonlinear functions [46]. In the research of biodiesel and engines, nonlinear regression can be used to predict complex dynamic changes during engine operation. For example, Figure 1 shows the variation of braking thermal efficiency with power output [47]. This enables researchers to describe engine performance under different conditions more accurately. Multiple regression is an extension of linear regression that allows for simultaneous analysis of the effects of multiple independent variables on a single dependent variable. In the study of biodiesel and engine performance, multiple regression is widely used to simultaneously consider the comprehensive impact of multiple factors (such as fuel composition, engine operating conditions, and environmental conditions) on the output results. For example, predicting the physical properties such as viscosity, density, and flash point of biodiesel based on raw materials [48]; predicting biodiesel performance based on fatty acid composition [49]; When predicting engine emissions (such as CO, NO_x, and HC), multiple regression models can integrate multiple input variables to establish a comprehensive prediction model [50].

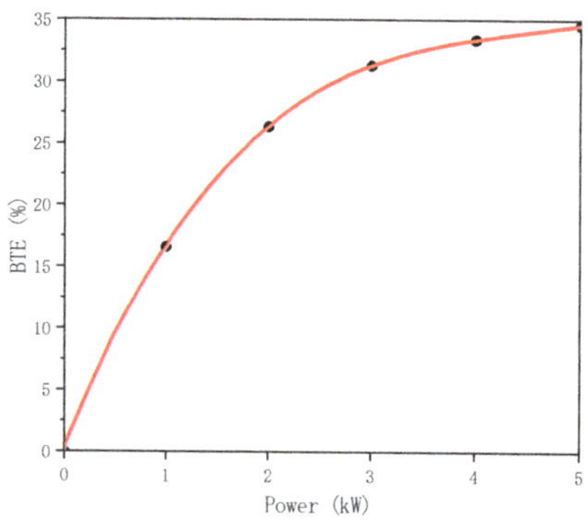

Figure 1. Variation of brake thermal efficiency with power output. The figure is recreated using data from reference [47].

Statistical modeling and regression methods have significant application value in the study of biodiesel and engine performance. They can provide clear explanations of variable relationships, which are easy to understand and implement. However, these methods typically assume that the model structure is known and there are no significant nonlinear or interactive effects in the data, which may limit their accuracy and applicability in practical applications. Therefore, when researchers use these methods, they usually need to perform sufficient preprocessing and exploratory analysis on the data to ensure the rationality and reliability of the model.

3.2. Machine Learning and Artificial Intelligence Methods

ML and artificial intelligence methods are increasingly important tools for predicting biodiesel properties and engine performance. This is due to their powerful nonlinear modeling capabilities, automatic feature learning capabilities, and advantages in processing high-dimensional data [51,52]. These methods can not only significantly improve prediction accuracy but also reduce the number of experiments and optimize the design process. The following are several common ML and artificial intelligence methods and their applications in predicting biodiesel properties and engine performance.

An ANN is a computational model that simulates biological neural networks and can learn complex nonlinear relationships between inputs and outputs through a large amount of training data. An ANN is particularly suitable for processing high-dimensional data and nonlinear systems, demonstrating extremely high accuracy in predicting the performance, such as brake thermal efficiency (BTE), brake-specific fuel consumption (BSFC), and emission characteristics of biodiesel blends [53–55]. Figure 2 shows the structure of the ANN used to study biodiesel performance and its performance in diesel engines. Research has shown that ANNs can accurately predict the performance and emissions of engines using different biodiesel blends by capturing the subtle impact of fuel composition changes on engine behavior. For example, when studying the effects of certain biodiesel additives on engine performance, ANN models can accurately predict how changes in additive concentration affect the combustion efficiency and emission levels of the engine by learning experimental data [56]. This high-precision predictive ability makes an ANN a valuable tool in biodiesel research.

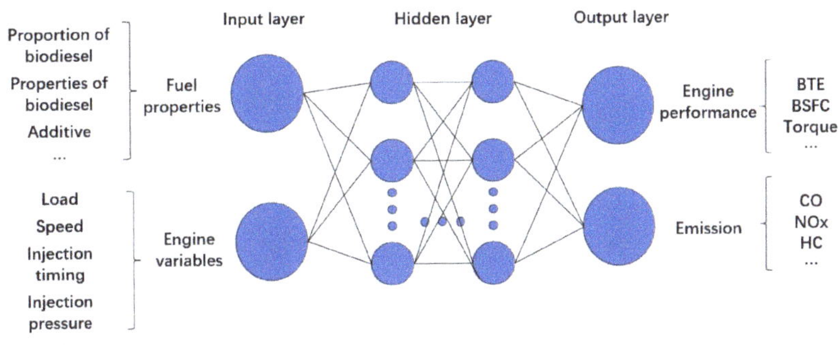

Figure 2. The structure of the ANN used to study biodiesel performance and its performance in diesel engines.

SVM is a supervised learning model based on statistical learning theory, commonly used for classification and regression problems. The advantage of SVM in predicting the properties and engine performance of biodiesel lies in its powerful nonlinear processing capability and efficiency in processing high-dimensional data. SVM maps input data to a high-dimensional space through kernel functions, enabling it to identify and capture hidden patterns and relationships in complex biodiesel-blended fuel datasets [57,58]. Figure 3 illustrates the application of SVM in predicting biodiesel yield. The SVM has shown significant performance in optimizing the production process and composition of biodiesel [59,60]. When studying engine emission characteristics, SVM can handle complex combinations of input variables, such as different biodiesel components, engine operating conditions, etc., to provide accurate emission predictions [61].

A decision tree is a predictive model that establishes relationships between input variables and target outputs by recursively splitting the data [62]. Random forest is an ensemble method of decision trees that enhances the stability and accuracy of the model by constructing multiple decision trees and combining their prediction results. In the prediction of biodiesel properties and engine performance, random forest can automatically identify and select the most important feature variables, thereby improving prediction performance. For example, when predicting engine combustion efficiency or emissions, random forests can identify the factors that have the greatest impact on target output from multiple fuel components and engine operating variables, thereby improving the accuracy and reliability of predictions [63,64].

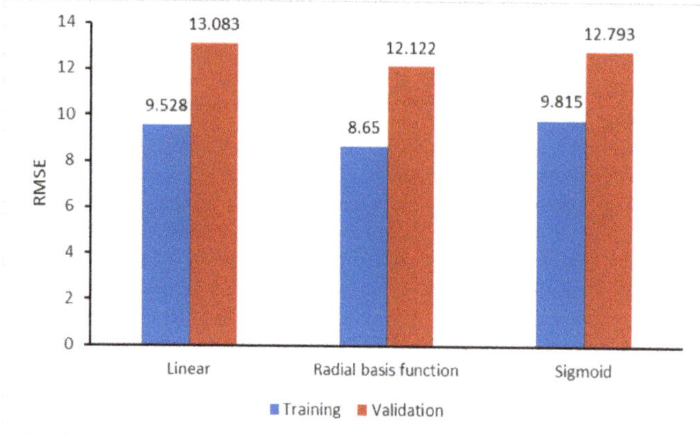

Figure 3. RMSE value comparisons for biodiesel yield by SVM regression with three kernels. The figure is recreated using data from reference [59].

ML and artificial intelligence methods offer significant advantages in predicting the properties and engine performance of biodiesel. They are capable of handling complex nonlinear relationships, automatically extracting data features, and providing high-precision prediction results [65]. However, these methods also face challenges, such as the large amount of data required for model training, high data quality requirements, and the opacity of the models that may limit interpretability. To overcome these challenges, researchers are exploring hybrid models that combine ML with other methods to improve model interpretability while ensuring high-precision predictions [66,67].

3.3. *Evolutionary Algorithms and Optimization Methods*

Evolutionary algorithms and optimization methods are powerful tools developed based on the principles of natural selection and genetics and are widely used in solving complex problems and multi-objective optimization [68]. These methods are suitable for the prediction and optimization of biodiesel properties and engine performance because they can search for the optimal solution in a wide range of search spaces, deal with complex multi-objective optimization problems, and do not depend on the specific mathematical model of the problem. The following are some common evolutionary algorithms and optimization methods in recent years and their applications in biodiesel and engine research.

A genetic algorithm (GA) is an evolutionary algorithm that simulates the process of natural selection. It optimizes candidate solutions in the search space through selection, crossover, mutation, and other operations [69]. GA is widely used in biodiesel research, such as predicting the properties of biodiesel or optimizing the biodiesel production process [70,71]. In the prediction of biodiesel properties and engine performance, GA is used to find the best fuel ratio to optimize combustion efficiency and reduce emissions [72]. GA's adaptability enables it to handle multi-objective optimization problems with complex nonlinear relationships and find the optimal fuel combinations and engine operating parameters.

Particle swarm optimization (PSO) is an optimization algorithm based on swarm intelligence that simulates the collective behavior of organisms, such as birds or fish, during foraging. PSO gradually approaches the global optimal solution through information sharing and cooperation among individuals. In biodiesel and engine research, PSO can be used to optimize multivariable systems, including fuel composition, engine tuning parameters. PSO is characterized by its fast convergence speed and simple calculation, which is especially suitable for continuous optimization problems [73]. Researchers can use PSO to find the optimal ratio of biodiesel-blended fuel to achieve the best combustion performance and the lowest emission level [74]. In addition, PSO can also be used to optimize the multi-objective performance of the engine, such as fuel economy and power output.

Genetic programming (GP) is an evolutionary computation method based on GA that is used to automatically generate computer programs. GP can be used to model complex nonlinear systems, predict the physical and chemical properties of biodiesel, and optimize the production process [75]. Figure 4 shows the GP tree model used for predicting kinematic viscosity [76]. In the prediction of biodiesel properties and engine performance, GP can generate nonlinear regression models or classifiers to predict engine performance and emissions under different fuel conditions. This method does not rely on the pre-defined model structure but automatically generates the optimal model through the evolution process. Researchers can use GP to generate a prediction model for simulating and optimizing engine operating parameters and fuel ratio to improve overall performance [77].

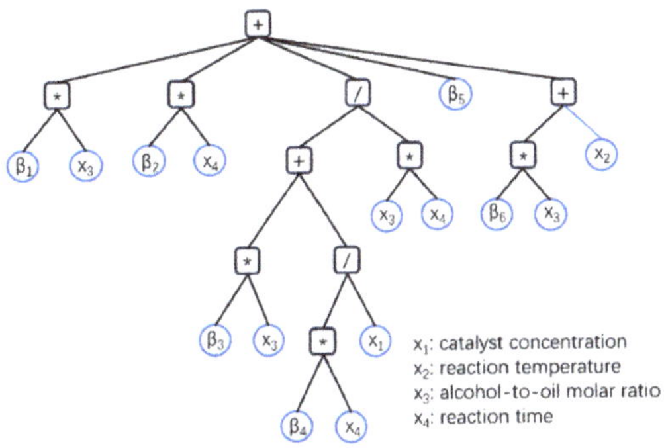

Figure 4. The GP tree model used for predicting kinematic viscosity. The figure is recreated based on the content of reference [76].

The application of evolutionary algorithms and optimization methods in biodiesel and engine research demonstrates great potential and flexibility. These methods can deal with complex multi-objective optimization problems and do not depend on the specific mathematical model of the problem, so they are widely applicable. However, these methods may face the problems of computational complexity and convergence speed when dealing with high-dimensional problems. In addition, the randomness and multiplicity of evolutionary algorithms may lead to the uncertainty of the results, which need to be addressed.

3.4. Scientific Modeling and Simulation Methods

Scientific modeling and simulation are key tools for studying complex physical and chemical processes. In the prediction and analysis of biodiesel and engine performance, these methods can simulate the actual combustion process and engine working state and provide in-depth theoretical support for optimizing design and improving performance [78]. Scientific modeling and simulation are not only limited to the analysis of experimental results but also can explore system behaviors under different conditions through virtual experiments so as to reduce the number and cost of actual experiments.

Computational fluid dynamics (CFD) is a scientific method that uses numerical analysis and data structures to simulate fluid flow. In biodiesel and engine research, CFD is widely used to simulate fluid flow, fuel injection, and combustion processes, as well as their impact on emissions in the combustion chamber [79]. CFD models can describe in detail the atomization, mixing, ignition, and combustion processes of fuel in the combustion chamber and predict engine performance and emission characteristics [80–82]. CFD simulation can provide the detailed distribution of temperature field, pressure field, and chemical reaction in the combustion process, which is very important for understanding the combustion behavior of biodiesel under different working conditions. For example, Gowrishankar, S. et al. [83] used CFD tools to compare and analyze the differences between conventional injection mode and premixed mode based on delayed injection. As shown in Table 2.

Table 2. CFD simulation results of two injection strategies [83].

Combustion Mode	Cylinder Pressure (Error %)	Heat Release Rate (Error %)	Fuel Distribution	Temperature Distribution
Conventional	1.50%	13%	Near piston bowl	Higher overall
Late injection	10%	15%	Throughout the combustion chamber	Lower temperatures

The chemical kinetic model is used to simulate the chemical reactions in the combustion process, including reaction rate, product distribution, heat release, etc. In biodiesel combustion, chemical kinetic models can help to study the decomposition, oxidation, and product formation processes of fuel molecules under high temperature and high pressure [84,85]. At present, the study of chemical kinetic models often involves the simplification of reaction mechanisms. Reaction mechanism simplification is a technique used to accelerate the calculation by simplifying the complex chemical reaction network, which is suitable for large-scale combustion simulation. In the combustion simulation of biodiesel, the chemical reaction network is usually very complex, including a large number of chemical substances and reaction steps. By simplifying the reaction mechanism, the reactions and substances that have little impact on the combustion behavior can be removed, so as to reduce the computational complexity and shorten the simulation time [86,87].

Scientific modeling and simulation methods offer powerful tools for the study of biodiesel combustion characteristics and engine performance. These methods can deeply reveal complex physical, chemical, and thermodynamic processes and provide high-fidelity simulation results. However, scientific modeling and simulation methods also face some challenges, such as the high demand for computing resources, the fact that the accuracy of the model depends on the accuracy of input parameters, and the high dependence on experimental data in some cases. To overcome these challenges, researchers usually combine experimental verification and optimization techniques to ensure the reliability and applicability of simulation results.

4. Recent Research Case Analysis

4.1. Properties of Biodiesel

As an alternative fuel, the physical and chemical properties of biodiesel have a direct impact on its performance in the engine. Understanding and predicting these properties not only helps optimize the production process of biodiesel but also improves engine combustion efficiency and reduces emissions. Related studies are shown in Table 3. This paper selects several key properties related to biodiesel combustion: density, viscosity, CN, and thermophysical properties. In the following chapters, we will discuss these properties in detail and analyze their impact on the performance of biodiesel.

Table 3. Summary of research on predicting the properties of biodiesel.

Ref.	Algorithm/Model Type	Input Parameters	Output Parameters	Fuel Composition
[75]	GP	Fatty acid composition	Kinematic viscosity, flash point, cold flow properties, CN and iodine numbers	Various biodiesel compositions
[88]	RSM	Ethanol, diesel, biodiesel blend ratios	Density, viscosity	Ethanol–diesel–biodiesel blends
[89]	Empirical Correlations	Biodiesel blend ratios	Density, viscosity	Ternary biodiesel blends
[90]	ANN, LSSVM	Intermolecular interactions, temperature	Density	Various biofuels
[91]	GPR	Molecular weight, carbon number, double bonds, fatty acid types	Kinematic viscosity, cloud point, pour point, iodine value	Various biodiesel compositions
[92]	ANN, Random Tree	Biodiesel composition, fatty acid profile	Kinematic viscosity, density, flash point, oxidation stability, acid value, and calorific values	Hybrid biodiesel fuels
[93]	Empirical Correlation	Fatty acid composition	CN	Various biodiesel compositions
[94]	LSSVM, GA, PSO, HGAPSO	FAME composition	CN	Various biodiesel compositions
[95]	ANN, SVM	FAME composition	CN	Various biodiesel compositions
[96]	ANN	FAME composition	CN	Various biodiesel compositions
[97]	ANFIS, PSO	Fatty acid composition	CN	Various biodiesel and diesel oils
[98]	Grey Wolf Optimizer, Experimental	Fuel Composition, Temperature	Calorific value	Nicotiana Tabacum L. oil methyl ester
[38]	Experimental Analysis	Temperature, Pressure, Fuel Composition	Thermal conductivity	Methyl pentanoate, methyl octanoate, methyl decanoate
[99]	Experimental Analysis	Temperature, Pressure, Fuel Blend Ratio	Density, viscosity, specific heat	Rapeseed and soybean oil methyl ester blends
[100]	Empirical Correlation, PC-SAFT	Temperature, Pressure, Fuel Blend Ratio	Density, viscosity, specific heat, isothermal compress-ibility	Waste cooking oil biodiesel, 1-butanol
[101]	ANN	Temperature, Composition	Solid fraction, cold flow properties	Ethylic biodiesel blends
[49]	Multiple Regression	Fatty acid composition	CN, density, kinematic viscosity, Heating value	Various biodiesel compositions
[102]	Regression Model	Storage time, antioxidant concentration, saponification value, acid value	Viscosity	Jatropha and Pongamia biodiesel
[103]	Empirical Models	Chemical structure, composition	Flash point	HC, biodiesel, petroleum fractions
[104]	Multivariate Regression	Near-infrared spectroscopy data	Cold filter plugging point, kinematic viscosity	Biodiesel and blends
[105]	ML	Temperature, pressure, composition	Density, viscosity	Biodiesel and diesel blends

Table 3. *Cont.*

Ref.	Algorithm/Model Type	Input Parameters	Output Parameters	Fuel Composition
[106]	ANN, Empirical Models	Temperature, blend ratio	Viscosity	Various biodiesel blends
[107]	Multiple Regression	Fatty acid composition	Iodine value, saponification value	Various biodiesel compositions
[108]	Experimental, Modeling	Temperature, Pressure	Thermal conductivity	Methyl butyrate, methyl caproate

4.1.1. Density and Kinematic Viscosity

The density and kinematic viscosity of biodiesel are important physical parameters that determine its combustion characteristics and engine performance. These properties directly affect the fuel injection, atomization, and combustion processes. Therefore, accurate prediction of these physical properties is very important for optimizing the formulation of biodiesel. In the research of this field, the traditional empirical formula can estimate these attributes to a certain extent, but with the progress of technology, researchers are increasingly turning to computational modeling and data analysis technology.

Existing models usually use the FAMEs component of biodiesel as input to predict multiple physical or chemical properties, including density and kinematic viscosity.

Researchers are increasingly focusing on predicting the properties of mixed fuels rather than studying single categories of biodiesel. For example, Razzaq et al. [88] studied the density and viscosity model of ethanol–diesel–biodiesel ternary mixture. It is pointed out that the physical properties of these mixtures can be effectively predicted by combining experimental data with regression analysis. This method shows the importance of combining experimental data with modeling technology.

Mujtaba et al. [89] evaluated the performance of three regression methods (linear regression, polynomial regression, and exponential regression) in predicting the density and viscosity of biodiesel–diesel mixtures containing additives. The study found that the exponential regression model performed well under specific conditions, but its versatility and accuracy still need to be further improved in the face of different biodiesel mixtures. This finding highlights the applicability differences of different regression methods in specific application scenarios.

In the application of ANN/ML, Nabipour et al. [90] tested four models to estimate the density of biofuels based on intermolecular interactions and the van der Waals radius of atoms, a least squares support vector machine (LSSVM), a radial basis function artificial neural network, a multilayer perceptron artificial neural network, and an adaptive neuro-fuzzy inference system (ANFIS). Among them, LSSVM performs best. This method provides a highly accurate density estimation model than the traditional method by analyzing the interaction of different fatty acids in biodiesel. By comparing various models, this study shows the superior performance of LSSVM in dealing with complex biofuel attribute prediction.

Pustokhina [91] successfully predicted the viscosity and other properties of biodiesel using a Gaussian process regression (GPR) model combined with a variety of kernel functions. The results show that the Matérn kernel function performs best in viscosity prediction, the determination coefficient R^2 reaches 0.992, and the root mean square error (RMSE) is 0.157, which indicates that the GPR model has high robustness and accuracy in dealing with nonlinear problems. This shows that nonlinear modeling technology has a wide application potential in predicting the complex properties of biodiesel. Giwa et al. [92] used ANN and a random tree algorithm to predict the density and viscosity of two or more kinds of oil–synthetic biodiesel blends. The research shows that the two kinds of prediction models based on FAME components have high prediction accuracy and perform well in processing complex data.

Additionally, Alviso et al. [75] used a GP model to predict the physical and chemical properties of biodiesel, including density and viscosity. The results show that GP can effectively capture the complex relationship between fatty acid composition and density, viscosity, and is also a reliable prediction tool. The introduction of GP technology provides a new solution for dealing with complex nonlinear relationships.

4.1.2. CN

CN of biodiesel is the key index to measure its ignition performance. High CN means that the ignition delay time of fuel is short, and the combustion is more complete, thereby improving the engine performance and reducing pollution emissions. With the increasingly important role of biodiesel in replacing traditional diesel fuel, accurate prediction of its CN is very important for optimizing fuel performance.

Some researchers used a linear regression model and multivariate analysis to estimate CN. For example, Lin et al. [93] used the composition information of fatty acid methyl ester (FAME) to predict the CN through a regression analysis model. The results showed that there was a significant correlation between different fatty acid compositions and CN, and the regression model could effectively estimate the CN of different biodiesel.

With the development of ML and artificial intelligence technology, more complex models are gradually applied in this field to improve the accuracy of prediction.

Bemani et al. [94] used LSSVM combined with GA, PSO, and hybrid genetic particle swarm optimization to predict CN, showing the potential of these evolutionary algorithms in optimizing models. The author believes that the combination of the LSSVM algorithm and GA, PSO, or hybrid genetic particle swarm optimization can be used as an accurate estimation model of the CN of biodiesel fuel. These methods show the advantages of the combination of evolutionary algorithms and ML technology and can achieve better prediction results on complex data sets.

Ghiasi et al. [95] compared the prediction performance of LSSVM and extra tree (ET). The results show that ET has higher reliability and stability in predicting CN.

Rahaju et al. [96] used a cascade neural network to predict the CN of 63 kinds of biodiesel. After training with 10 different algorithms, the study found that the Levenberg-Marquardt algorithm has the highest prediction accuracy, with a determination coefficient (R^2) of 0.9245 and a RMSE of 3.1541.

In addition, Noushabadi et al. [97] studied the prediction performance of the hybrid model based on PSO and ANFIS for the CN of biodiesel. The model can effectively integrate FAMEs data and optimize model parameters through an evolutionary algorithm so as to improve the prediction accuracy.

4.1.3. Thermophysical Properties

The interaction between thermophysical properties has complex effects on the overall performance of biodiesel. Research has shown that biodiesel blends exhibit higher instability compared to pure diesel [3]. Therefore, when developing biodiesel formulations, not only these thermal properties should be considered separately, but also their synergy should be comprehensively analyzed to optimize fuel performance and ensure long-term stable operation of the engine. This overall analysis method can help researchers more comprehensively understand the performance of biodiesel in practical applications.

Samuel et al. [98] used the grey wolf optimizer (GWO) to predict the heating value of biodiesel. GWO is an optimization algorithm based on swarm intelligence, which optimizes the solution of the problem by simulating the hunting behavior of gray wolves. This algorithm is applied to the heating value prediction of tobacco oil methyl ester, and the effectiveness of the model is verified by experimental data. The results show that the GWO model can accurately predict the heating value of biodiesel, showing high prediction accuracy. This research highlights the potential of the GWO in predicting complex fuel attributes.

Zheng et al. [38] measured the thermal conductivity of three biodiesel compounds (methyl valerate, methyl octanoate, and methyl decanoate) by experimental method. Using the transient hot wire method, the researchers obtained the thermal conductivity data of these compounds at different temperatures and pressures and used these data to verify and improve the existing prediction model.

Giuliano Albo et al. [99] proposed a formula for indirectly calculating the specific heat capacity of biodiesel through sound velocity and density. However, the accuracy of the results, especially, depends on the accuracy of density. This method can be used as an alternative method when the specific heat capacity cannot be measured directly.

Ait Belale et al. [100] studied the thermophysical properties of the binary liquid mixture of waste edible oil biodiesel and 1-butanol and predicted and correlated these thermophysical properties through experimental data and models (Tait equation and PC-SAFT model). The Tait equation fits the experimental data well, but due to its empirical nature, it cannot explain the interaction between fluid molecules. In contrast, the PC-SAFT model has many parameters, but it can better explain the intermolecular interaction and shows good prediction ability in the whole concentration range.

A neural network also shows great potential in the prediction of complex thermophysical properties, especially when dealing with multivariable nonlinear relationships.

Magalhães et al. [101] used the multilayer perceptron feed-forward neural network to predict the CTE and solid fraction of ethyl–biodiesel mixture. The ANN model can effectively predict the thermal expansion behavior of fuel at different temperatures by learning the nonlinear relationship in the experimental data, and the results show good prediction accuracy.

4.2. Performance and Emissions of Biodiesel in Diesel Engines

The performance and emission characteristics of diesel engines are significantly affected by fuel types and properties. As an alternative fuel, biodiesel has unique advantages in diesel engine performance due to its high oxygen content and excellent combustion characteristics. However, the disadvantages of biodiesel, such as high viscosity, poor low-temperature fluidity, and poor oxidation stability, may lead to engine performance degradation or emissions increasing under specific operating conditions. Therefore, accurately predicting the performance of biodiesel in diesel engines is of great significance for promoting the widespread application of biodiesel. Related studies are shown in Table 4.

Table 4. Summary of research on the performance of biodiesel in diesel engines.

Ref.	Theme	Algorithm/Model Type	Input Parameters	Output Parameters	Fuel Composition
[54]	Engine performance and emissions	ANN, RSM	Engine load, biodiesel ratio, injection pressure	BTE, BSFC, EGT, NO_x, CO, HC, smoke	Biodiesel–diesel blends
[77]	Engine performance and emissions	GP	Fuel blend ratio, engine load	BTE, BSFC, NO_x, CO, smoke	Biodiesel–diesel blends with nanoparticles
[109]	Engine performance and emissions	ANN	Injection timing, fuel blend ratio, engine speed	BTE, BSFC, NO_x, CO, HC	Waste cooking oil biodiesel
[110]	Engine performance and emissions	ANN	Engine load, fuel type	BTE, CO, HC, NO_x, smoke	Waste cooking oil biodiesel blends
[111]	Engine performance and emissions	ANN, RSM	Engine load, biodiesel ratio	BTE, BSFC, NO_x, CO, smoke	Animal fat-derived biodiesel
[112]	Engine performance and emissions	ANN, GEP	Engine load, fuel injection parameters, blending ratio	BTE, BSFC, CO, NO_x, UHC	Linseed oil biodiesel blends

Table 4. *Cont.*

Ref.	Theme	Algorithm/Model Type	Input Parameters	Output Parameters	Fuel Composition
[113]	Engine performance and emissions	ANN	Engine load, fuel, additive, CN	BTE, HC, CO, NO_x, smoke	Palm biodiesel–diesel blends with nanoparticles and ethanol
[114]	Engine performance and emissions	Random Forest, SVM, ANN	Additive concentration, fuel blend, engine load	BTE, NO_x, CO, HC, Smoke	Biodiesel with diphenylamine and ceria nanoparticles
[115]	Engine performance and emissions	ANN, RSM	Engine load, fuel blend ratio, DMC additive concentration	BTE, BSFC, NO_x, CO, exergy efficiency	Diesel–Bael biodiesel with DMC additive
[116]	Engine performance and emissions	ANN	Engine load, quantum dot concentration, fuel blend ratio	BTE, BSFC, NO_x, CO, HC	Biodiesel–diesel blends with carbon-doped quantum dots
[117]	Engine performance and emissions	RSM	Engine load, fuel blend ratio	BTE, BSFC, NO_x, CO	Algal biodiesel–diesel blends
[118]	Engine performance and emissions	ANN, GA	Engine load, biodiesel blend ratio, injection pressure	BTE, BSFC, NO_x, CO, Smoke	Castor oil biodiesel blends
[119]	Engine performance and emissions	ANN	Engine load, alumina nano-catalyst concentration, fuel blend ratio	BTE, NO_x, CO, UHC, vibration	Diesel–biodiesel with alumina nano-catalyst
[120]	Engine performance and emissions	ANN, RSM	Engine load, producer gas flow rate, biodiesel ratio	BTE, BSFC, NO_x, CO, smoke	Biodiesel-producer gas dual-fuel
[121]	Engine performance and emissions	ANN, Multiple Regression, Taguchi	Engine load, biodiesel–alcohol blend ratio	BTE, BSFC, NO_x, CO, smoke	Biodiesel–alcohol blends
[122]	Engine performance and emissions	ANN, RSM	Engine load, decanol proportion, fuel blend ratio	BTE, BSFC, NO_x, CO, smoke	Palm biodiesel–decanol blends
[123]	Engine performance and emissions	ANFIS, GA, PSO	Engine load, fuel injection timing, fuel injection pressure, biodiesel blend	BTE, UHC, NO_x	Jojoba biodiesel blends
[57]	Combustion characteristics	ANN, SVM	Engine load, fuel type, injection timing	Ignition delay (ID)	Diesel and biodiesel blends
[124]	Combustion characteristics	Chemical kinetic model	Hydrogen concentration, biogas–air ratio	Ignition delay, combustion duration, flame speed	Biogas–hydrogen–air mixtures
[125]	Combustion characteristics	CFD	Engine load, fuel type, heat loss parameters	Heat release rate, cylinder pressure, exhaust gas temperature	Various biodiesel blends
[126]	Combustion characteristics, Emission	Chemical kinetic model	Engine load, air–fuel ratio, fuel type	Heat release rate, NO_x, CO	Diesel–natural gas dual fuel

Table 4. *Cont.*

Ref.	Theme	Algorithm/Model Type	Input Parameters	Output Parameters	Fuel Composition
[127]	Combustion characteristics	ANN, CFD	Engine load, fuel blend ratio, engine speed	Heat release rate, cylinder pressure, ignition delay	Diesel–tomato seed oil biodiesel blends
[128]	Combustion characteristics	ANN	Engine load, fuel type	cylinder pressure, heat release rate, ignition delay	Diesel–safflower biodiesel blends

When studying the performance and emission characteristics of compression ignition engines, selecting appropriate parameters is the key. The selected parameters need to be able to fully reflect the operating efficiency, fuel economy, and environmental impact of the engine. In this paper, the key parameters such as combustion characteristics, BTE, BSFC, torque, and emission (such as NO_x, CO, HC, and particulate emissions) are selected for discussion.

4.2.1. Combustion Characteristics

The combustion characteristics of biodiesel in diesel engines have an important impact on its dynamic performance and emission performance. Combustion characteristics include combustion rate, ignition delay, heat release rate, and cylinder pressure, which directly determine engine efficiency and emissions.

Combustion rate refers to the speed at which fuel is burned in unit time, which is usually expressed by the mass or volume change rate of fuel.

Ignition delay refers to the time between fuel injection into the combustion chamber and fuel combustion, which is an important performance of the fuel injection system and fuel spontaneous combustion characteristics. Heat release rate is the heat released from the fuel per unit time, which reflects the energy release rate of the combustion process. In-cylinder pressure refers to the pressure measured inside the engine cylinder, which is usually measured at different stages of the combustion process. The cylinder pressure reflects the gas state in the cylinder and has a direct impact on the working process and performance of the engine.

In previous research, many mature thermal mathematical models have been widely used in the field of combustion research. They are used to simulate combustion characteristics and heat loss in compression ignition engines. In some papers focusing on the analysis of experimental data, the existing models will be combined to support their conclusions [124]. For example, Alhikami et al. [129] studied the spray ignition characteristics of biodiesel and other fuels through constant volume combustion chamber experiments. They made a comparative analysis with the prediction results of the model in the literature to help explain the relationship between the difference in ignition delay and fuel characteristics.

With the wide application of biodiesel as an alternative fuel, based on the existing theoretical framework, improving the accuracy of biodiesel combustion characteristics prediction through experimental verification and model adjustment has become the focus of research. Kamta Legue et al. [125] analyzed the effect of heat loss model changes on biodiesel combustion characteristics in combination with experiments and simulations. Through model prediction, the fuel ratio can be optimized, the heat loss can be reduced, and the combustion efficiency can be improved.

Another idea is to simplify the complex chemical mechanism model and analyze the interaction and influence of various components in the chemical reaction through the chemical reaction kinetics analysis technology. This method is also common in predicting the combustion characteristics of other fuels. For example, Liu [126] developed a simplified multi-component combustion mechanism for predicting the combustion characteristics of diesel natural gas dual fuel engines. In this study, several simplified steps and methods were introduced into the mechanism model, such as direct relationship graph, error propagation extended direct relationship graph, and species-wide sensitivity analysis. Through cross-reaction analysis, the model can accurately predict the combustion rate of fuel mixture, and the main emissions generated. This method not only reduces computational cost but also provides reliable prediction results. It is a direction worthy of further exploration in the future of biodiesel-related research.

In addition, some researchers choose to combine ANN/ML technology with other technologies to predict the combustion characteristics of biodiesel.

Karami et al. [127] used ANN and CFD models to predict the combustion characteristics of diesel engines using a tomato seed oil–biodiesel mixture. The results show that the ANN model can effectively predict the ignition delay, combustion duration (CD), heat release rate (HRR), and cylinder pressure (CP) under different biodiesel mixing ratios, and it is highly consistent with the experimental data.

Tuan et al. [57] compared the performance of ANN and SVM models in predicting the ignition delay of diesel and biodiesel blends. The results show that SVM has better prediction ability and accuracy than the ANN model. The authors suggest that SVM can be used to predict the ignition delay of diesel and biodiesel engines to improve combustion efficiency.

Dharmalingam et al. [109] used the Bayesian regularization neural network model to optimize the stratified injection strategy of waste edible oil biodiesel in diesel engines to improve combustion performance. The model successfully predicted the combustion characteristics under different injection conditions and reduced the generation of emissions through optimization.

4.2.2. BTE and BSFC

The performance of biodiesel in diesel engines, especially BTE and BSFC, is an important index to evaluate fuel economy and engine performance. BTE reflects the efficiency of the engine in converting the chemical energy of fuel into mechanical energy, while BSFC measures the amount of fuel consumed by the engine in generating unit power. Biodiesel usually has high oxygen content, which can improve combustion efficiency, but its low energy density may lead to the increase of BSFC. Therefore, how to accurately predict and optimize the BTE and BSFC of biodiesel has become a research hotspot in the academic community.

Although the traditional experimental methods can obtain accurate data, due to their high cost and time consumption, more and more researchers turn to the modeling method based on ML. These methods can deal with the complex nonlinear relationship and accurately predict the BTE and BSFC of biodiesel under different operating conditions.

Many researchers have introduced ANN, ML, and other models in the study of biodiesel engine performance.

Patnaik et al. [110] used the ANN model to predict the BTE and BSFC of biodiesel-blended fuel made from waste edible oil. The researchers trained the ANN model through a large number of experimental data and obtained the engine performance under different loads and different fuel ratios. This study shows the advantages of an ANN model in complex nonlinear data processing, especially when dealing with fuel mixtures under transformation conditions it can provide high-precision prediction results. Similarly, Simsek et al. [111] explored the effect of ANNs and RSMs to predict the BTE and BSFC of animal fat biodiesel in diesel engines. Moreover, the fuel ratio was adjusted according to the predicted results, optimizing the fuel economy of the engine.

Sharma et al. [130] compared the performance of enhanced regression tree (BRT) and ANN in predicting BTE and BSFC of biogas biodiesel dual fuel engines. The use of the BRT model shows its powerful ability in small samples and high-dimensional data. Especially in practical engineering applications, the robustness of BRT makes it an effective alternative to ANN. However, due to the complexity and high computational requirements of BRT models, their application may be limited to some extent.

In addition, some authors chose to use the gene expression programming (GEP) model and compared it with ANN. Sharma et al. [112] compared the performance of GEP and ANN in predicting the performance of biodiesel engines (including BTE and BSFC). The GEP model generates prediction models through an evolutionary algorithm, showing high prediction accuracy. The GEP model can generate accurate and explanatory prediction models through its unique evolutionary algorithm. Although its computational complexity is high, it has significant advantages in finding the optimal prediction formula and processing complex nonlinear data.

With the development of biodiesel research, more and more researchers have begun to explore the performance of biodiesel in the presence of different additives. At the same time, the prediction of diesel engine performance when using biodiesel with additives has become a hot research field. At present, many models can achieve this prediction goal more accurately.

Sule et al. [113] discussed the effect of adding nano particles and ethanol to biodiesel diesel blend fuel on engine BTE and BSFC. Through the ANN model, the researchers predicted and optimized the effect of additives on fuel performance. This study shows the advantages of combining experiment and modeling, especially when predicting and optimizing complex fuel formulations containing multiple additives.

Similarly, Kumar et al. [114] used ML methods to predict the BTE and BSFC of biodiesel-blended fuel with diphenylamine antioxidant and ceria nanoparticles in diesel engines. The ML algorithm effectively captures the performance changes of different additives and fuel ratios.

Ghanbari et al. [77] used the GP model to predict the BTE of biodiesel diesel blended fuel with nanoparticles. The GP model also has certain interpretability, which can generate accurate predictions of different fuel ratios and provide detailed performance analysis. By simulating the natural selection process, the model generates and optimizes the fuel performance prediction formula, which provides data support for the addition of different nanoparticles. This study shows the potential of evolutionary computation in fuel performance prediction.

Some researchers choose to combine ANNs and RSMs to make full use of their advantages: an ANN is good at processing complex data, while an RSM is outstanding in optimization. This method can provide a more comprehensive optimization scheme and has high practical application value. Aydın [54] and others combined ANNs and RSMs to predict and optimize the BTE and BSFC of biodiesel diesel blend fuel. The ANN model is used to predict the fuel performance, and RSM is used to optimize the fuel ratio to improve BTE and reduce BSFC.

In the face of biodiesel containing additives, the optimization method combining ANN and RSM also performs well. Pitchaiah [115] and others predicted and optimized BTE and BSFC of diesel engines using Bael biodiesel and DMC additives through ANN and RSM models. The study combines energy and exergy analysis to comprehensively evaluate the fuel performance.

4.2.3. Torque

In a diesel engine, the torque performance of biodiesel is an important index to measure its power output capacity. The torque directly affects the acceleration performance and traction of the engine. Accurately predicting the torque output of biodiesel under different operating conditions is very important for optimizing engine performance.

Some researchers also mentioned the attribute of torque when studying the prediction model of engine performance. In the following research cases, the torque is predicted by ANN, RSM, and a multi-objective optimization algorithm.

Taheri Garavand et al. [116] applied ANN to predict the torque performance of a carbon-based biodiesel mixture doped with quantum dots in internal combustion engines. The research shows that the ANN model can well predict the torque output of the engine under different load and speed conditions, indicating that this model has significant advantages in capturing complex fuel combustion characteristics.

Umeuzuegbu et al. [117] used RSM to model and predict the torque of seaweed biodiesel diesel blend fuel under different engine loads and speeds. The results show that with the increase in load, the torque also increases, while the increase in biodiesel proportion slightly reduces the torque output. This result proves the effectiveness of RSM in capturing torque changes.

Khoobbakht et al. [118] used computational modeling and multi-objective optimization technology to predict the torque of biodiesel made from castor oil in the engine. The results show that castor oil biodiesel can provide higher torque output under high load conditions, and the model optimization results show that the torque performance can be further optimized by adjusting the fuel ratio and engine operating parameters.

4.2.4. Emissions

Co, NO_x, HC, PM, smoke opacity, and other parameters are often considered when studying the emissions of biodiesel [131,132]. Among them, NO_x is one of the main pollutants emitted by diesel engines, which is formed at high temperatures. The use of biodiesel usually leads to an increase in NO_x emissions, which is due to the high oxygen content in biodiesel, resulting in an increase in combustion temperature. Therefore, the generation and control of NO_x has become an important research focus in the study of biodiesel emissions [133]. The mixing ratio of the biofuel mixture needs to be optimized according to engine operating conditions to balance the amount of NO_x and other exhaust emissions [134].

CO and HC emissions are usually related to incomplete combustion. Since the combustion efficiency of biodiesel is typically higher than that of traditional diesel, these emissions are often lower [135]. Researchers often track these pollutants to assess the environmental benefits of biodiesel.

In actual biodiesel diesel engine experiments, researchers can usually obtain information including BTE, BSFC, and various emissions parameters at the same time. Therefore, while training models to predict BTE and BSFC, many studies also predict emission characteristics (such as CO and NO_x), as shown in Table 4. This comprehensive data acquisition makes the prediction model more comprehensive and effective in application. However, although the prediction of CO and NO_x has received extensive attention in the research, the research on HC emissions and smoke opacity is relatively less mentioned. This difference may be related to the low emission of HC in biodiesel combustion, but it is still necessary to further explore the prediction of HC in order to improve the accuracy of the overall emission model.

ANN and RSM are considered to be accurate and reliable ways to predict HC emissions and smoke opacity [54,111,113]. For example, Hosseini [119] and others used experimental data to train multilayer perceptron feedforward neural networks through a back propagation algorithm. In the prediction of emissions, the model shows high accuracy. When predicting HC emissions, the model achieved a correlation coefficient of 0.98, indicating a strong relationship between the predicted and actual values.

Singh et al. [136] proposed a hybrid model combining ANFIS and GA to predict diesel engine performance and emission parameters, including HC emissions. This method significantly improves the prediction accuracy of engine performance and emission parameters. Compared with the ANFIS model alone, the hybrid model shows higher prediction accuracy.

In addition, some researchers have trained models specifically on the data of emissions. For example, Zhang et al. [137] use supervised ML tools to handle complex emission prediction tasks. The specific models used include multi-output least squares support vector regression and two types of ANN models (cascaded feedforward neural network and multilayer perceptron neural network). The results show that mls-svr model performs best in predicting CO_2, CO, and NO_x emissions.

5. Conclusions

Biodiesel is increasingly becoming an important alternative to fossil fuels due to its environmental and economic benefits. The development of predictive models and optimization techniques is crucial for the widespread application of biodiesel.

This article reviews the prediction models and optimization techniques for the characteristics and engine performance of biodiesel in recent years, analyzes statistical modeling, ML, and evolutionary algorithms, and emphasizes their effectiveness in predicting biodiesel characteristics (such as viscosity, density, and CN) and their impact on engine performance indicators (such as BTE and BSFC).

These studies indicate that combining experimental data with advanced computing techniques to improve prediction accuracy can provide important reference opinions for optimizing biodiesel formulations and engine tuning. Although significant progress has been made in this type of research, there are still challenges, such as relying on the data quality and interpretability of complex models. Future research can focus on improving these models and exploring hybrid methods to enhance their predictability and applicability in biodiesel engine systems. Research on the performance prediction of biodiesel can be expanded to include biodiesel derived from non-traditional biomass sources such as microalgae and waste oils. Given that many countries are promoting policies for the use of biodiesel, future studies could focus on increasing the proportion of biodiesel in fuel blends to develop comprehensive predictive models. This would broaden the applicability of these models. Additionally, research can be conducted to predict engine performance using biodiesel under specific operating conditions, further enhancing the practical relevance of these models.

Author Contributions: Conceptualization, W.A.; methodology, W.A.; software, W.A.; validation, W.A. and H.M.C.; formal analysis, W.A.; investigation, W.A.; resources, H.M.C.; data curation, W.A.; writing—original draft preparation, W.A.; writing—review and editing, W.A. and H.M.C.; visualization, W.A.; supervision, H.M.C.; project administration, H.M.C.; funding acquisition, H.M.C. All authors have read and agreed to the published version of the manuscript.

Funding: This work was supported by the National Research Foundation of Korea (NRF) grant funded by the Korean government (NRF-2022H1A7A2A02000033).

Conflicts of Interest: The authors declare no conflicts of interest.

Nomenclature

ANFIS	Adaptive neuro-fuzzy inference system
ANN	Artificial neural network
BRT	Enhanced regression tree
BSFC	Brake specific fuel consumption
BTE	Brake thermal efficiency
CFD	Computational fluid dynamics
CN	Cetane number
CO	Carbon monoxide
CTE	Coefficient of thermal expansion
FAMEs	Fatty acid methyl esters
GA	Genetic algorithm
GEP	Gene expression programming
GP	Genetic programming
GPR	Gaussian process regression
HC	Hydrocarbons
LSSVM	Least square support vector machine
ML	Machine learning
NO_x	Nitrogen oxide
PM	Particulate matter
PSO	Particle swarm optimization

RMSE	Root mean square error
RSM	Response surface methodology
SVM	Support vector machine

References

1. Su, Y.; Zhang, P.; Su, Y. An Overview of Biofuels Policies and Industrialization in the Major Biofuel Producing Countries. *Renew. Sustain. Energy Rev.* **2015**, *50*, 991–1003. [CrossRef]
2. de Souza, T.A.Z.; Pinto, G.M.; Julio, A.A.V.; Coronado, C.J.R.; Perez-Herrera, R.; Siqueira, B.O.P.S.; da Costa, R.B.R.; Roberts, J.J.; Palacio, J.C.E. Biodiesel in South American Countries: A Review on Policies, Stages of Development and Imminent Competition with Hydrotreated Vegetable Oil. *Renew. Sustain. Energy Rev.* **2022**, *153*, 111755. [CrossRef]
3. Giakoumis, E.G.; Dimaratos, A.M.; Rakopoulos, C.D.; Rakopoulos, D.C. Combustion Instability during Starting of Turbocharged Diesel Engine Including Biofuel Effects. *J. Energy Eng.* **2017**, *143*, 4016047. [CrossRef]
4. Setiawan, I.C.; Setiyo, M. Renewable and Sustainable Green Diesel (D100) for Achieving Net Zero Emission in Indonesia Transportation Sector. *Automot. Exp.* **2022**, *5*, 1–2. [CrossRef]
5. Maawa, W.N.; Mamat, R.; Najafi, G.; De Goey, L.P.H. Performance, Combustion, and Emission Characteristics of a CI Engine Fueled with Emulsified Diesel-Biodiesel Blends at Different Water Contents. *Fuel* **2020**, *267*, 117265. [CrossRef]
6. Giakoumis, E.G.; Rakopoulos, C.D.; Dimaratos, A.M.; Rakopoulos, D.C. Combustion Noise Radiation during the Acceleration of a Turbocharged Diesel Engine Operating with Biodiesel or N-Butanol Diesel Fuel Blends. *Proc. Inst. Mech. Eng. Part D J. Automob. Eng.* **2012**, *226*, 971–986. [CrossRef]
7. Khan, I.U. Analysis of Biodiesel and Fatty Acids Using State-of-the-Art Methods from Non-Edible Plants Seed Oil; Nicotiana Tobaccum and Olea Ferruginia. *Process Saf. Environ. Prot.* **2024**, *186*, 25–36. [CrossRef]
8. Krishnasamy, A.; Bukkarapu, K.R. A Comprehensive Review of Biodiesel Property Prediction Models for Combustion Modeling Studies. *Fuel* **2021**, *302*, 121085. [CrossRef]
9. Palani, Y.; Devarajan, C.; Manickam, D.; Thanikodi, S. Performance and Emission Characteristics of Biodiesel-Blend in Diesel Engine: A Review. *Environ. Eng. Res.* **2022**, *27*, 200338. [CrossRef]
10. Bukkarapu, K.R.; Krishnasamy, A. A Critical Review on Available Models to Predict Engine Fuel Properties of Biodiesel. *Renew. Sustain. Energy Rev.* **2022**, *155*, 111925. [CrossRef]
11. Suvarna, M.; Jahirul, M.I.; Aaron-Yeap, W.H.; Augustine, C.V.; Umesh, A.; Rasul, M.G.; Günay, M.E.; Yildirim, R.; Janaun, J. Predicting Biodiesel Properties and Its Optimal Fatty Acid Profile via Explainable Machine Learning. *Renew. Energy* **2022**, *189*, 245–258. [CrossRef]
12. Xiao, H.; Wang, W.; Bao, H.; Li, F.; Zhou, L. Biodiesel-Diesel Blend Optimized via Leave-One Cross-Validation Based on Kinematic Viscosity, Calorific Value, and Flash Point. *Ind. Crops Prod.* **2023**, *191*, 115914. [CrossRef]
13. Yusoff, M.N.A.M.; Zulkifli, N.W.M.; Sukiman, N.L.; Chyuan, O.H.; Hassan, M.H.; Hasnul, M.H.; Zulkifli, M.S.A.; Abbas, M.M.; Zakaria, M.Z. Sustainability of Palm Biodiesel in Transportation: A Review on Biofuel Standard, Policy and International Collaboration Between Malaysia and Colombia. *Bioenerg. Res.* **2021**, *14*, 43–60. [CrossRef] [PubMed]
14. Rakopoulos, C.D.; Antonopoulos, K.A.; Rakopoulos, D.C.; Hountalas, D.T.; Giakoumis, E.G. Comparative Performance and Emissions Study of a Direct Injection Diesel Engine Using Blends of Diesel Fuel with Vegetable Oils or Bio-Diesels of Various Origins. *Energy Convers. Manag.* **2006**, *47*, 3272–3287. [CrossRef]
15. Gebremariam, S.N.; Marchetti, J.M. Economics of Biodiesel Production: Review. *Energy Convers. Manag.* **2018**, *168*, 74–84. [CrossRef]
16. Mahlia, T.M.I.; Syazmi, Z.A.H.S.; Mofijur, M.; Abas, A.E.P.; Bilad, M.R.; Ong, H.C.; Silitonga, A.S. Patent Landscape Review on Biodiesel Production: Technology Updates. *Renew. Sustain. Energy Rev.* **2020**, *118*, 109526. [CrossRef]
17. Avinash, A.; Sasikumar, P.; Pugazhendhi, A. Analysis of the Limiting Factors for Large Scale Microalgal Cultivation: A Promising Future for Renewable and Sustainable Biofuel Industry. *Renew. Sustain. Energy Rev.* **2020**, *134*, 110250. [CrossRef]
18. Foteinis, S.; Chatzisymeon, E.; Litinas, A.; Tsoutsos, T. Used-Cooking-Oil Biodiesel: Life Cycle Assessment and Comparison with First- and Third-Generation Biofuel. *Renew. Energy* **2020**, *153*, 588–600. [CrossRef]
19. Hajjari, M.; Tabatabaei, M.; Aghbashlo, M.; Ghanavati, H. A Review on the Prospects of Sustainable Biodiesel Production: A Global Scenario with an Emphasis on Waste-Oil Biodiesel Utilization. *Renew. Sustain. Energy Rev.* **2017**, *72*, 445–464. [CrossRef]
20. Fonseca, J.M.; Teleken, J.G.; De Cinque Almeida, V.; Da Silva, C. Biodiesel from Waste Frying Oils: Methods of Production and Purification. *Energy Convers. Manag.* **2019**, *184*, 205–218. [CrossRef]
21. Wang, W.; Gowdagiri, S.; Oehlschlaeger, M.A. The High-Temperature Autoignition of Biodiesels and Biodiesel Components. *Combust. Flame* **2014**, *161*, 3014–3021. [CrossRef]
22. Giakoumis, E.G.; Sarakatsanis, C.K. Estimation of Biodiesel Cetane Number, Density, Kinematic Viscosity and Heating Values from Its Fatty Acid Weight Composition. *Fuel* **2018**, *222*, 574–585. [CrossRef]
23. Giakoumis, E.G. A Statistical Investigation of Biodiesel Physical and Chemical Properties, and Their Correlation with the Degree of Unsaturation. *Renew. Energy* **2013**, *50*, 858–878. [CrossRef]
24. Knothe, G. Improving Biodiesel Fuel Properties by Modifying Fatty Ester Composition. *Energy Environ. Sci.* **2009**, *2*, 759–766. [CrossRef]

25. Li, F.; Liu, Z.; Ni, Z.; Wang, H. Effect of Biodiesel Components on Its Lubrication Performance. *J. Mater. Res. Technol.* **2019**, *8*, 3681–3687. [CrossRef]
26. Giakoumis, E.G.; Rakopoulos, C.D.; Dimaratos, A.M.; Rakopoulos, D.C. Exhaust Emissions of Diesel Engines Operating under Transient Conditions with Biodiesel Fuel Blends. *Prog. Energy Combust. Sci.* **2012**, *38*, 691–715. [CrossRef]
27. Verma, S.; Upadhyay, R.; Shankar, R.; Pandey, S.P. Performance and Emission Characteristics of Micro-Algae Biodiesel with Butanol and TiO_2 Nano-Additive over Diesel Engine. *Sustain. Energy Technol. Assess.* **2023**, *55*, 102975. [CrossRef]
28. Mishra, S.; Bukkarapu, K.R.; Krishnasamy, A. A Composition Based Approach to Predict Density, Viscosity and Surface Tension of Biodiesel Fuels. *Fuel* **2021**, *285*, 119056. [CrossRef]
29. Fathurrahman, N.A.; Ginanjar, K.; Devitasari, R.D.; Maslahat, M.; Anggarani, R.; Aisyah, L.; Soemanto, A.; Solikhah, M.D.; Thahar, A.; Wibowo, E.; et al. Long-Term Storage Stability of Incorporated Hydrotreated Vegetable Oil (HVO) in Biodiesel-Diesel Blends at Highland and Coastal Areas. *Fuel Commun.* **2024**, *18*, 100107. [CrossRef]
30. Dunn, R.O. Effect of Temperature on the Oil Stability Index (OSI) of Biodiesel. *Energy Fuels* **2008**, *22*, 657–662. [CrossRef]
31. Lanjekar, R.D.; Deshmukh, D. A Review of the Effect of the Composition of Biodiesel on NOx Emission, Oxidative Stability and Cold Flow Properties. *Renew. Sustain. Energy Rev.* **2016**, *54*, 1401–1411. [CrossRef]
32. Alptekin, E.; Canakci, M. Determination of the Density and the Viscosities of Biodiesel–Diesel Fuel Blends. *Renew. Energy* **2008**, *33*, 2623–2630. [CrossRef]
33. Ramírez-Verduzco, L.F.; García-Flores, B.E.; Rodríguez-Rodríguez, J.E.; del Rayo Jaramillo-Jacob, A. Prediction of the Density and Viscosity in Biodiesel Blends at Various Temperatures. *Fuel* **2011**, *90*, 1751–1761. [CrossRef]
34. Ramírez Verduzco, L.F. Density and Viscosity of Biodiesel as a Function of Temperature: Empirical Models. *Renew. Sustain. Energy Rev.* **2013**, *19*, 652–665. [CrossRef]
35. Krisnangkura, K.; Yimsuwan, T.; Pairintra, R. An Empirical Approach in Predicting Biodiesel Viscosity at Various Temperatures. *Fuel* **2006**, *85*, 107–113. [CrossRef]
36. Giakoumis, E.G.; Sarakatsanis, C.K. A Comparative Assessment of Biodiesel Cetane Number Predictive Correlations Based on Fatty Acid Composition. *Energies* **2019**, *12*, 422. [CrossRef]
37. Fassinou, W.F.; Sako, A.; Fofana, A.; Koua, K.B.; Toure, S. Fatty Acids Composition as a Means to Estimate the High Heating Value (HHV) of Vegetable Oils and Biodiesel Fuels. *Energy* **2010**, *35*, 4949–4954. [CrossRef]
38. Zheng, X.; Qu, D.; Bao, Y.; Qin, G.; Liu, Y.; Luo, Q. Experimental Studies of Thermal Conductivity of Three Biodiesel Compounds: Methyl Pentanoate, Methyl Octanoate, and Methyl Decanoate. *J. Chem. Eng. Data* **2022**, *67*, 45–53. [CrossRef]
39. Santos, D.Q.; de Lima, A.L.; de Lima, A.P.; Neto, W.B.; Fabris, J.D. Thermal Expansion Coefficient and Algebraic Models to Correct Values of Specific Mass as a Function of Temperature for Corn Biodiesel. *Fuel* **2013**, *106*, 646–650. [CrossRef]
40. Veza, I.; Spraggon, M.; Fattah, I.M.R.; Idris, M. Response Surface Methodology (RSM) for Optimizing Engine Performance and Emissions Fueled with Biofuel: Review of RSM for Sustainability Energy Transition. *Results Eng.* **2023**, *18*, 101213. [CrossRef]
41. Singh Pali, H.; Sharma, A.; Kumar, N.; Singh, Y. Biodiesel Yield and Properties Optimization from Kusum Oil by RSM. *Fuel* **2021**, *291*, 120218. [CrossRef]
42. Dubey, A.; Prasad, R.S.; Kumar Singh, J.; Nayyar, A. Optimization of Diesel Engine Performance and Emissions with Biodiesel-Diesel Blends and EGR Using Response Surface Methodology (RSM). *Clean. Eng. Technol.* **2022**, *8*, 100509. [CrossRef]
43. Özgür, C. Optimization of Biodiesel Yield and Diesel Engine Performance from Waste Cooking Oil by Response Surface Method (RSM). *Pet. Sci. Technol.* **2021**, *39*, 683–703. [CrossRef]
44. Saravanan, S.; Rajesh Kumar, B.; Varadharajan, A.; Rana, D.; Sethuramasamyraja, B.; Lakshmi Narayana rao, G. Optimization of DI Diesel Engine Parameters Fueled with Iso-Butanol/Diesel Blends—Response Surface Methodology Approach. *Fuel* **2017**, *203*, 658–670. [CrossRef]
45. Kumar, S. Comparison of Linear Regression and Artificial Neural Network Technique for Prediction of a Soybean Biodiesel Yield. *Energy Sources Part A Recovery Util. Environ. Eff.* **2020**, *42*, 1425–1435. [CrossRef]
46. Archontoulis, S.V.; Miguez, F.E. Nonlinear Regression Models and Applications in Agricultural Research. *Agron. J.* **2015**, *107*, 786–798. [CrossRef]
47. Maheshwari, N.; Balaji, C.; Ramesh, A. A Nonlinear Regression Based Multi-Objective Optimization of Parameters Based on Experimental Data from an IC Engine Fueled with Biodiesel Blends. *Biomass Bioenergy* **2011**, *35*, 2171–2183. [CrossRef]
48. Mairizal, A.Q.; Awad, S.; Priadi, C.R.; Hartono, D.M.; Moersidik, S.S.; Tazerout, M.; Andres, Y. Experimental Study on the Effects of Feedstock on the Properties of Biodiesel Using Multiple Linear Regressions. *Renew. Energy* **2020**, *145*, 375–381. [CrossRef]
49. Kumbhar, V.; Pandey, A.; Sonawane, C.R.; El-Shafay, A.S.; Panchal, H.; Chamkha, A.J. Statistical Analysis on Prediction of Biodiesel Properties from Its Fatty Acid Composition. *Case Stud. Therm. Eng.* **2022**, *30*, 101775. [CrossRef]
50. Alahmer, A.; Alahmer, H.; Handam, A.; Rezk, H. Environmental Assessment of a Diesel Engine Fueled with Various Biodiesel Blends: Polynomial Regression and Grey Wolf Optimization. *Sustainability* **2022**, *14*, 1367. [CrossRef]
51. Aghbashlo, M.; Peng, W.; Tabatabaei, M.; Kalogirou, S.A.; Soltanian, S.; Hosseinzadeh-Bandbafha, H.; Mahian, O.; Lam, S.S. Machine Learning Technology in Biodiesel Research: A Review. *Prog. Energy Combust. Sci.* **2021**, *85*, 100904. [CrossRef]
52. Ishola, N.B.; Epelle, E.I.; Betiku, E. Machine Learning Approaches to Modeling and Optimization of Biodiesel Production Systems: State of Art and Future Outlook. *Energy Convers. Manag. X* **2024**, *23*, 100669. [CrossRef]

53. Subramanian, K.; Sathiyagnanam, A.P.; Damodharan, D.; Sivashanmugam, N. Artificial Neural Network Based Prediction of a Direct Injected Diesel Engine Performance and Emission Characteristics Powered with Biodiesel. *Mater. Today Proc.* **2021**, *43*, 1049–1056. [CrossRef]
54. Aydın, M.; Uslu, S.; Bahattin Çelik, M. Performance and Emission Prediction of a Compression Ignition Engine Fueled with Biodiesel-Diesel Blends: A Combined Application of ANN and RSM Based Optimization. *Fuel* **2020**, *269*, 117472. [CrossRef]
55. Tuan Hoang, A.; Nižetić, S.; Chyuan Ong, H.; Tarelko, W.; Viet Pham, V.; Hieu Le, T.; Quang Chau, M.; Phuong Nguyen, X. A Review on Application of Artificial Neural Network (ANN) for Performance and Emission Characteristics of Diesel Engine Fueled with Biodiesel-Based Fuels. *Sustain. Energy Technol. Assess.* **2021**, *47*, 101416. [CrossRef]
56. Karimmaslak, H.; Najafi, B.; Band, S.S.; Ardabili, S.; Haghighat-Shoar, F.; Mosavi, A. Optimization of Performance and Emission of Compression Ignition Engine Fueled with Propylene Glycol and Biodiesel–Diesel Blends Using Artificial Intelligence Method of ANN-GA-RSM. *Eng. Appl. Comput. Fluid Mech.* **2021**, *15*, 413–425. [CrossRef]
57. Tuan, N.V.; Minh, D.Q.; Khoa, N.X.; Lim, O. A Study to Predict Ignition Delay of an Engine Using Diesel and Biodiesel Fuel Based on the ANN and SVM Machine Learning Methods. *ACS Omega* **2023**, *8*, 9995–10005. [CrossRef]
58. Bukkarapu, K.R.; Krishnasamy, A. Biodiesel Composition Based Machine Learning Approaches to Predict Engine Fuel Properties. *Proc. Inst. Mech. Eng. Part D J. Automob. Eng.* **2024**, *238*, 1844–1860. [CrossRef]
59. Jin, X.; Li, S.; Ye, H.; Wang, J.; Wu, Y.; Zhang, D.; Ma, H.; Sun, F.; Pugazhendhi, A.; Xia, C. Investigation and Optimization of Biodiesel Production Based on Multiple Machine Learning Technologies. *Fuel* **2023**, *348*, 128546. [CrossRef]
60. Bukkarapu, K.R.; Krishnasamy, A. Investigations on the Applicability of Machine Learning Algorithms to Optimize Biodiesel Composition for Improved Engine Fuel Properties. *Int. J. Engine Res.* **2024**, *25*, 1299–1314. [CrossRef]
61. Raj Bukkarapu, K.; Krishnasamy, A. Support Vector Regression Approach to Optimize the Biodiesel Composition for Improved Engine Performance and Lower Exhaust Emissions. *Fuel* **2023**, *348*, 128604. [CrossRef]
62. Almohana, A.I.; Almojil, S.F.; Kamal, M.A.; Alali, A.F.; Kamal, M.; Alkhatib, S.E.; Felemban, B.F.; Algarni, M. Theoretical Investigation on Optimization of Biodiesel Production Using Waste Cooking Oil: Machine Learning Modeling and Experimental Validation. *Energy Rep.* **2022**, *8*, 11938–11951. [CrossRef]
63. Lionus Leo, G.M.; Jayabal, R.; Srinivasan, D.; Chrispin Das, M.; Ganesh, M.; Gavaskar, T. Predicting the Performance and Emissions of an HCCI-DI Engine Powered by Waste Cooking Oil Biodiesel with Al_2O_3 and $FeCl_3$ Nano Additives and Gasoline Injection—A Random Forest Machine Learning Approach. *Fuel* **2024**, *357*, 129914. [CrossRef]
64. Kumar, D.; Chhibber, V.K.; Singh, A. Emissions Prediction of Cashew Nut Shell Liquid Biodiesel Using Machine Learning. *Natl. Acad. Sci. Lett.* **2022**, *45*, 397–400. [CrossRef]
65. Moayedi, H.; Aghel, B.; Foong, L.K.; Bui, D.T. Feature Validity during Machine Learning Paradigms for Predicting Biodiesel Purity. *Fuel* **2020**, *262*, 116498. [CrossRef]
66. Rajkumar, S.; Das, A.; Thangaraja, J. Integration of Artificial Neural Network, Multi-Objective Genetic Algorithm and Phenomenological Combustion Modelling for Effective Operation of Biodiesel Blends in an Automotive Engine. *Energy* **2022**, *239*, 121889. [CrossRef]
67. Baghban, A.; Kardani, M.N.; Mohammadi, A.H. Improved Estimation of Cetane Number of Fatty Acid Methyl Esters (FAMEs) Based Biodiesels Using TLBO-NN and PSO-NN Models. *Fuel* **2018**, *232*, 620–631. [CrossRef]
68. Slowik, A.; Kwasnicka, H. Evolutionary Algorithms and Their Applications to Engineering Problems. *Neural Comput. Applic* **2020**, *32*, 12363–12379. [CrossRef]
69. Alam, T.; Qamar, S.; Dixit, A.; Benaida, M. Genetic Algorithm: Reviews, Implementations, and Applications. *Int. J. Eng. Pedagog.* **2020**, *10*, 57–77. [CrossRef]
70. Kolakoti, A.; Jha, P.; Mosa, P.R.; Mahapatro, M.; Kotaru, T.G. Optimization and Modelling of Mahua Oil Biodiesel Using RSM and Genetic Algorithm Techniques. *Math. Models Eng.* **2020**, *6*, 134–146. [CrossRef]
71. Onukwuli, D.O.; Esonye, C.; Ofoefule, A.U.; Eyisi, R. Comparative Analysis of the Application of Artificial Neural Network-Genetic Algorithm and Response Surface Methods-Desirability Function for Predicting the Optimal Conditions for Biodiesel Synthesis from Chrysophyllum Albidum Seed Oil. *J. Taiwan Inst. Chem. Eng.* **2021**, *125*, 153–167. [CrossRef]
72. Shirneshan, A.; Bagherzadeh, S.A.; Najafi, G.; Mamat, R.; Mazlan, M. Optimization and Investigation the Effects of Using Biodiesel-Ethanol Blends on the Performance and Emission Characteristics of a Diesel Engine by Genetic Algorithm. *Fuel* **2021**, *289*, 119753. [CrossRef]
73. Zhang, Q.; Ogren, R.M.; Kong, S.-C. A Comparative Study of Biodiesel Engine Performance Optimization Using Enhanced Hybrid PSO–GA and Basic GA. *Appl. Energy* **2016**, *165*, 676–684. [CrossRef]
74. Bagal, D.K.; Patra, A.K.; Jeet, S.; Barua, A.; Pattanaik, A.K.; Patnaik, D. MCDM Optimization of Karanja Biodiesel Powered CI Engine to Improve Performance Characteristics Using Super Hybrid Taguchi-Coupled WASPAS-GA, SA, PSO Method. In *Next Generation Materials and Processing Technologies*; Bag, S., Paul, C.P., Baruah, M., Eds.; Springer: Singapore, 2021; pp. 491–503.
75. Alviso, D.; Artana, G.; Duriez, T. Prediction of Biodiesel Physico-Chemical Properties from Its Fatty Acid Composition Using Genetic Programming. *Fuel* **2020**, *264*, 116844. [CrossRef]
76. Kumar, V.; Kalita, K.; Madhu, S.; Ragavendran, U.; Gao, X.-Z. A Hybrid Genetic Programming–Gray Wolf Optimizer Approach for Process Optimization of Biodiesel Production. *Processes* **2021**, *9*, 442. [CrossRef]

77. Ghanbari, M.; Najafi, G.; Ghobadian, B.; Yusaf, T.; Carlucci, A.P.; Kiani Deh Kiani, M. Performance and Emission Characteristics of a CI Engine Using Nano Particles Additives in Biodiesel-Diesel Blends and Modeling with GP Approach. *Fuel* **2017**, *202*, 699–716. [CrossRef]
78. Zandie, M.; Ng, H.K.; Gan, S.; Muhamad Said, M.F.; Cheng, X. Review of the Advances in Integrated Chemical Kinetics-Computational Fluid Dynamics Combustion Modelling Studies of Gasoline-Biodiesel Mixtures. *Transp. Eng.* **2022**, *7*, 100102. [CrossRef]
79. Maksom, M.S.; Nasir, N.F.; Asmuin, N.; Rahman, M.F.A.; Khairulfuaad, R. Biodiesel Composition Effects on Density and Viscosity of Diesel-Biodiesel Blend: A CFD Study. *CFD Lett.* **2020**, *12*, 100–109. [CrossRef]
80. Dixit, S.; Kumar, A.; Kumar, S.; Waghmare, N.; Thakur, H.C.; Khan, S. CFD Analysis of Biodiesel Blends and Combustion Using Ansys Fluent. *Mater. Today Proc.* **2020**, *26*, 665–670. [CrossRef]
81. Zandie, M.; Ng, H.K.; Gan, S.; Muhamad Said, M.F.; Cheng, X. A Comprehensive CFD Study of the Spray Combustion, Soot Formation and Emissions of Ternary Mixtures of Diesel, Biodiesel and Gasoline under Compression Ignition Engine-Relevant Conditions. *Energy* **2022**, *260*, 125191. [CrossRef]
82. Zandie, M.; Ng, H.K.; Muhamad Said, M.F.; Cheng, X.; Gan, S. Performance of a Compression Ignition Engine Fuelled with Diesel-Palm Biodiesel-Gasoline Mixtures: CFD and Multi Parameter Optimisation Studies. *Energy* **2023**, *274*, 127346. [CrossRef]
83. Gowrishankar, S.; Krishnasamy, A. CFD Analysis of Combustion and Emission Characteristics of Biodiesel under Conventional and Late-Injection Based Premixed Combustion Conditions. *Fuel* **2023**, *351*, 129021. [CrossRef]
84. Thomas, J.J.; Manojkumar, C.V.; Sabu, V.R.; Nagarajan, G. Development and Validation of a Reduced Chemical Kinetic Model for Used Vegetable Oil Biodiesel/1-Hexanol Blend for Engine Application. *Fuel* **2020**, *273*, 117780. [CrossRef]
85. Jung, J.W.; Lim, Y.C.; Suh, H.K. A Study on the Mechanism Reduction and Evaluation of Biodiesel with the Change of Mechanism Reduction Factors. *Proc. Inst. Mech. Eng. Part D J. Automob. Eng.* **2020**, *234*, 3398–3413. [CrossRef]
86. Zandie, M.; Ng, H.K.; Gan, S.; Muhamad Said, M.F.; Cheng, X. Development of a Reduced Multi-Component Chemical Kinetic Mechanism for the Combustion Modelling of Diesel-Biodiesel-Gasoline Mixtures. *Transp. Eng.* **2022**, *7*, 100101. [CrossRef]
87. Zhang, L.; Ren, X.; Lan, Z. A Reduced Reaction Mechanism of Biodiesel Surrogates with Low Temperature Chemistry for Multidimensional Engine Simulation. *Combust. Flame* **2020**, *212*, 377–387. [CrossRef]
88. Razzaq, L.; Farooq, M.; Mujtaba, M.A.; Sher, F.; Farhan, M.; Hassan, M.T.; Soudagar, M.E.M.; Atabani, A.E.; Kalam, M.A.; Imran, M. Modeling Viscosity and Density of Ethanol-Diesel-Biodiesel Ternary Blends for Sustainable Environment. *Sustainability* **2020**, *12*, 5186. [CrossRef]
89. Mujtaba, M.A.; Kalam, M.A.; Masjuki, H.H.; Razzaq, L.; Khan, H.M.; Soudagar, M.E.M.; Gul, M.; Ahmed, W.; Raju, V.D.; Kumar, R.; et al. Development of Empirical Correlations for Density and Viscosity Estimation of Ternary Biodiesel Blends. *Renew. Energy* **2021**, *179*, 1447–1457. [CrossRef]
90. Nabipour, N.; Daneshfar, R.; Rezvanjou, O.; Mohammadi-Khanaposhtani, M.; Baghban, A.; Xiong, Q.; Li, L.K.B.; Habibzadeh, S.; Doranehgard, M.H. Estimating Biofuel Density via a Soft Computing Approach Based on Intermolecular Interactions. *Renew. Energy* **2020**, *152*, 1086–1098. [CrossRef]
91. Pustokhina, I.; Seraj, A.; Hafsan, H.; Mostafavi, S.M.; Alizadeh, S.M. Developing a Robust Model Based on the Gaussian Process Regression Approach to Predict Biodiesel Properties. *Int. J. Chem. Eng.* **2021**, *2021*, 5650499. [CrossRef]
92. Giwa, S.O.; Taziwa, R.T.; Sharifpur, M. Dependence of Composition-Based Approaches on Hybrid Biodiesel Fuel Properties Prediction Using Artificial Neural Network and Random Tree Algorithms. *Renew. Energy* **2023**, *218*, 119324. [CrossRef]
93. Lin, C.-Y.; Wu, X.-E. Determination of Cetane Number from Fatty Acid Compositions and Structures of Biodiesel. *Processes* **2022**, *10*, 1502. [CrossRef]
94. Bemani, A.; Xiong, Q.; Baghban, A.; Habibzadeh, S.; Mohammadi, A.H.; Doranehgard, M.H. Modeling of Cetane Number of Biodiesel from Fatty Acid Methyl Ester (FAME) Information Using GA-, PSO-, and HGAPSO- LSSVM Models. *Renew. Energy* **2020**, *150*, 924–934. [CrossRef]
95. Ghiasi, M.M.; Mohammadzadeh, O.; Zendehboudi, S. Reliable Connectionist Tools to Determine Biodiesel Cetane Number Based on Fatty Acids Methyl Esters Content. *Energy Convers. Manag.* **2022**, *264*, 115601. [CrossRef]
96. Rahaju, S.M.N.; Hananto, A.L.; Paristiawan, P.A.; Mohammed, A.T.; Opia, A.C.; Idris, M. Comparison of Various Prediction Model for Biodiesel Cetane Number Using Cascade-Forward Neural Network. *Automot. Exp.* **2023**, *6*, 4–13. [CrossRef]
97. Noushabadi, A.S.; Dashti, A.; Raji, M.; Zarei, A.; Mohammadi, A.H. Estimation of Cetane Numbers of Biodiesel and Diesel Oils Using Regression and PSO-ANFIS Models. *Renew. Energy* **2020**, *158*, 465–473. [CrossRef]
98. Samuel, O.D.; Kaveh, M.; Verma, T.N.; Okewale, A.O.; Oyedepo, S.O.; Abam, F.; Nwaokocha, C.N.; Abbas, M.; Enweremadu, C.C.; Khalife, E.; et al. Grey Wolf Optimizer for Enhancing *Nicotiana tabacum* L. Oil Methyl Ester and Prediction Model for Calorific Values. *Case Stud. Therm. Eng.* **2022**, *35*, 102095. [CrossRef]
99. Giuliano Albo, P.A.; Lago, S.; Wolf, H.; Pagel, R.; Glen, N.; Clerck, M.; Ballereau, P. Density, Viscosity and Specific Heat Capacity of Diesel Blends with Rapeseed and Soybean Oil Methyl Ester. *Biomass Bioenergy* **2017**, *96*, 87–95. [CrossRef]
100. Ait Belale, R.; M'hamdi Alaoui, F.E.; Chhiti, Y.; Sahibeddine, A.; Munoz Rujas, N.; Aguilar, F. Study on the Thermophysical Properties of Waste Cooking Oil Biodiesel Fuel Blends with 1-Butanol. *Fuel* **2021**, *287*, 119540. [CrossRef]
101. Magalhães, A.M.S.; Brentan, B.M.; Meirelles, A.J.A.; Maximo, G.J. Thermal Properties of Ethylic Biodiesel Blends and Solid Fraction Prediction Using Artificial Neural Networks. *Fluid Phase Equilibria* **2023**, *574*, 113885. [CrossRef]

102. Kiran, A.V.; Jayapriya, J.; Ravi, M. Evaluation and Predictive Model Development of Oxidative Stability of Biodiesel on Storage. *Chem. Eng. Commun.* **2016**, *203*, 676–682. [CrossRef]
103. Santos, S.M.; Nascimento, D.C.; Costa, M.C.; Neto, A.M.B.; Fregolente, L.V. Flash Point Prediction: Reviewing Empirical Models for Hydrocarbons, Petroleum Fraction, Biodiesel, and Blends. *Fuel* **2020**, *263*, 116375. [CrossRef]
104. Cunha, C.L.; Torres, A.R.; Luna, A.S. Multivariate Regression Models Obtained from Near-Infrared Spectroscopy Data for Prediction of the Physical Properties of Biodiesel and Its Blends. *Fuel* **2020**, *261*, 116344. [CrossRef]
105. Hoang, A.T. Prediction of the Density and Viscosity of Biodiesel and the Influence of Biodiesel Properties on a Diesel Engine Fuel Supply System. *J. Mar. Eng. Technol.* **2021**, *20*, 299–311. [CrossRef]
106. Zheng, Y.; Shadloo, M.S.; Nasiri, H.; Maleki, A.; Karimipour, A.; Tlili, I. Prediction of Viscosity of Biodiesel Blends Using Various Artificial Model and Comparison with Empirical Correlations. *Renew. Energy* **2020**, *153*, 1296–1306. [CrossRef]
107. Gopinath, A.; Puhan, S.; Nagarajan, G. Theoretical Modeling of Iodine Value and Saponification Value of Biodiesel Fuels from Their Fatty Acid Composition. *Renew. Energy* **2009**, *34*, 1806–1811. [CrossRef]
108. Zheng, X.; Bao, Y.; Qu, D.; Liu, Y.; Qin, G. Measurement and Modeling of Thermal Conductivity for Short Chain Methyl Esters: Methyl Butyrate and Methyl Caproate. *J. Chem. Thermodyn.* **2021**, *159*, 106486. [CrossRef]
109. Dharmalingam, B.; Annamalai, S.; Areeya, S.; Rattanaporn, K.; Katam, K.; Show, P.-L.; Sriariyanun, M. Bayesian Regularization Neural Network-Based Machine Learning Approach on Optimization of CRDI-Split Injection with Waste Cooking Oil Biodiesel to Improve Diesel Engine Performance. *Energies* **2023**, *16*, 2805. [CrossRef]
110. Patnaik, S.; Khatri, N.; Rene, E.R. Artificial Neural Networks-Based Performance and Emission Characteristics Prediction of Compression Ignition Engines Powered by Blends of Biodiesel Derived from Waste Cooking Oil. *Fuel* **2024**, *370*, 131806. [CrossRef]
111. Simsek, S.; Uslu, S.; Simsek, H. Proportional Impact Prediction Model of Animal Waste Fat-Derived Biodiesel by ANN and RSM Technique for Diesel Engine. *Energy* **2022**, *239*, 122389. [CrossRef]
112. Sharma, P. Artificial Intelligence-based Model Prediction of Biodiesel-fueled Engine Performance and Emission Characteristics: A Comparative Evaluation of Gene Expression Programming and Artificial Neural Network. *Heat Trans* **2021**, *50*, 5563–5587. [CrossRef]
113. Sule, A.; Abdul Latiff, Z.; Azman Abas, M.; Rozi Mohammed Perang, M.; Veza, I.; Faizullizam Roslan, M. Investigation of Dual Impact of Nanoparticles-Ethanol as Additive to Biodiesel-Diesel Fuel on an Engine Using Artificial Neural Network Prediction Model. *Mater. Today Proc.* **2023**, S2214785323050629. [CrossRef]
114. Kumar, V.; Choudhary, A.K. Prediction of the Performance and Emission Characteristics of Diesel Engine Using Diphenylamine Antioxidant and Ceria Nanoparticle Additives with Biodiesel Based on Machine Learning. *Energy* **2024**, *301*, 131746. [CrossRef]
115. Pitchaiah, S.; Juchelková, D.; Sathyamurthy, R.; Atabani, A.E. Prediction and Performance Optimisation of a DI CI Engine Fuelled Diesel–Bael Biodiesel Blends with DMC Additive Using RSM and ANN: Energy and Exergy Analysis. *Energy Convers. Manag.* **2023**, *292*, 117386. [CrossRef]
116. Taheri-Garavand, A.; Heidari-Maleni, A.; Mesri-Gundoshmian, T.; Samuel, O.D. Application of Artificial Neural Networks for the Prediction of Performance and Exhaust Emissions in IC Engine Using Biodiesel-Diesel Blends Containing Quantum Dot Based on Carbon Doped. *Energy Convers. Manag. X* **2022**, *16*, 100304. [CrossRef]
117. Khoobbakht, M.; Soleymani, M.; Kheiralipour, K.; Karimi, M. Predicting Performance Characteristics of an Engine Fueled by Algal Biodiesel-Diesel Using Response Surface Methodology. *Renew. Energy Res. Appl.* **2024**, *5*, 269–279. [CrossRef]
118. Umeuzuegbu, J.C.; Okiy, S.; Nwobi-Okoye, C.C.; Onukwuli, O.D. Computational Modeling and Multi-Objective Optimization of Engine Performance of Biodiesel Made with Castor Oil. *Heliyon* **2021**, *7*, e06516. [CrossRef]
119. Hosseini, S.H.; Taghizadeh-Alisaraei, A.; Ghobadian, B.; Abbaszadeh-Mayvan, A. Artificial Neural Network Modeling of Performance, Emission, and Vibration of a CI Engine Using Alumina Nano-Catalyst Added to Diesel-Biodiesel Blends. *Renew. Energy* **2020**, *149*, 951–961. [CrossRef]
120. Sharma, P.; Sharma, A.K.; Balakrishnan, D.; Manivannan, A.; Chia, W.Y.; Awasthi, M.K.; Show, P.L. Model-Prediction and Optimization of the Performance of a Biodiesel—Producer Gas Powered Dual-Fuel Engine. *Fuel* **2023**, *348*, 128405. [CrossRef]
121. Agrawal, T.; Gautam, R.; Agrawal, S.; Singh, V.; Kumar, M.; Kumar, S. Optimization of Engine Performance Parameters and Exhaust Emissions in Compression Ignition Engine Fueled with Biodiesel-Alcohol Blends Using Taguchi Method, Multiple Regression and Artificial Neural Network. *Sustain. Futures* **2020**, *2*, 100039. [CrossRef]
122. Kumar, A.N.; Kishore, P.S.; Raju, K.B.; Ashok, B.; Vignesh, R.; Jeevanantham, A.K.; Nanthagopal, K.; Tamilvanan, A. Decanol Proportional Effect Prediction Model as Additive in Palm Biodiesel Using ANN and RSM Technique for Diesel Engine. *Energy* **2020**, *213*, 119072. [CrossRef]
123. Kumar Singh, N.; Singh, Y.; Sharma, A.; Kumar, S. Diesel Engine Performance and Emission Analysis Running on Jojoba Biodiesel Using Intelligent Hybrid Prediction Techniques. *Fuel* **2020**, *279*, 118571. [CrossRef]
124. Benaissa, S.; Adouane, B.; Ali, S.M.; Mohammad, A. Effect of Hydrogen Addition on the Combustion Characteristics of Premixed Biogas/Hydrogen-Air Mixtures. *Int. J. Hydrogen Energy* **2021**, *46*, 18661–18677. [CrossRef]
125. Kamta Legue, D.R.; Ayissi, Z.M.; Babikir, M.H.; Obounou, M.; Ekobena Fouda, H.P. Experimental and Simulation of Diesel Engine Fueled with Biodiesel with Variations in Heat Loss Model. *Energies* **2021**, *14*, 1622. [CrossRef]
126. Liu, Z.; Yang, L.; Song, E.; Wang, J.; Zare, A.; Bodisco, T.A.; Brown, R.J. Development of a Reduced Multi-Component Combustion Mechanism for a Diesel/Natural Gas Dual Fuel Engine by Cross-Reaction Analysis. *Fuel* **2021**, *293*, 120388. [CrossRef]

127. Karami, R.; Rasul, M.G.; Khan, M.M.K.; Mahdi Salahi, M.; Anwar, M. Experimental and Computational Analysis of Combustion Characteristics of a Diesel Engine Fueled with Diesel-Tomato Seed Oil Biodiesel Blends. *Fuel* **2021**, *285*, 119243. [CrossRef]
128. Işcan, B. ANN Modeling for Justification of Thermodynamic Analysis of Experimental Applications on Combustion Parameters of a Diesel Engine Using Diesel and Safflower Biodiesel Fuels. *Fuel* **2020**, *279*, 118391. [CrossRef]
129. Alhikami, A.F.; Yao, C.-E.; Wang, W.-C. A Study of the Spray Ignition Characteristics of Hydro-Processed Renewable Diesel, Petroleum Diesel, and Biodiesel Using a Constant Volume Combustion Chamber. *Combust. Flame* **2021**, *223*, 55–64. [CrossRef]
130. Sharma, P.; Sahoo, B.B. Precise Prediction of Performance and Emission of a Waste Derived Biogas–Biodiesel Powered Dual–Fuel Engine Using Modern Ensemble Boosted Regression Tree: A Critique to Artificial Neural Network. *Fuel* **2022**, *321*, 124131. [CrossRef]
131. Rakopoulos, C.D.; Rakopoulos, D.C.; Hountalas, D.T.; Giakoumis, E.G.; Andritsakis, E.C. Performance and Emissions of Bus Engine Using Blends of Diesel Fuel with Bio-Diesel of Sunflower or Cottonseed Oils Derived from Greek Feedstock. *Fuel* **2008**, *87*, 147–157. [CrossRef]
132. Rakopoulos, C.D.; Dimaratos, A.M.; Giakoumis, E.G.; Rakopoulos, D.C. Investigating the Emissions during Acceleration of a Turbocharged Diesel Engine Operating with Bio-Diesel or n-Butanol Diesel Fuel Blends. *Energy* **2010**, *35*, 5173–5184. [CrossRef]
133. Giakoumis, E.G.; Rakopoulos, C.D.; Rakopoulos, D.C. Assessment of NOx Emissions during Transient Diesel Engine Operation with Biodiesel Blends. *J. Energy Eng.* **2014**, *140*, A4014004. [CrossRef]
134. Rakopoulos, D.C.; Rakopoulos, C.D.; Giakoumis, E.G. Impact of Properties of Vegetable Oil, Bio-Diesel, Ethanol and n -Butanol on the Combustion and Emissions of Turbocharged HDDI Diesel Engine Operating under Steady and Transient Conditions. *Fuel* **2015**, *156*, 1–19. [CrossRef]
135. Giakoumis, E.G. A Statistical Investigation of Biodiesel Effects on Regulated Exhaust Emissions during Transient Cycles. *Appl. Energy* **2012**, *98*, 273–291. [CrossRef]
136. Singh, N.K.; Singh, Y.; Sharma, A.; Rahim, E.A. Prediction of Performance and Emission Parameters of Kusum Biodiesel Based Diesel Engine Using Neuro-Fuzzy Techniques Combined with Genetic Algorithm. *Fuel* **2020**, *280*, 118629. [CrossRef]
137. Zhang, L.; Zhu, G.; Chao, Y.; Chen, L.; Ghanbari, A. Simultaneous Prediction of CO_2, CO, and NOx Emissions of Biodiesel-Hydrogen Blend Combustion in Compression Ignition Engines by Supervised Machine Learning Tools. *Energy* **2023**, *282*, 128972. [CrossRef]

energies

MDPI

Article

Numerical Analysis of Diesel Engine Combustion and Performance with Single-Component Surrogate Fuel

Mehedi Hassan Pranta and Haeng Muk Cho *

Department of Mechanical Engineering, Kongju National University, Cheonan 31080, Republic of Korea; mhp432@gmail.com
* Correspondence: hmcho@kongju.ac.kr; Tel.: +82-(10)-87113252

Abstract: Compression ignition engines are widely recognized for their reliability and efficiency, remaining essential for transportation and power generation despite the transition toward sustainable energy solutions. This study employs ANSYS Forte to analyze the combustion and performance characteristics of a direct-injection, single-cylinder, four-stroke engine fueled with an n-heptane-based diesel surrogate. The investigation considers varying SOI timings (−32.5°, −27.5°, −22.5°, and −17.5° BTDC) and EGR rates (0%, 15%, 30%, 45%, and 60%). The simulation incorporates the RNG k-ε turbulence model, the power-law combustion model, and the KH-RT spray breakup model. The results indicate that the optimal peak pressure and temperature occur at an SOI of −22.5° BTDC with 0% EGR. Advancing SOI enhances oxidation, reducing NOx and CO emissions but increasing UHC due to delayed fuel–air mixing. Higher EGR rates lower in-cylinder pressure, temperature, HRR, and NOx emissions while elevating CO and UHC levels due to oxygen depletion and incomplete combustion. These findings highlight the trade-offs between combustion efficiency and emissions, emphasizing the need for optimized SOI and EGR strategies to achieve balanced engine performance.

Keywords: IC engine; engine performance; exhaust emission; numerical analysis; FORTE

1. Introduction

Internal combustion engines have long been the dominant power source in the transportation sector, with hydrocarbons serving as the primary fuel for over a century. The increasing demand for heavy-duty vehicles is expected to drive diesel consumption at a higher rate than gasoline. However, advancements in engine efficiency—achieved through downsizing, turbocharging, and reductions in vehicle weight—are anticipated to mitigate the growth in gasoline consumption [1]. The rapid expansion of the automotive industry and the rising number of vehicles have significantly contributed to elevated exhaust gas emissions, exacerbating environmental concerns and accelerating the depletion of fossil fuel reserves [2,3]. In response to stringent emissions regulations and the growing emphasis on ecological sustainability, the engine industry has intensified research into alternative energy sources for next-generation powertrains [4,5].

Combustion characteristics and emission profiles are largely dictated by engine design and operational parameters, such as pressure dynamics, fuel spray models, and injection timing, which play a pivotal role in overall engine performance [6,7]. To address future demands for high-efficiency and low-emission combustion systems, researchers have explored various advanced combustion strategies. Among these, low-temperature combustion (LTC) has emerged as a promising approach, integrating several modern com-

https://doi.org/10.3390/en18051082

bustion techniques to achieve significant NOx reductions. However, LTC requires a high activation energy threshold, posing challenges in its practical implementation [8,9].

In the automotive sector, exhaust gas recirculation (EGR) has gained popularity due to its ability to significantly lower engine emissions, especially NOx emissions, a dangerous environmental pollutant that must be minimized [10,11]. Part of the engine's exhaust gasses is diverted and reintroduced into the intake mechanism to properly dilute the entering air–fuel mixture [12]. Instead of depending exclusively on aftertreatment systems, this strategy seeks to reduce pollutant production during the combustion process, which has been investigated as an alternative to conventional post-treatment techniques [13]. Recirculation of exhaust gasses has been widely recognized as an effective strategy for reducing engine emissions. By introducing recirculated exhaust gasses into the intake air, the overall oxygen concentration is reduced, leading to a lower peak combustion temperature [12]. Since NOx formation is highly temperature-dependent, this reduction in peak temperature significantly inhibits its production [13]. The primary objective of the EGR system is not only to decrease combustion temperatures but also to ensure they remain below 1500 °C (1750 K), a critical threshold beyond which NOx formation occurs through a chain reaction mechanism. By maintaining combustion temperatures below this limit, EGR effectively mitigates NOx emissions while contributing to improved environmental performance. Moreover, adding ineffective exhaust gasses to the blend of air and fuel can produce more uniform combustion surroundings, which may help lower particulate matter and other controlled pollutants that are generally linked to concentrated regions of higher temperatures and substantial air–fuel proportions [12,14]. The EGR has been widely recognized for its ability to enhance engine efficiency while significantly reducing NOx emissions. As the EGR rate increases, NOx emissions decrease due to lower combustion temperatures, contributing to improved environmental performance [11,15,16]. However, several studies have reported that EGR reduces the oxygen concentration in the intake air, leading to incomplete combustion and a marginal decline in brake power and brake thermal efficiency [17]. Additionally, the increased presence of inert gasses within the combustion chamber elevates specific fuel consumption [18]. Nevertheless, recent research indicates that optimizing EGR rates and incorporating alternative fuels can effectively counteract efficiency losses while maintaining the NOx reduction benefits [19].

The combustion and performance characteristics of an engine are profoundly influenced by the timing of fuel injection, commonly referred to as the start of injection (SOI). Modifying the timing of both main and pilot injections significantly affects engine efficiency as well as emissions of carbon monoxide (CO), nitrogen oxides (NOx), and unburned hydrocarbons (HCs), as it alters multiple aspects of combustion dynamics [20]. One of the primary advantages of direct fuel injection over conventional port-fuel injection systems is the precise control it offers over injection timing, enabling more effective regulation of fuel–air mixture formation and combustion processes [21]. Additionally, variations in injection angles have been shown to influence combustion temperature and in-cylinder pressure, further impacting emissions of HC, NOx, and CO [22].

Direct injection technology provides several key benefits, including enhanced volumetric efficiency, reduced heat losses, and the ability to operate at higher compression ratios, all of which contribute to improved engine performance and fuel efficiency. As a result, the optimization of fuel injection strategies has become a critical research area for both academics and original equipment manufacturers (OEMs). Particular attention has been given to leveraging direct injection for advanced combustion techniques, such as stratified charge direct injection, which aim to maximize efficiency while minimizing pollutant formation [23]. The precise timing of fuel injection plays a crucial role in shaping combustion behavior by influencing fuel spray dispersion, turbulence interactions, and in-cylinder air motion, all of which directly impact combustion efficiency and emissions [23].

A comprehensive understanding of in-cylinder flow dynamics and fuel–air mixture formation is essential for optimizing combustion characteristics and minimizing emissions [24]. While blending alternative fuels with diesel or gasoline can effectively reduce emissions, it often results in a slight compromise in brake power output [25]. Diesel and gasoline fuels consist of hundreds of chemical compounds, making it impractical to model their combustion behavior with a complete reaction mechanism (RM) that accounts for all chemical species. This challenge extends to diesel substitutes as well. Consequently, surrogate fuels—simplified fuel models that emulate the key physicochemical properties of real fuels—are widely used to facilitate combustion research [24].

Single-component surrogates such as *n*-dodecane and n-heptane are well-characterized in terms of their physicochemical properties, making them valuable in studies related to spray atomization, chemical kinetic modeling, and combustion simulation [26]. In numerous combustion investigations, an n-heptane-based surrogate has been widely used to reproduce the characteristics of diesel fuel and has generated a great deal of scientific interest. Diesel fuel combustion's varied characteristics can lead to unfavorable engine performance and adverse environmental consequences, including loud knocking and high emissions of particulate matter and nitrogen oxides [27]. An n-heptane-based surrogate is widely accepted for diesel engine simulations due to its well-defined combustion characteristics, simplified chemical kinetics, and computational efficiency [28]. Diesel is a complex mixture of hydrocarbons, making multi-component surrogates more representative but computationally demanding. n-heptane (C_7H_{16}) provides a practical alternative by closely approximating key diesel combustion properties such as autoignition delay and ignition temperature [29]. With a cetane number of approximately 56, it effectively replicates ignition behavior, making it useful for autoignition and combustion studies [30]. Furthermore, its well-documented chemical reaction mechanisms enable reduced kinetic models, allowing for faster, more accurate numerical simulations compared to multi-component surrogates, which require solving intricate reaction pathways [31]. Despite these advantages, n-heptane has notable limitations, primarily its inability to capture the aromatic and cycloalkane components of real diesel, leading to discrepancies in soot formation and emissions predictions [32]. Additionally, unlike diesel fuels, which exhibit a distribution of ignition delays due to their diverse hydrocarbon composition, n-heptane follows a single ignition delay behavior, making multi-component surrogates more suitable for advanced combustion strategies like low-temperature combustion [33]. While these limitations restrict its applicability for precise emissions modeling, n-heptane remains highly valuable in engineering applications where computational efficiency and fundamental ignition studies are prioritized. It is widely employed in computational fluid dynamics (CFD) simulations and chemical kinetics modeling due to its well-established reaction mechanisms and its ability to approximate ignition delay and flame propagation under high-pressure diesel engine conditions [30]. However, for real-world diesel performance optimization and emissions analysis, more complex surrogates incorporating aromatics and cycloalkanes may be preferable [34]. Despite this, n-heptane remains a practical and widely accepted surrogate for diesel engine simulations, particularly when prioritizing computational efficiency and fundamental combustion studies. This numerical study aims to analyze the behavior of a diesel engine under varying SOI timing and EGR conditions while ensuring computational efficiency by using n-heptane-based surrogate fuel.

The study considers EGR levels from 0% to 60% to comprehensively evaluate its impact on combustion characteristics and emissions. While excessive EGR (>45%) can introduce combustion instability due to oxygen dilution and lower in-cylinder temperatures, investigating a wide range of EGR rates provides valuable insights into the transition from stable to unstable combustion regimes [35,36]. Additionally, high EGR rates significantly influence HC and CO emissions due to incomplete oxidation, making it essential to quantify the trade-offs between NOx reduction and combustion efficiency [37]. By extending the analysis up to 60% EGR, this study aims to establish critical limits beyond which combustion performance deteriorates, contributing to a more comprehensive understanding of the practical feasibility of high EGR applications in diesel engines.

CFD simulation has been used to continue the analysis since numerical analysis has become a valuable tool for studying and developing engine control techniques because of its greater flexibility and cheaper cost. It plays a crucial role in combustion research by providing insights into complex processes such as fluid flow, heat transfer, and chemical reactions [38]. The simulation used ANSYS 2022 R1 Forte, one of the most excellent tools for analyzing engine efficiency and emissions, to make optimization decisions [39]. This application can be used to analyze several engine data, including temperature, fuel consumption, in-cylinder pressure, and emissions of HC, NOx, and CO. This software simulates internal combustion engine combustion procedures by combining liquid fuel spray, turbulent gas dynamics, and comprehensive chemical kinetics in a highly effective way [40].

2. Materials and Methods

2.1. Simulation Setup

The operation of an IC engine was simulated using the FORTE 2022 R1 software, developed by ANSYS. This software integrates CFD for modeling liquid fuel spray, turbulent gas dynamics, and transport phenomena with an industry-standard chemical kinetics solver (CHEMKIN files). The inclusion of CHEMKIN ensures accurate resolution of chemical reaction mechanisms and thermodynamic properties, enhancing the fidelity of combustion simulations.

In this study, a Cummins N-14 single-cylinder diesel engine was utilized. The engine configuration consisted of one exhaust valve and two intake valves, with direct injection implemented in a quiescent combustion chamber. The detailed engine specifications are presented in Table 1 [41].

Table 1. Engine specification.

Parameters	Unit	Description
Engine type		Cummins N-14 DI diesel
Number of cylinders		1
Swirl ratio		0.5
Engine RPM		1200
Bore	cm	13.97
Stroke	cm	15.24
Displacement	L	2.34
Connecting rod length	cm	30.48
Compression ratio		16.0
Fuel injector type		Common rail
Number of nozzles		8, equally spaced

Diesel fuel combustion was modeled using a simplified *n*-heptane chemical mechanism. The chemical kinetics were represented using a standard ANSYS Chemkin Pro chemistry set, incorporating a semi-detailed *n*-heptane mechanism comprising 36 species and 74 reactions [42]. The key chemical properties of this *n*-heptane-based surrogate fuel are summarized in Table 2 [43].

Table 2. Chemical properties.

Fuel Properties	Unit	Value
Density at 298 K	kg/m^3	684
Boiling point	K	371
Ignition temperature	K	496
Lower heating value	KJ/kg	44,926
Heat of vaporization	KJ/mol	31.77
Viscosity	mPa·s	0.42
Cetane number		56

The lower density of n-heptane (684 kg/m^3) compared to conventional diesel (850 kg/m^3) poses significant challenges for high-pressure common rail systems, such as those in Cummins N-14 DI engines operating at ~1000 bar, particularly in maintaining consistent volumetric energy delivery and equivalence ratios. To address these challenges, this study employed ANSYS Forte simulations with meticulously calibrated injection parameters, including pressure and duration, to ensure accurate fuel mass delivery and combustion modeling. While the reduced density of n-heptane could theoretically increase leakage risks in diesel-optimized fuel systems, the simulations assumed idealized sealing conditions, a well-established approach in surrogate fuel studies that effectively isolates combustion dynamics from mechanical inefficiencies. This methodology enables a robust representation of spray development, mixture formation, and combustion phasing, providing critical insights into the adaptability of common rail systems to alternative fuel properties without compromising analytical fidelity.

2.2. Governing Equations

Internal combustion engines utilize both the air–fuel mixture before combustion and the burned byproducts after combustion as working fluids. To simulate the turbulent reacting flows in diesel engines, ANSYS Forte was employed, incorporating turbulent flows governed by the Navier–Stokes Equation with chemical blending and reaction. The Eulerian-Lagrangian formulation method was applied to model reactive spray, a key characteristic of diesel engine combustion [44]. The RNG k-ε turbulence model was selected due to its capability to accurately depict turbulent transport in highly sheared circulatory flows at a lower computational cost [45]. This model enhances estimations of swirling and recirculating flows for spray and combustion simulations by incorporating an additional component into the dissipation rate equation, improving precision in highly strained flows [46]. However, while the RNG k-ε model is more accurate than the standard k-ε model, it struggles with identifying significant anisotropic turbulence and compressibility effects, which can be critical in transient diesel combustion scenarios. Alternative turbulence models, such as Large Eddy Simulation (LES) and the Reynolds Stress Model (RSM), have demonstrated superior accuracy in analyzing complex flow systems, though at a significantly higher computational cost [47].

To accurately capture flame propagation dynamics, the G-equation combustion model was utilized, while the KH-RT droplet breakup model was applied to simulate the atomization and breakup of diesel fuel spray, ensuring a realistic representation of spray-induced turbulence. Additionally, the Han–Reitz heat transfer model was implemented to enhance the accuracy of in-cylinder heat transfer predictions and thermal loss estimations [48]. To strike a balance between computational efficiency and predictive accuracy, a simplified chemical kinetics model based on an n-heptane single-fuel surrogate was used, incorporating 36 species and 74 reactions. This approach significantly reduced computational costs while preserving critical combustion characteristics. However, the simplification of reaction pathways may introduce discrepancies in ignition delay, flame speed, and pollutant formation, particularly for emissions such as NOx and soot [49]. A more detailed multi-step reaction mechanism, although computationally demanding, could improve forecasting accuracy by accounting for intermediate species and complex oxidation pathways [50].

2.2.1. Eulerian Phase

The fuel's vapor phase and the surrounding air are both included in the Eulerian phase. The Navier–Stokes equations of mass, momentum, and energy conservation represent the motion of the Eulerian phase. Equation (1) is applied for mass conservation:

$$\frac{\partial(\rho v)}{\partial t} + \frac{\partial(\rho v u_i)}{\partial x_i} = S_m, \tag{1}$$

where ρv is the steam density and u_i represents the Eulerian velocity. S_m does droplet evaporation derive the source. This phrase makes the link between the Eulerian and Lagrangian phases possible. Using the term origin, the mass dissipated throughout the liquid phase's evaporation can be added to the Eulerian phase.

Equation (2) is utilized for the momentum conservation:

$$\frac{\partial(\rho u)}{\partial t} + \frac{\partial(\rho u u_i)}{\partial x_i} = -\frac{\partial p}{\partial x_i} + \frac{\partial \tau_{ij}}{\partial x_i} + S, \tag{2}$$

The stress tensor τ_{ij} is given by Equation (3):

$$\tau_{ij} = 2\mu S_{ij} - \frac{2}{3}\mu S_{kk}\delta_{ij} = \mu\left(\frac{\partial u_j}{\partial x_i} + \frac{\partial u_i}{\partial x_j} - \frac{2}{3}\frac{\partial u_k}{\partial x_k}\delta_{ij}\right), \tag{3}$$

where δ_{ij} represent as the Kronecker symbol.

Equation (4) is used for energy conservation:

$$\frac{\partial(\rho v e_t)}{\partial t} + \frac{\partial(\rho v e_t u_j)}{\partial x_j} = -\frac{\partial(p u_j)}{\partial x_j} + \frac{\partial(\tau_{ij} u_i)}{\partial x_j} - \frac{\partial q_i}{\partial x_j} + S_e, \tag{4}$$

Heat flux q_i and e_t are calculated through Equations (5) and (6).

$$e_t = \frac{p}{\gamma_v - 1} + \frac{1}{2}u_i u_i, \tag{5}$$

$$q_i = -k_v \frac{\partial T}{\partial x_j}, \tag{6}$$

where γ_v is an adiabatic index of gas and k_v represents the thermal conductivity index of gas. S_e is the inverse of heat change due to evaporation and conduction.

2.2.2. Lagrangian Phase

The liquid phase of the fuel is contained in the Lagrangian phase. Transfer parameters allow the two phases to be linked together, with this phase imposed on top of the Eulerian phase.

$$\frac{du_l}{dt} = F_T(u_v - u_l) + F_{MV} + \frac{g(\rho_l - \rho_v)}{\rho_l}, \tag{7}$$

Equation (7) represents the conservation of momentum where F_T is the drag force, F_{MV} is virtual mass. These can be solved using Equations (8) and (9)

$$F_T = \frac{18\mu_v C_D Re_r}{24} \cdot \frac{1}{\rho_l d_l^2}, \tag{8}$$

$$F_{MV} = \frac{1}{2}\frac{\rho_r}{\rho_l} \cdot \frac{d}{dt}(u_v - u_l), \tag{9}$$

2.2.3. Standard k-ε Model

A semi-empirical method, the k-ε model assumes that the flow is entirely turbulent and particle viscosity has little effect. This framework determines the turbulent kinetic energy k and its dissipation rate ε separately using two transport equations [44]. The model works well in several turbulent scenarios and is stable and computationally effective. Equations (10) and (11) represent the transport equations of k and ε.

$$\frac{\partial\overline{\rho}\widetilde{k}}{\partial t} + \nabla\cdot\left(\overline{\rho}\widetilde{u}\widetilde{k}\right) = -\frac{2}{3}\overline{\rho}\widetilde{u}\nabla\cdot\widetilde{u} + (\overline{\sigma} - \Gamma) : \nabla\widetilde{u} + \nabla\cdot\left(\frac{\mu + \mu_T}{Pr_k}\nabla\widetilde{k}\right) - \overline{\rho}\widetilde{\varepsilon} + \overline{\dot{W}}^s \tag{10}$$

$$\frac{\partial\overline{\rho}\widetilde{\varepsilon}}{\partial t} + \nabla\cdot\left(\overline{\rho}\widetilde{u}\widetilde{\varepsilon}\right) = -\left(\frac{2}{3}C_{\varepsilon1} - C_{\varepsilon3}\right)\overline{\rho}\widetilde{\varepsilon}\nabla\cdot\widetilde{u} + \nabla\cdot\left(\frac{v^+ v_T}{Pr_\varepsilon}\nabla\widetilde{\varepsilon}\right) + \frac{\widetilde{\varepsilon}}{\widetilde{k}}\left[C_{\varepsilon1}(\overline{\sigma} - \Gamma) : \nabla\widetilde{u} - C_{\varepsilon2}\overline{\rho}\widetilde{\varepsilon} + C_s\overline{\dot{W}}^s\right] \tag{11}$$

Here, Pr_k, Pr_ε, $C_{\varepsilon1}$, $C_{\varepsilon2}$, $C_{\varepsilon3}$ are constants of the model and $\overline{\dot{W}}^s$ represents the negative rate of the turbulent eddies of dispersing spray droplets.

2.2.4. RNG k-ε Model

The Reynolds Average Navier–Stokes (RANS) is a time-averaged equation and can be expressed as in Equation (12) [45].

$$\frac{\partial\overline{\rho}\widetilde{\varepsilon}}{\partial t} + \nabla\cdot\left(\overline{\rho}\widetilde{u}\widetilde{\varepsilon}\right) = -\left(\tfrac{2}{3}C_{\varepsilon1} - C_{\varepsilon3}\right)\overline{\rho}\widetilde{\varepsilon}\nabla\cdot\widetilde{u} + \nabla\cdot\left(\tfrac{v^+ v_T}{Pr_\varepsilon}\nabla\widetilde{\varepsilon}\right) + \tfrac{\widetilde{\varepsilon}}{\widetilde{k}}\left[C_{\varepsilon1}(\overline{\sigma} - \Gamma) : \nabla\widetilde{u} - C_{\varepsilon2}\overline{\rho}\widetilde{\varepsilon} + C_s\overline{\dot{W}}^s\right] - \overline{\rho}Rc \tag{12}$$

R can be defined from the following Equation (13)

$$R = \frac{C_\mu\eta^3\left(1 - \frac{\eta}{\eta_0}\right)\widetilde{\varepsilon}^2}{(1 + \beta\eta^3)\widetilde{k}}, \tag{13}$$

where η can be derived from Equation (14)

$$\eta = S\frac{\widetilde{k}}{\widetilde{\varepsilon}}, \tag{14}$$

Scalar quantity S can be derived from Equation (15)

$$S = \sqrt{2\overline{S} : \overline{S}}, \tag{15}$$

$\overline{S}$ represents the tensor of the mean strain rate, which can be derived from Equation (16)

$$\overline{S} = 0.5\left(\nabla\widetilde{u} + \nabla\widetilde{u}^T\right), \tag{16}$$

In the ANSYS Forte model, constant values Pr_k, Pr_ε, $C_{\varepsilon 1}$, $C_{\varepsilon 2}$, $C_{\varepsilon 3}$ were modified to account for the compressibility effect [45]. Equations (17)–(19) represent the constants' modification.

$$C_{\varepsilon_3} = \frac{-1 + 2C_{\varepsilon_2} - 3m(n-1) + (-1)^{\delta}\sqrt{6}C_\mu C_\eta \eta}{3}, \quad (17)$$

where $m = 0.5$, $n = 1.4$ for an ideal gas

$$R = \frac{\eta\left(1 - \frac{\eta}{\eta_0}\right)}{(1 + \beta\eta^3)}, \quad (18)$$

With

$$\delta = \begin{cases} 1\ if\ \nabla\tilde{u} < 0 \\ 0\ if\ \nabla\tilde{u} > 0 \end{cases}, \quad (19)$$

With this method, the value of C_ε fluctuates between −0.9 and 1.72621, and ANSYS Forte is automatically calculated based on the model's other constant specifications and flow circumstances. Table 3 shows the values of constants in both the standard k-ε model and the RNG k-ε model.

Table 3. Constant values in standard k-ε model and RNG k-ε model.

Model	C_μ	$C_{\varepsilon 1}$	C_{ε_2}	C_{ε_3}	$1/Pr_k$	$1/Pr_\varepsilon$	η_0	β
Standard k-ε model	0.09	1.44	1.92	−1.0	1.0	0.769	-	-
RNG k-ε model	0.0845	1.42	1.68	-	1.39	1.39	4.38	0.012

2.2.5. Turbulent Flame Speeds

The flame front and flame position are tracked using a Favre-averaged G-equation. Throughout combustion, the unburned zone's flame front has a G-equation of zero-G (x, t) < 0, but the burnt zone has a G (x, t) > 0 [51]. Equations (20)–(23) can be written as follows to calculate the flame speed [52,53].

$$\frac{\partial\tilde{G}}{\partial t} + \left(\tilde{\vec{u}} - \vec{u}_{vertex}\right) \cdot \nabla\tilde{G} = \frac{\bar{\rho}_u}{\bar{\rho}_b} S_T^O \cdot \left|\nabla\tilde{G}\right| - D_T\tilde{\kappa}\left|\nabla\tilde{G}\right|, \quad (20)$$

$$\frac{\partial\widetilde{G''^2}}{\partial t} + \tilde{\vec{u}} \cdot \nabla\widetilde{G''^2} = \bar{\bar{\nabla}}\left(\frac{\bar{\rho}_u}{\bar{\rho}_b} D_T \bar{\bar{\nabla}}\widetilde{G''^2}\right) + 2D_T\left(\nabla\tilde{G}\right)^2 - C_S\frac{\tilde{\varepsilon}}{\tilde{k}}\widetilde{G''^2}, \quad (21)$$

Here, $\bar{\bar{\nabla}}$ represents the tangential gradient operator, $\vec{u}$ fluid velocity, $\vec{u}_{vertex}$ is the moving vertex velocity, $\bar{\rho}_u$ and $\bar{\rho}_b$, respectively, denote the average densities of the unburned and burned blends. $\tilde{k}$ is the Favre mean flame front curvature and D_T denotes turbulent diffusivity. In the G-equation, S_T^O, the turbulent flame can be derived from the following equation.

$$\frac{S_T^O}{S_L^O} = 1 + I_P\left\{-\frac{a_4 b_3^2 l_I}{2b_1 l_F} + \left[\left(\frac{a_4 b_3^2 l_1}{2b_1 l_F}\right)^2 + a_4 b_3^2 \frac{\acute{u} l_I}{S_T^O l_F}\right]^{1/2}\right\}, \quad (22)$$

where C_S, a_4, b_1,b_3 are model constants, laminar flame thickness, and the turbulence integral length scale are, respectively, shown by l_I, and l_F. The laminar flame speed S_L^O can be derived from the following equation.

$$S_L^O = S_{L,\ ref}^O \left(\frac{T_u}{T_{u,ref}}\right)^{\alpha} \left(\frac{p}{p_{ref}}\right)^{\beta} F_{dil}, \quad (23)$$

Here, ref is denoted as the reference condition, which is 1 atm and 298 K. F_{dil} represents factor for the diluent's effect.

2.2.6. Injection Model

The spray division is captured using the KH-RT spray model [51]. Equations (24) and (25) represent the KH-RT spray model, where A_{KH} is the wavelength and Ω_{KH} is the rate of increase in the fastest-growing wave.

$$A_{KH} = \frac{9.02 r_0 \left(1 + 0.45\sqrt{Z}\right)\left(1 + 0.4T_a^{0.7}\right)}{\left(1 + 0.865We^{1.67}\right)^{0.7}}, \quad (24)$$

$$\Omega_{KH} = \frac{0.34 + 0.38We^{1.5}}{(1 + Z)(1 + 1.4T^{0.6})}\sqrt{\frac{\sigma}{\rho l r^3}}, \quad (25)$$

2.2.7. Exhaust Gas Recirculation

The exhaust gas recirculation (EGR) percentage can be calculated using Equation (26) [54].

$$\%of\ EGR = \frac{m_{EGR}}{m_a + m_f}, \quad (26)$$

where m_{EGR}, m_a, and m_f respectively represent the mass of EGR, mass of air, and mass of fuel.

Balanced stoichiometric combustion reaction for n-heptane with air is shown in Equation (27).

$$C_7H_{16} + 11(O_2 + 3.76N_2) = 7CO_2 + 8H_2O + 41.36N_2 \quad (27)$$

Molar mass of C_7H_{16} is 100 g/mol.

Equation (28) shows the stoichiometric air–fuel ratio and for n-heptane it is 15.1.

$$AFR = \frac{Mass\ of\ air}{Mass\ of\ fuel} \quad (28)$$

Mass of air, m_a is 15.1 m_f, then total mass of air and fuel is 16.1 m_f.

Using Equation (26), mass of EGR can be calculated. On the other hand, the mass of EGR is the sum of the masses of CO_2, H_2O, O_2, and N_2. Consequently, the value of mass of fuel, m_f can be obtained.

2.3. Computational Domain and Meshing

In ANSYS Forte, the Sector Mesh Generator is employed to construct the computational geometry required for simulations, optimizing mesh efficiency and reducing computational costs. Given the combustion chamber's design, which features eight fuel injection holes, the computational domain is divided into eight periodic sectors to enhance simulation accuracy while minimizing computational demand. The 45-degree sector model is generated based on key engine parameters, including bore, stroke, crevice width, and squish. Additionally, the engine bowl geometry is utilized to define the combustion chamber topology, ensuring an accurate representation of in-cylinder flow and combustion dynamics. Figure 1 provides a visualization of the 3D combustion chamber geometry and the mesh layout of the computational domain.

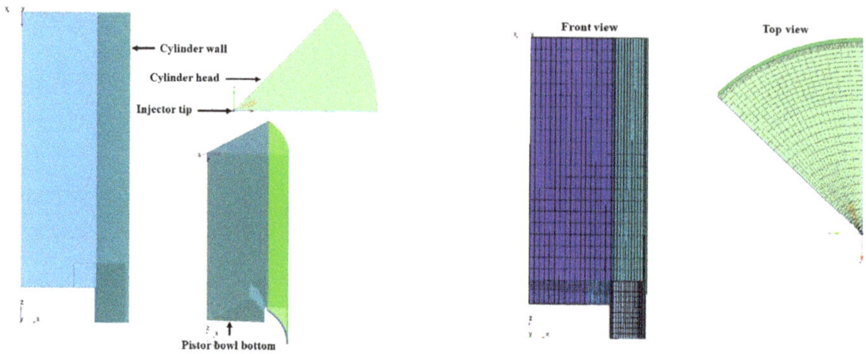

(a) (b)

Figure 1. (**a**) 3D geometry of the combustion chamber; (**b**) mesh layout of the computational domain.

Ansys Forte software employs an immersed boundary approach to generate real-time Cartesian volume meshes, which are automatically updated at each piston position based on specified criteria and refinement parameters. When no mesh refinement is applied, the global mesh size determines the background mesh resolution. Mesh refinement enhances surface definition and computational accuracy in critical regions, adapting dynamically to parameters such as velocity and temperature. Although finer meshes improve accuracy, they demand higher computational resources, necessitating a balance between precision and efficiency. To evaluate mesh sensitivity, six different grid resolutions were tested under identical boundaries and initial conditions. Table 4 presents a comparison of the cell numbers and their corresponding maximum pressure and temperature.

Table 4. Cell numbers and their corresponding maximum pressure and temperature.

Cell Numbers	Maximum Pressure (MPa)	Maximum Temperature (K)
1386	8.34	1215
8262	8.46	1236
15,560	8.48	1237
17,619	8.51	1249
21,784	8.51	1240
32,671	8.5	1238

Considering the computational time with 17,619 cells, it has considerable capacity for ensuring grid-independent solutions.

2.4. Boundary Conditions and Numerical Parameters

Boundary conditions play a crucial role in defining the model's scope and ensuring the accurate representation of physical phenomena in computational simulations. They explain the effects of these models before going into detail about new behaviors and connecting models from various physics theoretical frameworks. The boundary conditions employed in this study are detailed in Table 5. To match with the injection requirement for the cylinder displacement and proper air–fuel mixing, 53.5 mg of fuel was injected into the combustion chamber [55].

Table 5. Boundary conditions.

Variables	Unit	Description
Engine speed	rpm	1200
Piston temperature	K	500
Cylinder wall temperature	K	420
Cylinder head temperature	K	470
SOI	deg	−17.5, −22.5, −27.5, −32.5
Injection duration	deg	7.75
Injected fuel mass	mg	53.5
Heat transfer model		Han-Reitz
Droplet breakup model		Kelvin Helmholtz-Rayleigh Taylor (KH-RT)
Combustion model		G-equation

Additionally, various numerical parameters essential for the simulation setup are presented in Table 6.

Table 6. Numerical parameters.

Variables	Description
Turbulence model	RNG k-ε
Epsilon coefficient 1	1.42
Epsilon coefficient 2	1.68
Viscosity coefficient	0.085
RNG beta parameter	0.012
Epsilon spray coefficient	1.5
Droplet breakup model	KH-RT
Size constant of KH breakup	1.0
Time constant of KH breakup	40
Size constant of RT breakup	0.15
Time constant of RT breakup	1.0
RT distance constant	1.9
Droplet size distribution	Uniform
Reaction mechanism	n-haptane
No. of species	36
No. of reactions	74

2.5. Model Validation

The numerical simulation was performed using ANSYS Forte, and the model was validated to ensure the reliability and accuracy of the obtained results. For validation, the simulated data from the current study was compared with findings from existing literature. Figure 2 illustrates the simulation outcomes for the n-heptane combustion process, providing a comparative analysis of in-cylinder pressure between the present study and previously published data, which demonstrate sufficient consistency. Approximately 2.47% [56], 1.18% [57], 2.12% [58], and 6.46% [24] variation in maximum pressure have been identified compared to the current study, indicating that the current model can be used for additional analysis.

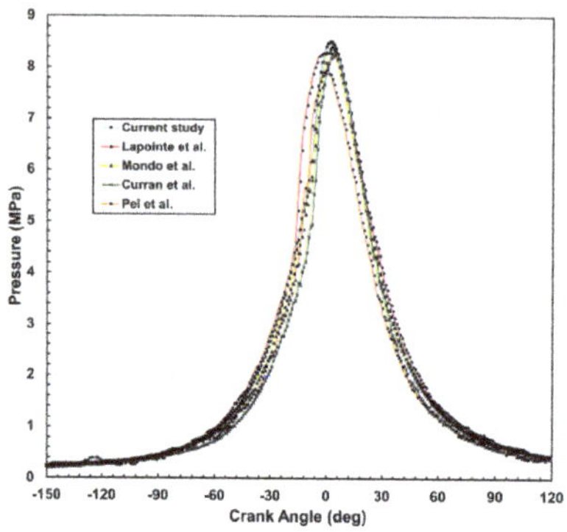

Figure 2. Comparison of in-cylinder pressure in the CI engine between the current study, Lapointe et al. [56], Mondo et al. [57], Curran et al. [58], and Pei et al. [24].

3. Result and Discussion

This study provides a comprehensive evaluation of engine performance and combustion characteristics for an n-heptane-based diesel surrogate, focusing on variations in fuel injection timing and EGR. The SOI is a critical parameter, as it directly influences combustion initiation, emissions, and overall engine performance. Optimizing injection timing can enhance these characteristics, leading to a more efficient combustion process.

EGR is a widely adopted technique for reducing NOx emissions by recirculating a portion of exhaust gasses into the intake air before combustion. The study examines key in-cylinder parameters, including pressure, temperature, heat release rate, CO emissions, unburned hydrocarbons, and NOx emission index (EINOx), across different crankshaft angles. These parameters were analyzed under varying SOI and EGR conditions to assess their impact on engine behavior.

For model validation against the existing literature, the SOI and EGR values were set at $-22.5°$ crank angle before the top dead center (BTDC) and 0%, respectively, in the current study. These conditions serve as the reference case for further analysis involving different SOI and EGR levels.

3.1. In-Cylinder Pressure and Temperature

In an internal combustion engine, in-cylinder pressure and temperature vary with crank angle due to changes in cylinder volume, combustion, and heat transfer. Figures 3 and 4 illustrate the variation in in-cylinder pressure in terms of SOI and EGR across crank angles. At 0% EGR and an SOI of $-22.5°$ BTDC, the peak pressure reaches 8.51 MPa, but it decreases by 3.1%, 1%, and 1.31% when the SOI is shifted to $-17.5°$, $-27.5°$, and $-32.5°$, respectively. This trend highlights the role of combustion phasing, where intermediate SOI values ($-22.5°$ to $-27.5°$ BTDC) achieve higher peak pressures compared to earlier ($-32.5°$ BTDC) or later ($-17.5°$ BTDC) injections. Earlier SOI provides extended time for fuel–air mixing, potentially enhancing charge homogeneity, yet excessively early injection may lead to wall-wetting and mixture cooling, reducing combustion efficiency. Conversely, delayed SOI limits mixture formation and flame propagation, leading to incomplete combustion and lower peak pressures. The optimal SOI balances mixture preparation and combustion phasing, maximizing pressure near top dead center (TDC) to improve engine work output.

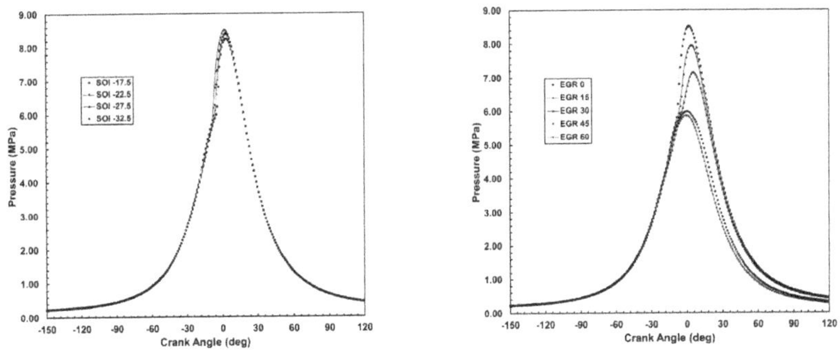

(a) (b)

Figure 3. In-cylinder pressure distribution at: (**a**) 0% EGR; (**b**) SOI −22.5°.

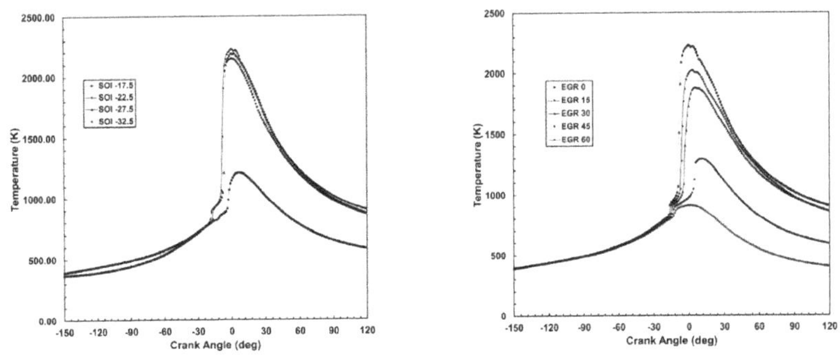

(a) (b)

Figure 4. In-cylinder temperature distribution at: (**a**) 0% EGR; (**b**) SOI −22.5°.

This variation in in-cylinder pressure highlights the combined effects of SOI timing and EGR on combustion characteristics and engine performance. Reduced air–fuel mixing time at SOI −17.5° and earlier combustion during the compression stroke at −27.5° and −32.5° influence pressure development, as a portion of combustion energy counteracts the upward piston motion, reducing peak pressure despite longer mixing durations. At an SOI of −22.5° BTDC, increasing EGR from 0% to 60% results in a proportional pressure reduction of 6.73%, 15.9%, 29.69%, and 31.09% at EGR levels of 15%, 30%, 45%, and 60%, respectively. This inverse relationship between EGR rate and peak pressure stems from both physical and chemical effects. EGR introduces inert exhaust gasses into the intake charge, diluting the air–fuel mixture and increasing the specific heat capacity of the charge. This leads to a lower combustion temperature, as a portion of the heat released during combustion is absorbed by the inert gasses rather than contributing to pressure rise. Additionally, higher EGR rates reduce the availability of fresh air and oxygen, leading to incomplete combustion and further suppressing pressure buildup.

The in-cylinder temperature distribution is significantly influenced by SOI timing and EGR, mirroring pressure trends. At 0% EGR, peak temperature reaches 2228.12 K at an SOI of −22.5° BTDC, decreasing by 45.54% (1213.34 K), 1.65% (2191.44 K), and 3.6% (2147.83 K) at −17.5°, −27.5°, and −32.5°, respectively. Intermediate SOI timings (−22.5° to −27.5° BTDC) yield higher peak temperatures compared to earlier (−32.5° BTDC) or later (−17.5° BTDC) injections due to the balance between fuel–air mixing and combustion phasing. While earlier SOI allows extended mixing, excessive advances may induce over-mixing and cooling effects, slightly lowering peak temperature. Conversely, delayed SOI limits mixture preparation, resulting in less complete combustion and lower temperatures. Furthermore, increasing EGR from 0% to 60% progressively reduces peak temperature by 9.26% (2021.83 K), 15.71% (1878.12 K), 42.06% (1290.97 K), and 59.12% (910.75 K) at EGR levels of 15%, 30%, 45%, and 60%, respectively. This reduction is attributed to the dilution of the air–fuel mixture and the increased specific heat capacity of the charge, which absorbs more combustion heat, leading to lower peak temperatures. While EGR effectively suppresses NOx emissions by reducing combustion temperature, excessive EGR can negatively impact combustion efficiency and power output due to slower burn rates and reduced energy release.

3.2. Heat Release Rate

The HRR indicates the combustion rate, and Figure 5 shows the fluctuation in heat release rate along the crank angle at various SOIs and EGR percentages. The combustion process is divided into two phases: the first phase shows a slight increase in HRR due to fuel–air mixing, where the dispersion of turbulent combustion rises with an increase in the air–fuel mixture, while in the second phase, the HRR reaches its peak due to a higher air–fuel equivalent ratio.

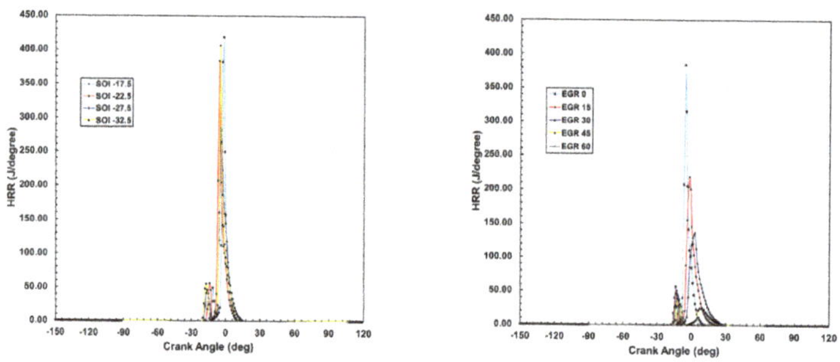

(a) (b)

Figure 5. The heat release rate variation at: (**a**) 0% EGR; (**b**) SOI −22.5°.

The HRR at SOI −22.5° was 384.14 J/degree, increasing by 8.9% and 5.74% at SOI −17.5° and −32.5°, respectively, but decreasing by 25.79% at −27.5°. At −17.5°, the higher HRR is attributed to enhanced fuel accumulation within a shorter ignition delay, causing combustion to begin abruptly but inadequately. Despite the higher HRR, it does not result in higher pressure or temperature since combustion starts later in the expansion stroke. At −27.5°, HRR is reduced due to premature heat losses and a more extended combustion period. Partial recovery of HRR is observed at −32.5° due to improved premixed combustion, though losses persist. Lower HRR limits peak pressure and temperature, but improved mixing at −32.5° partially increases HRR. However, while varying EGR percentages from 0 to 15, 30, 45, and 60, HRR rates gradually decreased by 42.93%, 64.32%, 89.83%, and 91.93%.

3.3. CO Mass Fraction

The variation in carbon monoxide (CO) emissions is strongly influenced by SOI timing and EGR, as shown in Figure 6. CO emissions primarily result from incomplete combustion due to insufficient oxygen or a rich air–fuel mixture, with peak CO formation occurring near combustion events. Later SOI timings (−17.5° and −22.5° BTDC) exhibit higher CO emissions compared to earlier SOI timings (−27.5° and −32.5° BTDC) due to reduced mixing time, leading to locally fuel-rich zones where incomplete oxidation occurs. Advancing SOI improves fuel–air mixing, enhancing combustion efficiency and lowering CO emissions by 14.62% and 7.43% at −27.5° and −32.5° BTDC, respectively.

However, excessive SOI advancement can impact overall combustion efficiency, reducing peak temperature and pressure. Additionally, increasing EGR levels affects CO emissions due to temperature-dependent oxidation. At 15% and 30% EGR, CO emissions rise by 3.36% and 5.87%, respectively, due to lower oxygen availability and incomplete combustion. However, at higher EGR rates of 45% and 60%, significant temperature reductions suppress CO formation, leading to emission reductions of 3.53% and 40.45%, respectively.

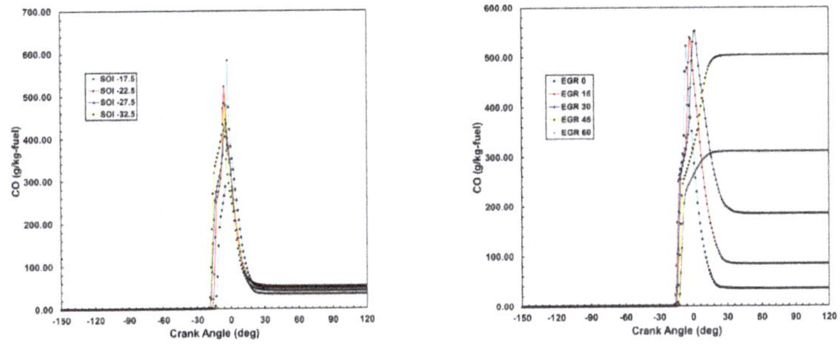

(a) (b)

Figure 6. Variation in CO mass fraction across CA: (**a**) 0% EGR; (**b**) SOI −22.5°.

3.4. Unburned Hydrocarbon

The formation of unburned hydrocarbons (UHCs) in exhaust gasses is strongly influenced by the homogeneity of the air–fuel mixture, which plays a critical role in flame propagation. The presence of squish volumes and crevices contributes to UHC generation, as the flame tends to be extinguished in these near-wall regions.

At an SOI of −22.5°, UHC emissions were recorded at 937.75 g/kg-fuel. As the SOI was retarded to −17.5°, the ignition delay decreased, leading to improved fuel consumption and a 7.16% reduction in UHC emissions. However, when the SOI was advanced to −27.5° and −32.5°, UHC emissions increased by 3.36% and 5.38%, respectively. This increase can be attributed to prolonged ignition delay, which limits effective fuel–air mixing and results in incomplete combustion.

At a fixed SOI of −22.5°, increasing the EGR rate to 15%, 30%, 45%, and 60% led to a progressive reduction in oxygen concentration within the combustion chamber. This resulted in a deteriorated air–fuel mixture, causing UHC emissions to rise by 2.03%, 3.09%, 3.53%, and 4.24%, respectively. Figure 7 illustrates the impact of varying SOI and increasing EGR rates on air–fuel mixture uniformity, ignition delay, and subsequent UHC emissions.

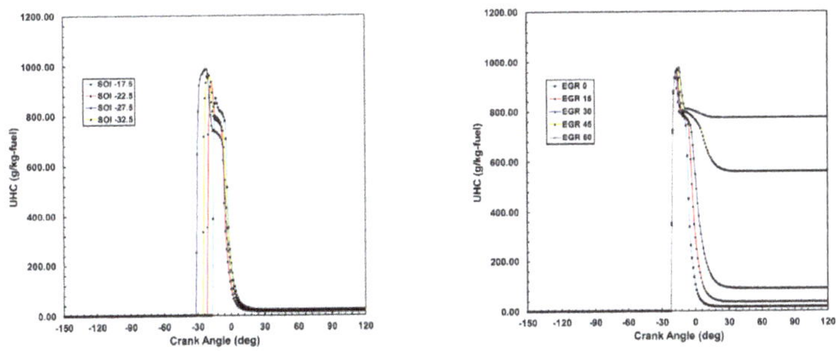

(a) (b)

Figure 7. Unburned hydrocarbon emission depending on: (**a**) 0% EGR; (**b**) SOI −22.5°.

3.5. Emission Indexed NOx (EINOx)

Figure 8 illustrates the EINOx distribution for various SOI and EGR rates, highlighting the significant impact of increased EGR on NOx formation. As the EGR rate increases, the reintroduction of exhaust gasses into the combustion chamber reduces oxygen availability and lowers combustion temperatures, thereby suppressing thermal NOx formation. The results indicate that at EGR 0%, NOx emissions are 1.27 g/kg-fuel, which decreases substantially with increasing EGR. At EGR 15% and 30%, NOx emissions drop to 0.15 g/kg-fuel and 0.03 g/kg-fuel, respectively, and are nearly eliminated at EGR 45% and 60%, reaching 0.00 g/kg-fuel. Additionally, advancing the SOI further contributes to NOx reduction, as earlier combustion results in greater heat loss to the cylinder walls, limiting peak flame temperatures. While higher EGR rates effectively mitigate NOx emissions, excessive EGR can lead to combustion inefficiencies, necessitating careful optimization for practical applications.

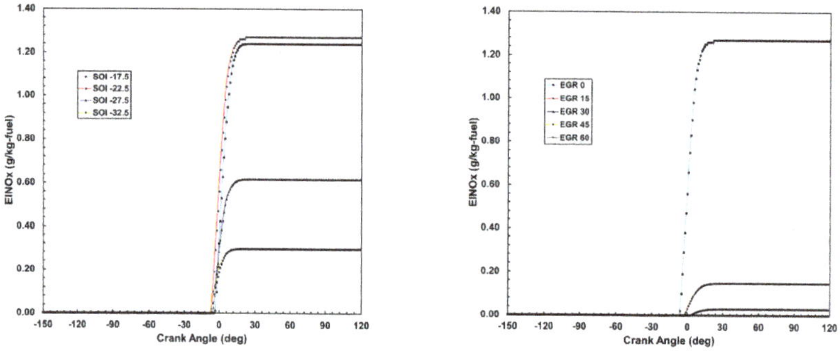

(a) (b)

Figure 8. EINOx distribution at: (**a**) 0% EGR; (**b**) SOI −22.5°.

3.6. Fuel Vapor Mass Fraction Contour

Figure 9 visually represents the n-heptane surrogate, illustrating fuel injection through fuel vapor mass fraction contours at various crank angles (CAs). During the intake stroke, the temperature remains nearly constant but gradually increases throughout the compression stroke. After −20° BTDC, as fuel begins to be injected into the cylinder, the temperature rises rapidly, reaching a maximum of 2228.12 K. This contour corresponds to fuel injection at an SOI of −22.5° and a 0% EGR rate.

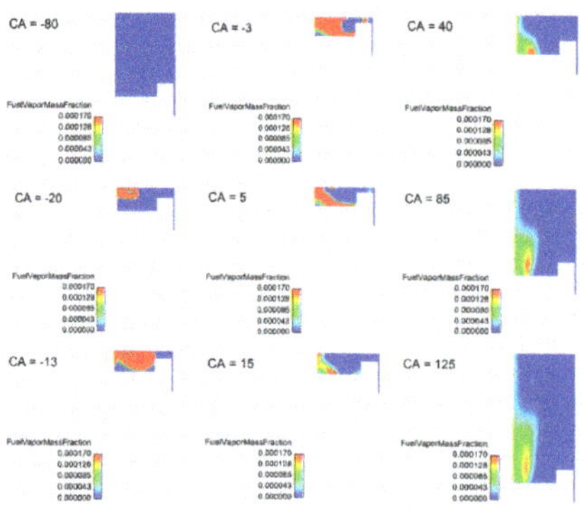

Figure 9. Fuel vapor mass fraction contour comparison at various CA.

4. Conclusions

Computational fluid dynamics is a powerful tool for evaluating combustion characteristics and pollutant emissions in internal combustion engines. In this study, a CFD simulation was conducted to analyze the combustion and emission behavior of a single-cylinder direct injection diesel engine operating at 1200 RPM, fueled with an n-heptane-based surrogate. The k-ε RNG turbulence model was employed to predict combustion dynamics and emissions. The study investigated the effects of fuel injection timing at SOI values of −17.5°, −22.5°, −27.5°, and −32.5° BTDC, as well as the impact of EGR at rates of 0%, 15%, 30%, 45%, and 60%. The key findings are summarized as follows:

1. A fuel injection timing of −22.5° BTDC with 0% EGR results in increased in-cylinder pressure, temperature, HRR, and EINOx emissions.
2. Advancing SOI to −27.5° and −32.5° BTDC leads to earlier combustion energy release, which counteracts the piston's upward movement, affecting overall engine performance. While this can lower CO emissions due to improved fuel oxidation, it also increases UHC emissions due to prolonged ignition delay and incomplete combustion.
3. Advancing SOI significantly reduces NOx emissions by increasing heat transfer to the cylinder walls during the early combustion phase.
4. Higher EGR rates reduce fresh air intake, leading to lower in-cylinder pressure, HRR, temperature, and EINOx emissions. However, excessive EGR increases UHC and CO emissions due to diminished oxygen availability and incomplete combustion.

Author Contributions: Conceptualization, M.H.P. and H.M.C.; methodology, M.H.P.; software, M.H.P.; validation, M.H.P. and H.M.C.; formal analysis, M.H.P.; investigation, M.H.P.; resources, M.H.P.; data curation, M.H.P.; writing—original draft preparation, M.H.P.; writing—review and editing, M.H.P. and H.M.C.; visualization, M.H.P.; supervision, H.M.C.; project administration, H.M.C.; funding acquisition, H.M.C. All authors have read and agreed to the published version of the manuscript.

Funding: This research received no external funding.

Data Availability Statement: The original contributions presented in this study are included in the article. Further inquiries can be directed to the corresponding author.

Conflicts of Interest: The authors declare no conflicts of interest.

Abbreviations

The following abbreviations are used in this manuscript:

LTC	Low-Temperature Combustion
RM	Reaction Mechanism
RNG	Re-Normalization Group
HRR	Heat Release Rate
DI	Direct Injection
KH-RT	Kelvin-Helmholtz-Raleigh Taylor
OEMs	Original Equipment Manufacturers
CI	Compression Ignition
CO_2	Carbon Dioxide
HC	Hydrocarbon
CO	Carbon Monoxide
NOx	Nitric Oxides
IC	Internal Combustion

LES	Large Eddy Simulation
RSM	Reynolds Stress Model
AFR	Air Fuel Ratio
3D	Three Dimension
CA	Crank Angle
TDC	Top Dead Center
BTDC	Before Top Dead Center
UHC	Unburned Hydrocarbon
J/degree	Joule per degree
g/kg-fuel	Grams per kilogram of fuel
RANS	Reynolds Averaged Navier–Stokes
CFD	Computational Fluid Dynamics
K	Kelvin
SOI	Start of Injection
EGR	Exhaust Gas Recirculation
RPM	Revolution Per Minute

References

1. Kalghatgi, G. Is It Really the End of Internal Combustion Engines and Petroleum in Transport? *Appl. Energy* **2018**, *225*, 965–974. [CrossRef]
2. Lyu, P.; Wang (Slade), P.; Liu, Y.; Wang, Y. Review of the Studies on Emission Evaluation Approaches for Operating Vehicles. *J. Traffic Transp. Eng.* **2021**, *8*, 493–509. [CrossRef]
3. Zhang, L.; Long, R.; Chen, H.; Geng, J. A Review of China's Road Traffic Carbon Emissions. *J. Clean. Prod.* **2019**, *207*, 569–581. [CrossRef]
4. Mock, P. CO2 Emission Standards for Passenger Cars and Light-Commercial Vehicles in the European Union. Available online: https://theicct.org/publication/co2-emission-standards-for-passenger-cars-and-light-commercial-vehicles-in-the-european-union/ (accessed on 25 January 2025).
5. Duan, X.; Liu, J.; Tan, Y.; Luo, B.; Guo, G.; Wu, Z.; Liu, W.; Li, Y. Influence of Single Injection and Two-Stagnation Injection Strategy on Thermodynamic Process and Performance of a Turbocharged Direct-Injection Spark-Ignition Engine Fuelled with Ethanol and Gasoline Blend. *Appl. Energy* **2018**, *228*, 942–953. [CrossRef]
6. Gopinath, S.; Devan, P.K.; Sabarish, V.; Sabharish Babu, B.V.; Sakthivel, S.; Vignesh, P. Effect of Spray Characteristics Influences Combustion in DI Diesel Engine—A Review. *Mater. Today Proc.* **2020**, *33*, 52–65. [CrossRef]
7. Mansor, M.R.A.; Mohamad, T.I.; Sabah, O. Numerical Investigation on Combustion and Emissions in a Direct Injection Compression Ignition Engine Fuelled with Various Hydrogen–Methane–Diesel Blends at Different Intake Air Temperatures. *Energy Rep.* **2021**, *7*, 403–421. [CrossRef]
8. Kumar, R.; Zheng, M. Fuel Efficiency Improvements of Low Temperature Combustion Diesel Engines. In *Proceedings of the SAE World Congress & Exhibition*; SAE International: Warrendale, PA, USA, 2008.
9. Bendu, H.; Sivalingam, M. Experimental Investigation on the Effect of Charge Temperature on Ethanol Fueled HCCI Combustion Engine. *J. Mech. Sci. Technol.* **2016**, *30*, 4791–4799. [CrossRef]
10. D'Aniello, F.; Arsie, I.; Pianese, C.; Stola, F. Development of an Integrated Control Strategy for Engine and SCR System Based on Effective EGR Rate. *IFAC-PapersOnLine* **2020**, *53*, 14034–14039. [CrossRef]
11. Lopatin, O.P. The Effect of Operational Modes of Diesel Engines to Emissions of Nitrogen Oxides. *IOP Conf. Ser. Mater. Sci. Eng.* **2020**, *862*, 062087. [CrossRef]
12. Peng, B.; Feng, C.; Liu, S.; Zhang, R. Synthesis of CuO Catalyst Derived from HKUST-1 Temple for the Low-Temperature NH3-SCR Process. *Catal. Today* **2018**, *314*, 122–128. [CrossRef]
13. Rimkus, A.; Melaika, M.; Matijošius, J. Efficient and Ecological Indicators of CI Engine Fuelled with Different Diesel and LPG Mixtures. *Procedia Eng.* **2017**, *187*, 504–512. [CrossRef]
14. Khujamberdiev, R.; Cho, H. Impact of Biodiesel Blending on Emission Characteristics of One-Cylinder Engine Using Waste Swine Oil. *Energies* **2023**, *16*, 5489. [CrossRef]
15. Yasin, M.H.M.; Mamat, R.; Yusop, A.F.; Idris, D.M.N.D.; Yusaf, T.; Rasul, M.; Najafi, G. Study of a Diesel Engine Performance with Exhaust Gas Recirculation (EGR) System Fuelled with Palm Biodiesel. *Energy Procedia* **2017**, *110*, 26–31. [CrossRef]
16. Rajesh Kumar, B.; Saravanan, S. Effect of Exhaust Gas Recirculation (EGR) on Performance and Emissions of a Constant Speed Di Diesel Engine Fueled with Pentanol/Diesel Blends. *Fuel* **2015**, *160*, 217–226. [CrossRef]

17. Nalluri, P.; Premkumar, P.; Sastry, M.R.C. Thermodynamic Simulation Study on Compression Ignition Engine Operating with Oil Obtained via Catalytic Pyrolysis of Waste Polythene Covers. *Sustain. Chem. Clim. Action* **2024**, *4*, 100041. [CrossRef]
18. Fatkhurrozak, F.; Sanjaya, F.L.; Syarifudin, S. The Effect of Additional Diethyl Ether (DEE) and Jatropa on the Brake Power and the Fuel Consumption of Diesel Engine with an EGR System. *AIP Conf. Proc.* **2024**, *3070*, 40002. [CrossRef]
19. Sethin, A.; Oo, Y.M.; Thawornprasert, J.; Somnuk, K. Effects of Blended Diesel–Biodiesel Fuel on Emissions of a Common Rail Direct Injection Diesel Engine with Different Exhaust Gas Recirculation Rates. *ACS Omega* **2024**, *9*, 20906–20918. [CrossRef]
20. Vedula, R.T.; Song, R.; Stuecken, T.; Zhu, G.G.; Schock, H. Thermal Efficiency of a Dual-Mode Turbulent Jet Ignition Engine under Lean and near-Stoichiometric Operation. *Int. J. Engine Res.* **2017**, *18*, 1055–1066. [CrossRef]
21. Siadkowska, K.; Barański, G.; Sochaczewski, R.; Wendeker, M. Experimental Investigation on Indicated Pressure and Heat Release for Direct Hydrogen Injection in a Dual Fuel Diesel Engine. *Adv. Sci. Technol. Res. J.* **2022**, *16*, 54–66. [CrossRef]
22. Lu, Z.; Ma, M.; Wang, T.; Lu, T.; Wang, H.; Feng, Y.; Shi, L. Numerical Research of the In-Cylinder Natural Gas Stratification in a Natural Gas-Diesel Dual-Fuel Marine Engine. *Fuel* **2023**, *337*, 126861. [CrossRef]
23. Cardone, M.; Marialto, R.; Ianniello, R.; Lazzaro, M.; Di Blasio, G. Spray Analysis and Combustion Assessment of Diesel-LPG Fuel Blends in Compression Ignition Engine. *Fuels* **2020**, *2*, 1–15. [CrossRef]
24. Pei, Y.; Mehl, M.; Liu, W.; Lu, T.; Pitz, W.J.; Som, S. A Multicomponent Blend as a Diesel Fuel Surrogate for Compression Ignition Engine Applications. *J. Eng. Gas Turbines Power* **2015**, *137*, 111502. [CrossRef]
25. Khandal, S.V.; Banapurmath, N.R.; Gaitonde, V.N. Performance Studies on Homogeneous Charge Compression Ignition (HCCI) Engine Powered with Alternative Fuels. *Renew. Energy* **2019**, *132*, 683–693. [CrossRef]
26. Gheraissa, N.; Bouras, F.; Khaldi, F.; Hidouri, A.; Rehouma, F.; Dogga, A. A Comparative Study of the Combustion Supplied by Multi-Fuels: Computational Analysis. *Energy Rep.* **2021**, *7*, 3819–3832. [CrossRef]
27. Sathishkumar, S.; Ibrahim, M.M. Investigation on the Effect of Injection Schedule and EGR in Hydrogen Energy Share Using Common Rail Direct Injection Dual Fuel Engine. *Int. J. Hydrog. Energy* **2021**, *46*, 11494–11510. [CrossRef]
28. Parveg, A.S.M.S.; Ratner, A. Droplets Combustion Characteristics Comparison of Single Component and Multicomponent Diesel Surrogates With Petroleum-Based Commercial Diesel Fuel. In Proceedings of the ASME 2023 International Mechanical Engineering Congress and Exposition, New Orleans, LA, USA, 29 October–2 November 2023.
29. Salahi, M.M.; Mahmoudzadeh Andwari, A.; Kakoee, A.; Hyvonen, J.; Gharehghani, A.; Mikulski, M.; Lendormy, É. Novel Chemical Kinetics Mechanism for Robust Simulation of Multi-Component Fuel Blends in Engine Conditions. In *Proceedings of the Conference on Sustainable Mobility*; SAE International: Warrendale, PA, USA, 2024.
30. Dahham, R.Y.; Wei, H.; Zhang, R.; Li, J.; Shu, G.; Pan, J. Numerical Study for the Comparison between Direct Dual-Fuel Stratification and Reactivity-Controlled Compression Ignition of Ammonia-Based Engines. *Appl. Therm. Eng.* **2025**, *258*, 124624. [CrossRef]
31. Mahdi, M.; Amin, S.; Andwari, M.; Kakoee, A.; Golbaghi, K. New Chemical Kinetics Mechanism for Simulation of Natural Gas/Hydrogen/Diesel Multi-Fuel Combustion in Engines. In Proceedings of the Second SIMS EUROSIM Conference on Modelling and Simulation, SIMS EUROSIM 2024, Oulu, Finland, 10–12 September 2024; pp. 11–12. [CrossRef]
32. Rorimpandey, P.; Yip, H.L.; Srna, A.; Zhai, G.; Wehrfritz, A.; Kook, S.; Hawkes, E.R.; Chan, Q.N. Hydrogen-Diesel Dual-Fuel Direct-Injection (H2DDI) Combustion under Compression-Ignition Engine Conditions. *Int. J. Hydrogen Energy* **2023**, *48*, 766–783. [CrossRef]
33. Wang, X.; Bu, H.; Chen, H.; Liu, J.; Chen, Z.; Gao, J. Numerical Investigation of Diesel Spray Combustion Characteristics in the Ammonia/Air Atmosphere. *J. Energy Inst.* **2024**, *116*, 101718. [CrossRef]
34. Udaybhanu, G.; Jameel, A.G.A.; Roberts, W.L.; Mahendra Reddy, V. Development of Laminar Burning Velocity Prediction Model and Correlation of Iso-Octane Air Mixtures Using Artificial Neural Network. *Energy* **2024**, *307*, 132639. [CrossRef]
35. Elkelawy, M.; El Shenawy, E.A.; Mohamed, S.A.; Elarabi, M.M.; Alm-Eldin Bastawissi, H. Impacts of EGR on RCCI Engines Management: A Comprehensive Review. *Energy Convers. Manag. X* **2022**, *14*, 100216. [CrossRef]
36. Qi, D.; Ma, L.; Chen, R.; Jin, X.; Xie, M. Effects of EGR Rate on the Combustion and Emission Characteristics of Diesel-Palm Oil-Ethanol Ternary Blends Used in a CRDI Diesel Engine with Double Injection Strategy. *Appl. Therm. Eng.* **2021**, *199*, 117530. [CrossRef]
37. Öztürk, E.; Can, Ö. Effects of EGR, Injection Retardation and Ethanol Addition on Combustion, Performance and Emissions of a DI Diesel Engine Fueled with Canola Biodiesel/Diesel Fuel Blend. *Energy* **2022**, *244*, 123129. [CrossRef]
38. Sadeq, A.M. *Combustion Advancements: From Molecules to Future Challenges*; Amazon KDP: Seattle, WA, USA, 2023; ISBN 9798990783614.
39. Bayramoğlu, K.; Yılmaz, S. Emission and Performance Estimation in Hydrogen Injection Strategies on Diesel Engines. *Int. J. Hydrog. Energy* **2021**, *46*, 29732–29744. [CrossRef]
40. Rizvi, I.H.; Gupta, R. Numerical Investigation of Injection Parameters and Piston Bowl Geometries on Emission and Thermal Performance of DI Diesel Engine. *SN Appl. Sci.* **2021**, *3*, 626. [CrossRef]

41. Plengsa-Ard, C.; Kaewbumrung, M. CFD Modelling Wall Heat Transfer inside a Combustion Chamber Using ANSYS Forte. *IOP Conf. Ser. Mater. Sci. Eng.* **2018**, *297*, 012036. [CrossRef]
42. Rashid, N.M.H.N.A.; Hairuddin, A.A.; Rezali, K.A.M.; Masuri, S.U.; Mossa, A.A.M.A.; Jaafar, J.; Fitriyana, D.F. Computational Fluid Dynamics (CFD) Validation and Investigation the Effect of Piston Bowl Geometries Performance on Port Fuel Injection-Homogeneous Charge Compression Ignition (PFI-HCCI) Engines. *J. Adv. Res. Numer. Heat Transf.* **2024**, *18*, 30–48. [CrossRef]
43. Aggarwal, S.K.; Fu, X.; Wijeyakulasuriya, S. Effects of Fuel Reactivity and Injection Timing on Diesel Engine Combustion and Emissions. *Int. J. Green Energy* **2016**, *13*, 431–445. [CrossRef]
44. Nkol, F.P.N.; Freidy, E.J.; Banta, N.J.I.; Yotchou, G.V.T.; Abbe, C.V.N.; Mouangue, R.M. Simulating the Effect of Methanol and Spray Tilt Angle on Pollutant Emission of a Diesel Engine Using Different Turbulence Models. *Int. J. Heat Technol.* **2023**, *41*, 1105–1120. [CrossRef]
45. Han, Z.; Reitz, R.D. Turbulence Modeling of Internal Combustion Engines Using RNG κ-ε Models. *Combust. Sci. Technol.* **1995**, *106*, 267–295. [CrossRef]
46. Kumar, A.; Bharti, R.P. Evaluation of Rans-Based Turbulence Models for Isothermal Flow in a Realistic Can-Type Gas Turbine Combustor Application. *J. Comput. Sci.* **2024**, *81*, 102362. [CrossRef]
47. Sonachalam, M.; Manieniyan, V. Optimization of Critical Angle, Distance and Flow Rate of Secondary Fuel Injection in DI Diesel Engine Using Computational Fluid Dynamics. *SN Appl. Sci.* **2021**, *3*, 126. [CrossRef]
48. Abay, M. Computational Fluid Dynamics Analysis of Flow and Combustion of a Diesel Engine. *J. Therm. Eng.* **2017**, *4*, 1878–1895. [CrossRef]
49. Chakchak, S.; Hidouri, A.; Ghabi, A.; Chrigui, M.; Boushaki, T. Numerical Study of Turbulent Swirling Diffusion Flame Under Lean and Rich Conditions Using Turbulence Realizable K-Epsilon Model. *Combust. Sci. Technol.* **2023**, *195*, 1461–1482. [CrossRef]
50. Liu, Z.; Wijeyakulasuriya, S.; Mashayekh, A.; Chai, X. Investigation of Reynolds Stress Model for Complex Flow Using CONVERGE. In *Proceedings of the WCX SAE World Congress Experience*; SAE International: Warrendale, PA, USA, 2020.
51. Singh, S.; Reitz, R.D.; Musculus, M.P.B. Comparison of the Characteristic Time (CTC), Representative Interactive Flamelet (RIF), and Direct Integration with Detailed Chemistry Combustion Models against Optical Diagnostic Data for Multi-Mode Combustion in a Heavy-Duty DI Diesel Engine. In *Proceedings of the SAE 2006 World Congress & Exhibition*; SAE International: Warrendale, PA, USA, 2006.
52. Yoshikawa, T.; Reitz, R.D. Validation of a Grid Independent Spray Model and Fuel Chemistry Mechanism for Low Temperature Diesel Combustion. *Int. J. Spray Combust. Dyn.* **2009**, *1*, 283–316. [CrossRef]
53. Xu, C.; Zhong, A.; Wang, C.; Jiang, C.; Li, X.; Zhou, K.; Huang, Y. Combustion Characteristics and Laminar Flame Speed of Premixed Ethanol-Air Mixtures with Laser-Induced Spark Ignition. *Biofuels Eng.* **2017**, *2*, 63–72. [CrossRef]
54. Performance of a CI Engine Operating in Highly Premixed Late Injection and Low Temperature Combustion Mode to Achieve Cleaner Emission—A CFD Analysis. In Proceedings of the International Conference on Mechanical, Industrial and Energy Engineering (ICMIEE), Khulna, Bangladesh, 22–24 December 2022.
55. Torn, D. SIMULACION CFD DE LA INYECCI 'ON Y' COMBUSTION CON SISTEMA DUAL-FUEL EN UN MCI HEAVY-DUTY. 2020. Available online: http://hdl.handle.net/2117/340307 (accessed on 20 February 2025).
56. Lapointe, S.; Zhang, K.; McNenly, M.J. Reduced Chemical Model for Low and High-Temperature Oxidation of Fuel Blends Relevant to Internal Combustion Engines. *Proc. Combust. Inst.* **2019**, *37*, 789–796. [CrossRef]
57. Mondo, K.; Agrebi, S.; Hamdi, F.; Lakhal, F.; Sadiki, A.; Chrigui, M. Impact of Multi-Component Surrogates on the Performances, Pollutants, and Exergy of IC Engines. *Entropy* **2022**, *24*, 671. [CrossRef] [PubMed]
58. Curran, H.J.; Gaffuri, P.; Pitz, W.J.; Westbrook, C.K. A Comprehensive Modeling Study of N-Heptane Oxidation. *Combust. Flame* **1998**, *114*, 149–177. [CrossRef]

energies

MDPI

Article

Data-Based In-Cylinder Pressure Model with Cyclic Variations for Combustion Control: An RCCI Engine Application †

Maarten Vlaswinkel [1,*] and Frank Willems [1,2]

1 Control Systems Technology, Department of Mechanical Engineering, Eindhoven University of Technology, 5600 MB Eindhoven, The Netherlands
2 Powertrains Department, TNO Mobility & Built Environment, 5700 AT Helmond, The Netherlands; f.p.t.willems@tue.nl
* Correspondence: m.g.vlaswinkel@tue.nl
† This paper is an extended version of our paper published in IFAC Symposium on Advances in Automotive Control (AAC) 2022.

Abstract: Cylinder-pressure-based control is a key enabler for advanced pre-mixed combustion concepts. In addition to guaranteeing robust and safe operation, it allows for cylinder pressure and heat release shaping. This requires fast control-oriented combustion models. Over the years, mean-value models have been proposed that can predict combustion metrics (e.g., gross indicated mean effective pressure ($IMEP_g$), or the crank angle where 50% of the total heat is released (CA50)) or models that predict the full in-cylinder pressure. However, these models are not able to capture cycle-to-cycle variations. The inclusion of the cycle-to-cycle variations is important in the control design for combustion concepts, like reactivity-controlled compression ignition, that can suffer from large cycle-to-cycle variations. In this study, the in-cylinder pressure and cycle-to-cycle variations are modelled using a data-based approach. The in-cylinder conditions and fuel settings are the inputs to the model. The model combines principal component decomposition and Gaussian process regression. A detailed study is performed on the effects of the different hyperparameters and kernel choices. The approach is applicable to any combustion concept, but is most valuable for advance combustion concepts with large cycle-to-cycle variation. The potential of the proposed approach is successfully demonstrated for a reactivity-controlled compression ignition engine running on diesel and E85. The average prediction error of the mean in-cylinder pressure over a complete combustion cycle is 0.051 bar and of the corresponding mean cycle-to-cycle variation is 0.24 bar^2. This principal-component-decomposition-based approach is an important step towards in-cylinder pressure shaping. The use of Gaussian process regression provides important information on cycle-to-cycle variation and provides next-cycle control information on safety and performance criteria.

Keywords: internal combustion engine; combustion modelling; control-oriented modelling; eigenpressure; Gaussian process regression

1. Introduction

Concerns about global warming have resulted in dramatic reduction targets for CO_2 emissions from on-road applications. This has boosted interest in high-efficiency and low-carbon propulsion methods in the transportation sector. This has led to a trend towards electrification for personal mobility, but the go-to technology for heavy-duty applications has not yet been decided. High-efficiency and clean internal combustion engines together with sustainable fuels are expected to play a significant role in the future [1–3]. Advanced combustion concepts provide promising solutions to increase thermal efficiency. Concepts like homogeneous charge compression ignition, partial premixed combustion, and reactivity-controlled compression ignition (RCCI) have been proposed [4]. From these concepts, RCCI provides high thermal efficiency and fuel flexibility as well as controllability. RCCI uses a combination of a low- and high-reactivity fuels during combustion [5,6]. By changing the ratio between low- and high-reactivity fuels and their injection timing, it is possible to optimise combustion phasing, duration, and magnitude. However, continuous monitoring of the combustion process and regulation of this ratio and timing is required to guarantee robust and safe operation [7,8].

1.1. Control Challenges for Advanced Combustion Concepts

The introduced advanced combustion concepts rely on controlled auto-ignition of the in-cylinder mixture of air, residuals from previous combustion, and fuel. These concepts are sensitive to changes in operating conditions, such as intake temperature and intake air mixture. This can result in misfires as well as undesired large cyclic variations, which are associated with unstable combustion. Also, mechanical limits for safe operation can be violated. This can lead to engine damage.

Cylinder-pressure-based control (CPBC) is a key concept for guaranteeing safe and stable operation of these advanced combustion concepts [8]. Typically, the measured in-cylinder pressure is used in next-cycle combustion control strategies to minimise cyclic variations in key combustion metrics. Several CPBC strategies have already been proposed in the literature; an overview of applied combustion metrics and control approaches can be found in [9]. Traditionally, these methods aim to realise the desired engine load and combustion phasing by controlling gross indicated mean effective pressure ($IMEP_g$) and the crank angle (CA) where 50% of the heat is released (CA50), respectively. Combustion is considered to be stable in case the cyclic variance in $IMEP_g$ is below 5%. For engine safety, peak pressure ($\max(p)$) and peak pressure rise rate ($\max\left(\frac{dp}{d\theta}\right)$) are monitored.

Alternatively, CPBC opens the route to in-cylinder pressure and heat release shaping. More precisely, contrary to the traditional control of individual combustion metrics, this approach aims to control the entire in-cylinder pressure curve. Consequently, focus is on the realisation of ideal thermodynamic cycles, which are associated with maximal thermal efficiency [10]. This is a promising approach that can explicitly deal with in-cylinder-pressure-related safety constraints and that further enhances robustness of the controlled combustion process. However, for this approach, control development requires information of the entire pressure curve.

1.2. Control-Oriented Combustion Modelling

Models are becoming increasingly important in control development. In addition to their role as digital twins in simulations, they are used in control design, they can assist in control calibration, and they can be embedded in model-based controllers. In this work, we focus on the development of control-oriented models (COMs) for controller design and calibration. To support in-cylinder pressure shaping studies, the COM should describe the relevant combustion characteristics, including the relation between the in-cylinder mixture composition, intake manifold pressure, and temperature, as model inputs, and the full in-cylinder pressure curve. In case of advanced combustion concepts, a description of the cycle-to-cycle variations should also be available.

For the COMs, a distinction can be made between two types of models:

- Physics-based models, that use first-principle physical relations to capture combustion behaviour;
- Data-based models, that use black-box modelling methods, where measurements are used to create a mapping from input to output.

For combustion modelling, various models are found in the literature.

1.2.1. Physics-Based Combustion Models

To model important combustion metrics, e.g., $IMEP_g$ or CA50, basic physics-based models have been proposed [11–16]. These models provide a deterministic and dynamic view of the relationship between actuation and combustion metrics without determining the full in-cylinder pressure. To add new combustion metrics these models should be extended with new descriptions to capture the behaviour of these new metrics. This can be time-consuming and reduces the flexibility of these models during combustion control development.

To model the full in-cylinder pressure, more complex first-principle models have been proposed. These include a multi-zone model [17] and a fluid dynamic model [18]. The complexity of these models results in computation times that exceed the combustion time. Therefore, they are not suited as COMs. A reduction in computation time is achieved by using static, data-driven, deterministic regression models to capture the behaviour of important combustion metrics.

1.2.2. Data-Based Combustion Models

Various data-based combustion modelling approaches have been introduced. For example, a Gaussian process regression (GPR) model to map in-cylinder conditions to combustion metrics [19]; a state-space model identified using data to model combustion phasing and peak pressure rise rate [20]; or a frequency response function method to determine cylinder-individual behaviour [21]. These models are made to only provide information on the modelled combustion metrics. Therefore, the model has to be extended to include other metrics.

Capturing the full in-cylinder pressure using data, principal component decomposition (PCD) models have been proposed. These models consist of a weighted sum of principal components, where the weights are modelled using regression methods. A deterministic neural network to capture the behaviour of the weights has been proposed [22]. To include cycle-to-cycle variations in the model, a GPR model to capture the behaviour of the weights has been proposed [23]. Alternatively, a method that uses double Wiebe functions to model the full in-cylinder pressure curve has been used [24]. The parameters of the double Wiebe function are determined using measurement data. A random forest machine learning approach is applied to describe the change in the mean behaviour and cycle-to-cycle behaviour of these parameters. However, determining these parameters from a measured in-cylinder pressure curve can be difficult.

The use of the PCD of the in-cylinder pressure has already been proposed in several control and detection methods. This decomposition was used as input to a virtual emission sensor [25]. They were able to predict the air-to-fuel ratio and NO_x emissions quite accurately. Also, this decomposition was used for knock detection and avoidance [26,27]. They used the decomposition to derive a measure of proximity to engine knocking. This decomposition was used as an alternative method to maximise the thermal efficiency [10]. They used the decomposition to derive a measure of the closeness of a measured in-cylinder pressure to an idealised thermodynamic cycle.

1.3. Research Objective and Main Contributions

In this study, we will extend the work of Vlaswinkel et al. [23] by giving an extensive analysis on (1) the comparison of different kernels in the GPR approach with regards to prediction quality of important combustion metrics; (2) understanding the effects of modelling a correlated process as an uncorrelated Gaussian process; (3) using a data set with a wide range of operating conditions to show the effectiveness of the model.

This work is organised as follows. In Section 2, an overview is given of the experimental setup and the data sets used. Section 3 describes the data-based combustion model, including cycle-to-cycle variation. A detailed analysis of the effect on different hyperparameters is presented in Section 4. The prediction quality of the combustion model is demonstrated and validated in Section 5.

2. Single-Cylinder Engine Test Bench

In this section, we will give a description of the setup and the data sets used. A discussion is provided on the chosen inputs to the model and how these are determined.

2.1. System Description

In this study, a modified PACCAR MX13 engine is used, as shown in Figure 1. Cylinders 2 to 6 have been removed and only cylinder 1 is operational. To keep the engine running at a constant speed, the electric motor of the engine dynamometer provides the required torque. The focus is on RCCI combustion with a single injection of diesel to autoignite the well-mixed charge of E85, air, and recirculated exhaust gas. The injection of diesel does not ignite the mixture, but the ignition is caused by the increased temperature as a result of cylinder compression. Therefore, there is a clear temporal separation between the injection of diesel and combustion. The direct injection (DI) of diesel is handled by a Delphi DFI21 injector connected to a common rail. The E85 port fuel injection (PFI) is handled by a Bosch EV14 injector fitted into the intake channel set at 5 bar. Both the DI and PFI fuel mass flows are measured using a Siemens Sitrans FC Mass 2100 Coriolis mass flow meter coupled with Mass 6000 signal converters. Boosted intake air is supplied at 8 bar and the pressure and temperature are regulated using a pressure regulator and an electric heater, respectively. The exhaust gas recirculation (EGR) fraction is regulated by the EGR and back-pressure butterfly valves. The EGR flow is cooled down to approximately room temperature by a cooled stream of process water. The condensation tank collects the condensation from the EGR flow and is drained regularly. The expansion and mixing tank are both attached to a surge tank to dampen pressure fluctuations in the intake and exhaust manifold as a result of single-cylinder operation. The in-cylinder pressure is sampled at 0.2° CA with a Kistler 6125C uncooled pressure transducer and amplified with a Kistler 5011B. A Leine Linde RSI 503 encoder provides crank angle information at a 0.2° interval. A Bronkhorst IN-FLOW F-106BI-AFD-02-V digital mass flow meter is used to measure the mass of the intake air flow. Pressures and temperatures located at different locations in the air path are measured every combustion cycle using a Gems Sensors & Controls 3500 Series pressure transmitter and Type-K thermocouples, respectively. The concentration of CO_2 in the intake and exhaust flows are measured using an Horiba MEXA 7100 DEGR system. Table 1 lists the main specifications of the engine setup.

Table 1. Main specifications of the engine setup.

Parameter	Value
PFI fuel	E85
DI fuel	Diesel (EN590)
Compression ratio	17.2
Intake valve closure	−173° CA aTDC
Exhaust valve opening	146° CA aTDC
Engine speed	1200 rpm
Oil temperature	90 °C
Coolant temperature	87 °C

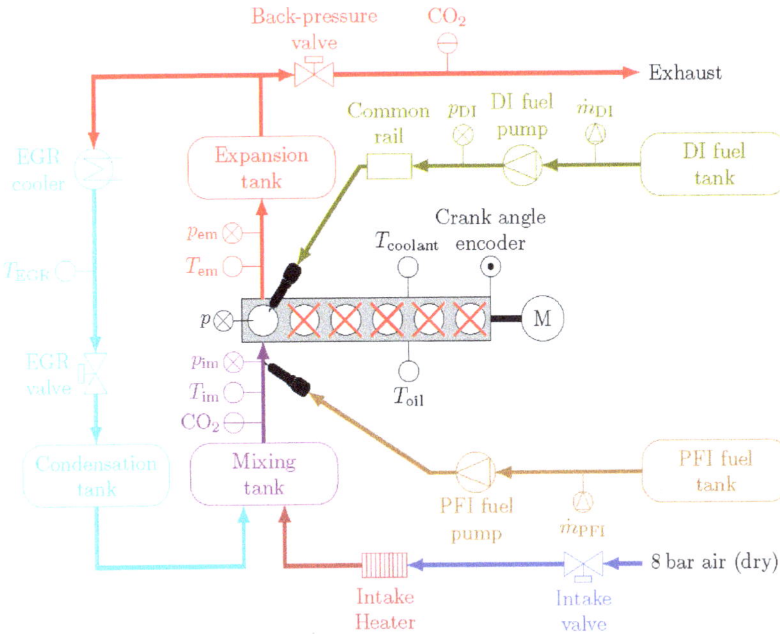

Figure 1. Schematic of the single-cylinder PACCAR MX13 engine equipped with exhaust gas recirculation (EGR), direct injection (DI), and port fuel injection (PFI).

2.2. Data Set for Model Training and Validation

The model relates in-cylinder conditions, determined at intake valve closing, to a resulting in-cylinder pressure. These conditions consist of a range of parameters related to engine speed, cylinder wall temperature, and mixture composition, pressure, and temperature. Since the engine is running at a single speed and at steady-state conditions the most relevant changes throughout the data set are a result of differences in mixture composition, pressure, and temperature. These can be described using intake and fuelling conditions. The chosen measurable parameters used to describe in-cylinder conditions are:

- Total injected energy

$$Q_{\text{total}} = m_{\text{PFI}}\text{LHV}_{\text{PFI}} + m_{\text{DI}}\text{LHV}_{\text{DI}}, \tag{1}$$

  where m_{PFI} and m_{DI} are the injected masses of PFI and DI fuels, and LHV_{PFI} and LHV_{DI} are the lower heating values of the PFI and DI fuels;
- Energy-based blend ratio

$$\text{BR} = \frac{m_{\text{PFI}}\text{LHV}_{\text{PFI}}}{Q_{\text{total}}}; \tag{2}$$

- Start of injection of the directly injected fuel SOI_{DI};
- Pressure at the intake manifold p_{im};
- Temperature at the intake manifold T_{im};
- EGR ratio

$$X_{\text{EGR}} = \frac{\text{CO}_{2,\text{in}}}{\text{CO}_{2,\text{out}}} \tag{3}$$

  with $\text{CO}_{2,\text{in}}$ and $\text{CO}_{2,\text{out}}$ the concentrations of CO_2 as a fraction of the volume flow at the intake and exhaust, respectively.

The variations in the in-cylinder conditions for the training data and validation data are shown in Figure 2. Figure 2a shows the distribution of each individual measure for the in-cylinder conditions. Figure 2b shows the joint distribution of the measures used for the in-cylinder conditions. The data set contains 95 different measurements consisting of $n_{\mathrm{cyc}} = 50$ consecutive cycles each. Both small and large cycle-to-cycle variations, and non-firing behaviour are present within the data set. In this work, each cycle is used and no averaging over the n_{cyc} in-cylinder conditions and in-cylinder pressure traces in a measurement is performed before analysis. The data set is randomly divided into a training set of $n_{\mathrm{train}} = 75$ measurements and a validation set of the remaining $n_{\mathrm{val}} = 20$ measurements.

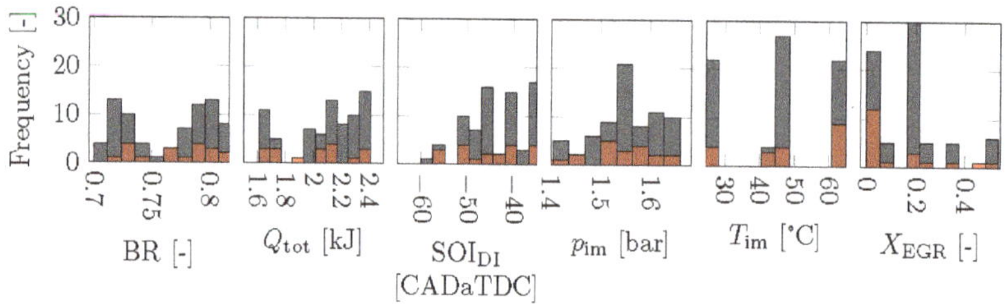

(**a**) Distribution of the individual measures of the in-cylinder conditions

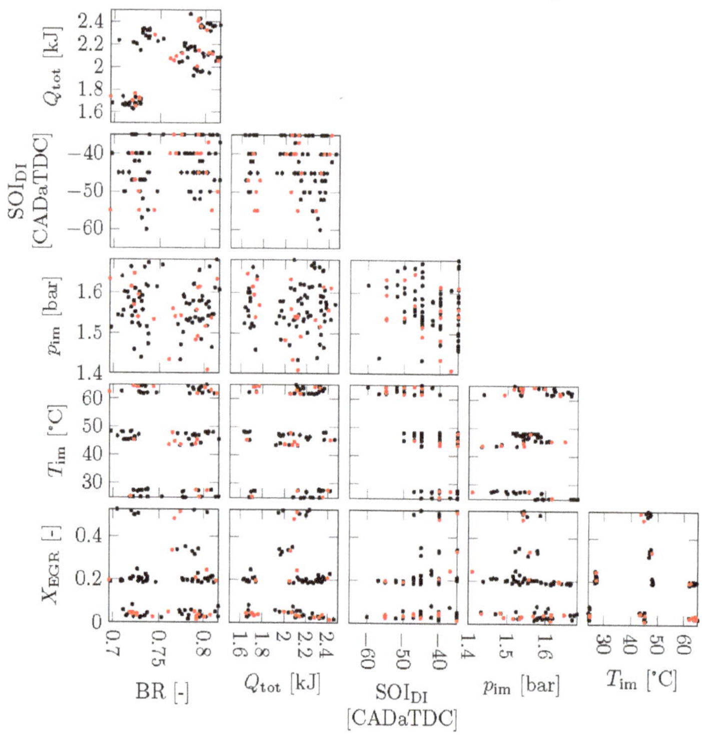

(**b**) Joint distribution between the measures of the in-cylinder conditions

Figure 2. Distribution of the in-cylinder conditions of the training (black) and validation data (red).

3. Combustion Model

In this section, the data-based approach to model the in-cylinder pressure is introduced. It is based on the method presented in Vlaswinkel et al. [23]. The approach combines principal component decomposition (PCD) and Gaussian process regression (GPR). To describe the in-cylinder pressure during the compression and power stroke, PCD is used to minimise the amount of information required by separating the influence of the in-cylinder conditions s_{ICC} and the crank angle θ into two different mappings. GPR gives the possibility to model the in-cylinder pressure and cycle-to-cycle variation at different in-cylinder conditions. In Vlaswinkel et al. [23], the entire combustion cycle, including compression and expansion during motoring, is captured using the PCD/GPR method. In this study, the compression and expansion effects are separated from the effects of the actual combustion. The compression and expansion effects are modelled using adiabatic compression and expansion, while the effects of the actual combustion are modelled using the PCD/GPR method.

3.1. Principal Component Decomposition of the In-Cylinder Pressure

The in-cylinder pressure $p(\theta, s^*_{\text{ICC}})$ at crank angle $\theta \in \{-180^\circ, -180^\circ + \Delta\text{CA}, \ldots, 180^\circ - \Delta\text{CA}, 180^\circ\}$, with ΔCA the crank angle resolution, is decomposed as

$$p(\theta, s^*_{\text{ICC}}) = p_{\text{mot}}(\theta, s^*_{\text{ICC}}) + w(s^*_{\text{ICC}})^\top f(\theta), \tag{4}$$

where $w(s^*_{\text{ICC}})$ is a vector of weights and $f(\theta)$ is the vector of principal components. In these vectors, the ith element is related to the ith principal component (PC). The in-cylinder condition $s^*_{\text{ICC}} \in \mathcal{S}^* \subset \mathcal{S}$ is in the set $\mathcal{S}^*$ containing all in-cylinder conditions present in the training set and the set $\mathcal{S}$ spanning the modelled operation domain. It is assumed that the in-cylinder pressure during the intake stroke is equal to p_{im}.

The PCs are computed using the eigenvalue method. The $n_{\text{train}} \cdot n_{\text{cyc}}$ in-cylinder pressures $p(\theta, s^*_{\text{ICC}})$ contained in the training set are used. The vector F_i is the ith unit eigenvector of the matrix $PP^\top$, where $P \in \mathbb{R}^{n_{\text{CA}} \times n_{\text{train}} n_{\text{cyc}}}$, with n_{CA} the number of crank angle values. The elements in matrix P are defined as

$$[P]_{ab} := p(\theta_a, s^*_{\text{ICC},b}) - p_{\text{mot}}(\theta_a, s^*_{\text{ICC},b}), \tag{5}$$

such that the ath row of P contains the values of the in-cylinder pressure at the ath crank angle for all $s^*_{\text{ICC}} \in \mathcal{S}^*$ and the bth column of P contains the full in-cylinder pressure at all $\theta \in \{-180^\circ, -180^\circ + \Delta\text{CA}, \ldots, 180^\circ - \Delta\text{CA}, 180^\circ\}$ for the bth s^*_{ICC}. The ith PC is defined as

$$f_i(\theta_a) = [F_i]_a. \tag{6}$$

The weight related to the ith PC is given by

$$w_i(s^*_{\text{ICC}}) = P(s^*_{\text{ICC}}) F_i, \tag{7}$$

where $[P(s^*_{\text{ICC}})]_a = p(\theta_a, s^*_{\text{ICC}}) - p_{\text{mot}}(\theta_a, s^*_{\text{ICC}})$. The training set generates a single set of PCs. These PCs are ordered by relevance, where $i = 1$ is the most relevant PC. The determination of the PCs and the required number of PCs will be considered later in this study.

3.2. Gaussian Process Regression to Capture Effects of In-Cylinder Conditions

GPR is used to estimate the behaviour of $w(s_{\text{ICC}})$ over the full operation domain $\mathcal{S}$. To include cycle-to-cycle variations, $w(s_{\text{ICC}})$ is described by a stochastic process as

$$w(s_{\text{ICC}}) := \mathcal{N}(\hat{w}(s_{\text{ICC}}), W(s_{\text{ICC}})) \tag{8}$$

with mean $\hat{w}(s_{\text{ICC}}) := \mathbb{E}[w(s_{\text{ICC}})]$ and variance $W(s_{\text{ICC}}) := \mathbb{E}[(w(s_{\text{ICC}}) - \hat{w}(s_{\text{ICC}}))(w(s_{\text{ICC}}) - \hat{w}(s_{\text{ICC}}))^\top]$. During this study, the correlation between the output variables $w_i(s_{\text{ICC}})$ and $w_j(s_{\text{ICC}})$ $\forall i, j \in \{1, 2, \ldots, n_{\text{PC}}\}$ with n_{PC}, the number of PCs, will be neglected (i.e., $W(s_{\text{ICC}})$ is a diagonal matrix), since most of the literature on GPR assumes the output variables to be uncorrelated. This might affect the quality of the prediction of the cycle-to-cycle variation.

To improve the accuracy of prediction and the determination of hyperparameters, normalised in-cylinder conditions $\bar{s}_{\text{ICC}}$ and weights $\bar{w}_i(s^*_{\text{ICC}})$ will be used. Scaling the in-cylinder condition uses the mean $\bar{\mu}_{s^*_{\text{ICC}},j}$ and standard deviation $\bar{\sigma}_{s^*_{\text{ICC}},j}$ of the jth in-cylinder condition variable over the full training set S^* as

$$\bar{s}_{\text{ICC},j} = \frac{s_{\text{ICC},j} - \bar{\mu}_{s^*_{\text{ICC}},j}}{\bar{\sigma}_{s^*_{\text{ICC}},j}}. \tag{9}$$

The scaling of the weights uses the mean $\bar{\mu}_{w^*_{\text{ICC}},i}$ and standard deviation $\bar{\sigma}_{w^*_{\text{ICC}},i}$ of the ith in-cylinder conditions variable over the full training set S^* as

$$\bar{w}_i(s^*_{\text{ICC}}) = \frac{w_i(s^*_{\text{ICC}}) - \bar{\mu}_{w_i}}{\bar{\sigma}_{w_i}}. \tag{10}$$

Following [28], the scaled expected value and scaled covariance matrix without correlation can be computed as

$$\hat{\bar{w}}_i(\bar{s}_{\text{ICC}}) = K(\bar{s}_{\text{ICC}}, \bar{s}^*_{\text{ICC}}, \phi)(K(\bar{s}^*_{\text{ICC}}, \bar{s}^*_{\text{ICC}}, \phi) + \varphi_{\text{n}} I)^{-1} \bar{w}_i(\bar{s}^*_{\text{ICC}}) \tag{11}$$

and

$$\begin{aligned} \bar{W}_{ii}(\bar{s}_{\text{ICC}}) = {} & K(\bar{s}_{\text{ICC}}, \bar{s}_{\text{ICC}}, \phi) - \\ & K(\bar{s}_{\text{ICC}}, \bar{s}^*_{\text{ICC}}, \phi)(K(\bar{s}^*_{\text{ICC}}, \bar{s}^*_{\text{ICC}}, \phi) + \varphi_{\text{n}} I)^{-1} K^{\mathsf{T}}(\bar{s}_{\text{ICC}}, \bar{s}^*_{\text{ICC}}, \phi), \end{aligned} \tag{12}$$

where $K(\cdot, \cdot, \phi)$ is the kernel and ϕ and φ_{n} are the kernel's hyperparameters. The selection of both elements will be discussed in the next section.

To optimise the set of hyperparameters ϕ and φ_{n} found in the kernels, the marginal log-likelihood is maximised for each PC separately. The marginal log-likelihood is often used in determining the hyperparameters in GPR and does not depend on the kernel type. It is given by

$$\ln(\text{Prob}(\bar{w}_i \,|\, \bar{s}^*_{\text{IVC}}, \phi)) = -\frac{1}{2} \bar{w}_i^{\mathsf{T}} K^{-1}_{\bar{s}^*_{\text{IVC}}} \bar{w}_i - \frac{1}{2} \ln(\det(K_{\bar{s}^*_{\text{IVC}}})) - \frac{n_{\text{exp}} n_{\text{cyc}}}{2} \ln(2\pi), \tag{13}$$

where $\bar{w}_i$ is a vector of the weights related to the ith PC at measured $\bar{s}^*_{\text{IVC}}$ in the training set and $K_{\bar{s}^*_{\text{IVC}}} := K(\bar{s}^*_{\text{IVC}}, \bar{s}^*_{\text{IVC}}, \phi) + \varphi_{\text{n}} I$.

Finally, the scaled expected value and scaled covariance matrix are descaled to complete the description of (8). The descaled expected value is given by

$$\hat{w}_i(\bar{s}_{\text{ICC}}) = \hat{\bar{w}}_i(\bar{s}_{\text{ICC}}) \bar{\sigma}_{w_i} + \bar{\mu}_{w_i} \tag{14}$$

and the descaled covariance matrix is given by

$$W_{ii}(\bar{s}_{\text{ICC}}) = \bar{W}_{ii}(\bar{s}_{\text{ICC}}) \bar{\sigma}_{w_i}. \tag{15}$$

3.3. *Reconstructing the In-Cylinder Pressure with Cycle-to-Cycle Variation*

The PCs $f(\theta)$ (Section 3.1) and the estimate behaviour of $w(s_{\text{ICC}})$ (Section 3.2) can be combined to reconstruct a predicted in-cylinder pressure $p(\theta, s_{\text{ICC}})$. Using (4), the mean and variance of the in-cylinder pressure can be described by

$$\hat{p}(\theta, s_{\text{ICC}}) = \mathbb{E}[p(\theta, s_{\text{ICC}})] = \hat{w}^{\mathsf{T}}(s_{\text{ICC}}) f(\theta) + f_{\text{mot}}(\theta, s_{\text{ICC}}) \tag{16}$$

and

$$\hat{\sigma}^2_p(\theta, s_{\text{ICC}}) = \mathbb{E}\Big[(p(\theta, s_{\text{ICC}}) - \mathbb{E}[p(\theta, s_{\text{ICC}})])^2\Big] = f^{\mathsf{T}}(\theta) W(s_{\text{ICC}}) f(\theta), \tag{17}$$

respectively.

4. Combustion Model Identification

The PCD and GPR require the selection of the number of PCs as well as the kernel type and hyperparameters. The training set is used to determine the PCs and values for the hyperparameters, while the validation set is used to determine the required amount of PCs n_{PC} and the best performing kernel type. For this selection, an assessment is made on the prediction accuracy of combustion metrics that are relevant for control [29]. To this end, the mean absolute error (MAE) is analysed, which is defined as

$$\text{MAE}(z) := \frac{1}{n_{\text{val}} n_{\text{cyc}}} \sum_{k=1}^{n_{\text{val}} n_{\text{cyc}}} \left| z_{k,\text{meas}} - z_{k,\text{model}} \right|, \tag{18}$$

where n_{val} is the number of validation measurements, and z_{meas} and z_{model} are the combustion metrics resulting from the measured in-cylinder pressure and modelled in-cylinder pressure, respectively. The following combustion metrics are studied:

- gross indicated mean effective pressure,

$$\text{IMEP}_{\text{g}} = \frac{1}{V_{\text{d}}} \int_{\theta=-180^\circ}^{\theta=180^\circ} p(\theta)\, dV(\theta) \tag{19}$$

  with displacement volume V_{d};
- peak pressure, $\max(p(\theta))$;
- peak pressure rise rate, $\max\left(\frac{dp}{d\theta}\right)$;
- crank angle where 50% of the total heat is released,

$$\text{CA50} = \left\{ \theta \,\middle|\, \frac{Q(\theta)}{\max(Q(\theta))} = 0.5 \right\} \tag{20}$$

  with the heat release given by [30]

$$Q(\theta) = \frac{1}{\kappa - 1} p(\theta) V(\theta) + \int_{\alpha=-180^\circ}^{\alpha=\theta} p(\alpha) \frac{dV}{d\alpha} d\alpha - \frac{1}{\kappa - 1} p(-180^\circ) V(-180^\circ); \tag{21}$$

- burn duration, CA75-CA25, with CA75 and CA25 computed in a similar fashion to CA50;
- burn ratio,

$$R_{\text{b}} = \frac{\text{CA75} - \text{CA50}}{\text{CA50} - \text{CA10}}. \tag{22}$$

4.1. Selection of Principal Components

The first hyperparameter is the number of PCs n_{PC}. The GPR formulation proposed in Section 3.2 is not used in this part of the discussion. Figure 3 shows the four most relevant PCs derived from the training data, as discussed in Section 3.1. This figure illustrates that adding more PCs will add more higher-frequency components to the in-cylinder pressure. Figure 4 shows the absolute error in the corresponding combustion metrics by comparing measurements and model results. The modelled, decomposed in-cylinder pressure is based on an increasing number of PCs, using (6) to compute the required weights. Each measured cycle in the validation set is analysed separately. The figure indicates the minimum, maximum, median, and first and third quartiles, while the crosses show outliers. It can be seen that the largest gain in improvement is made at lower numbers of PCs. From the used training and validation sets, it is concluded that having more than eight PCs gives a negligible improvement. Therefore, $n_{PC} = 8$ is used in this study.

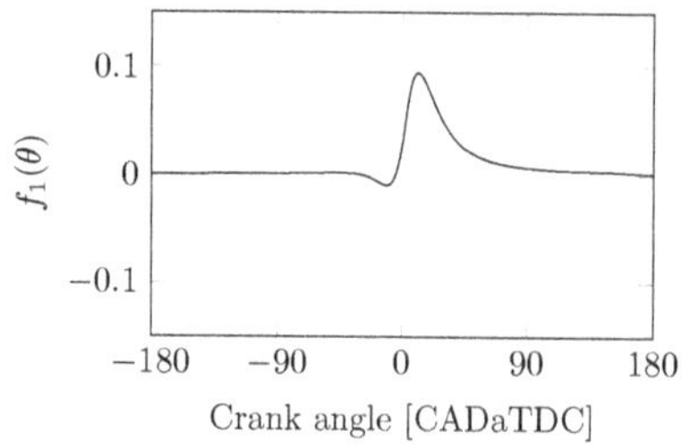

(**a**) Most relevant PC

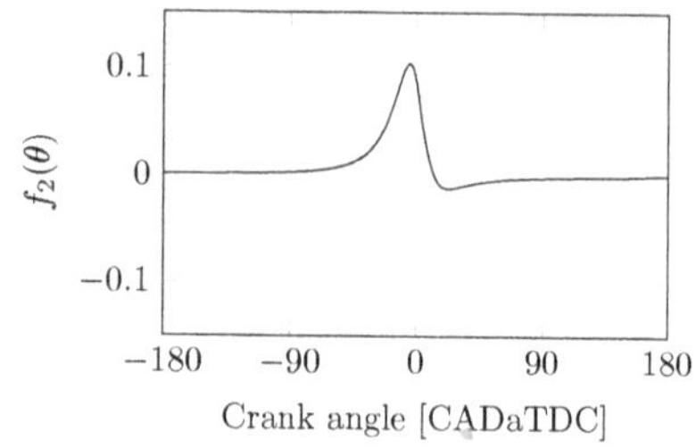

(**b**) Second most relevant PC

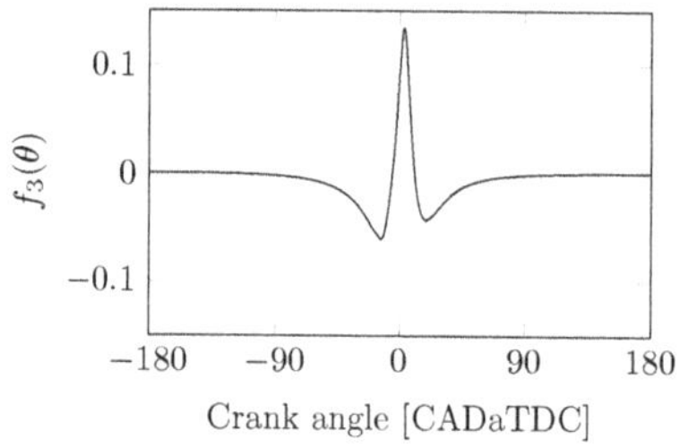

(**c**) Third most relevant PC

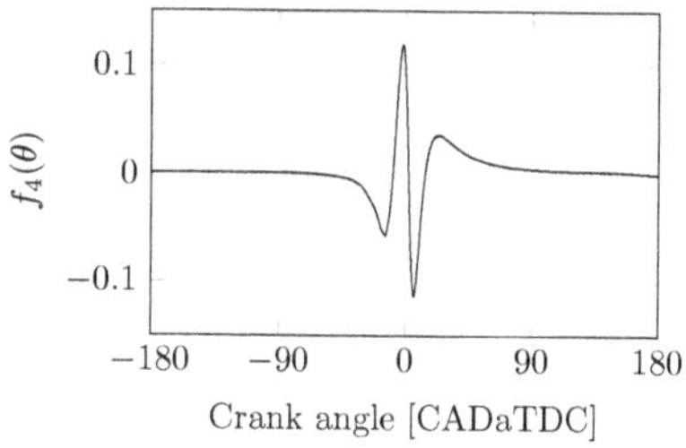

(**d**) Fourth most relevant PC

Figure 3. Four most relevant principal components (PCs) resulting from the used training data.

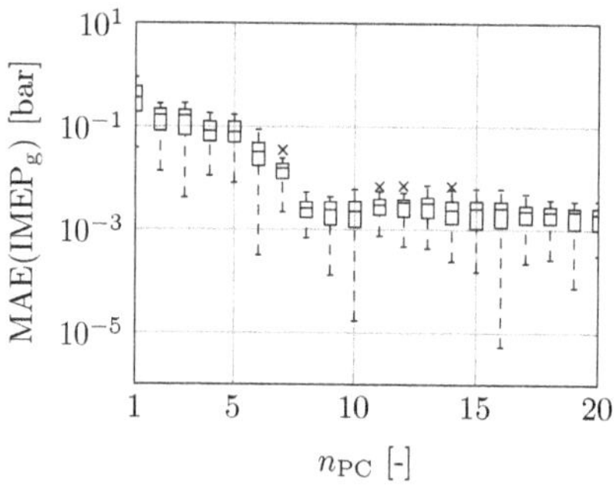

(**a**) Gross indicated mean effective pressure

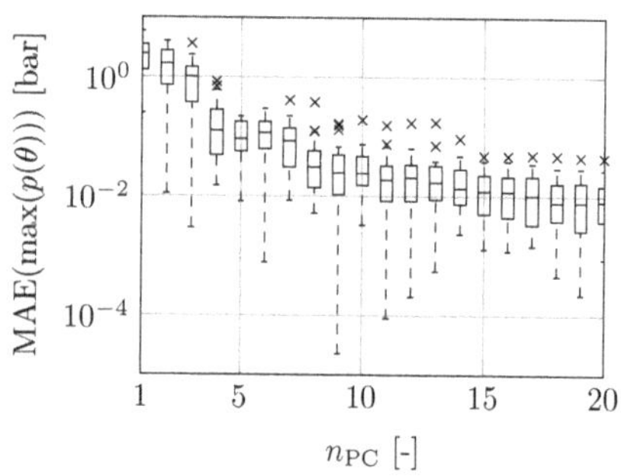

(**b**) In-cylinder peak pressure

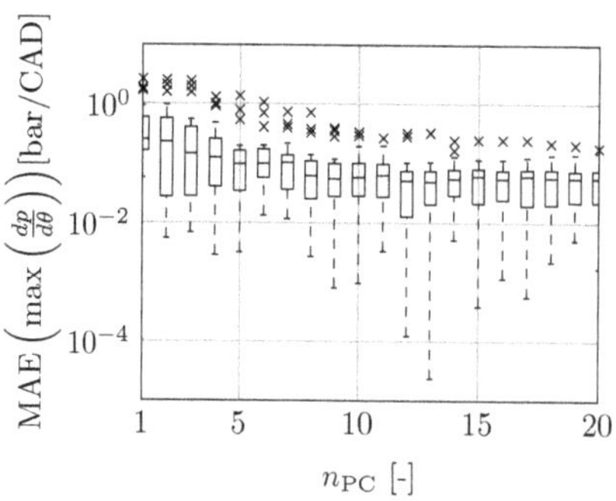

(**c**) In-cylinder peak pressure rise rate

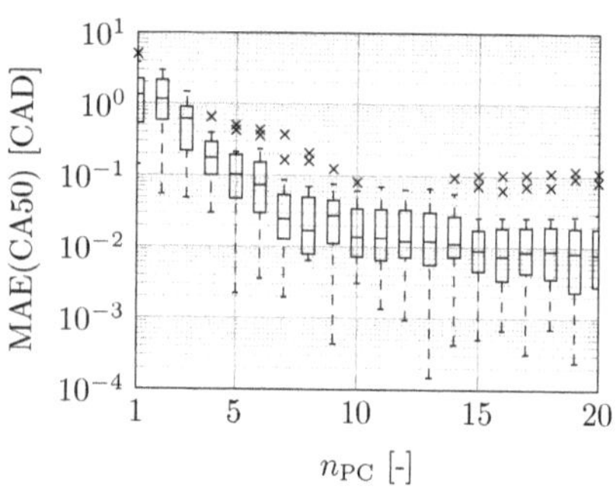

(**d**) Crank angle at 50% total heat release

Figure 4. *Cont.*

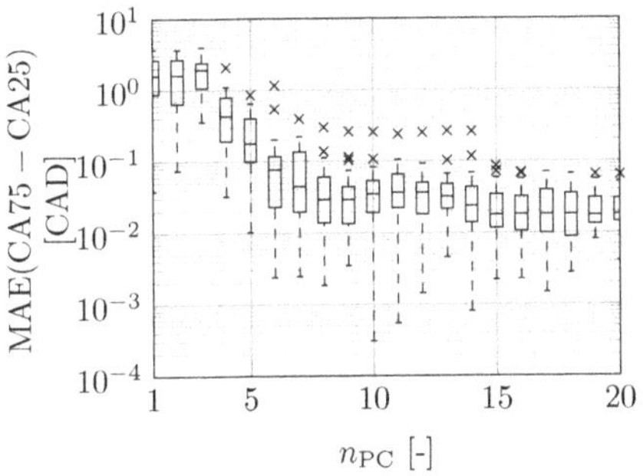

(e) Combustion duration

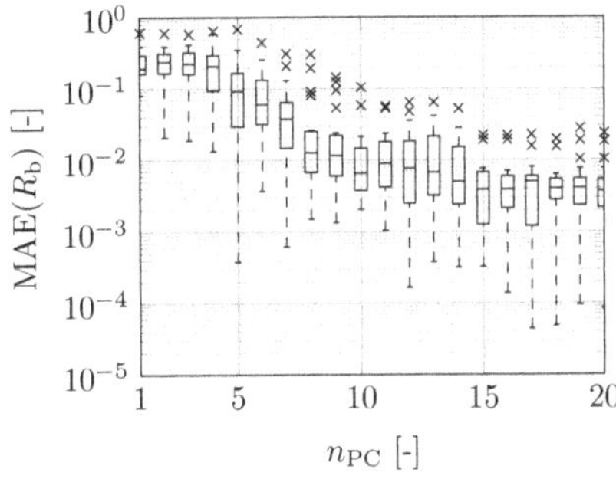

(f) Burn ratio

Figure 4. Prediction error of combustion metrics for the validation data set using different numbers of principal components n_{PC}. The box plot shows the minimum, maximum, median, and first and third quartiles, while the crosses show outliers.

4.2. Selection of Kernel

Another important aspect in the quality of the model lies in the chosen kernel. This describes the correlation between all measured $w(s^*_{ICC})$ and predicted means $\hat{w}(s_{ICC})$ and variance $W(s_{ICC})$. The kernel types compared in this study rely on the distance measure

$$r(\bar{s}_{ICC}, \bar{s}'_{ICC}) := \sqrt{(\bar{s}_{ICC} - \bar{s}'_{ICC})^{\top} \Phi_l^{-2} (\bar{s}_{ICC} - \bar{s}'_{ICC})}, \quad (23)$$

where $\bar{s}_{ICC}$ and $\bar{s}'_{ICC}$ are scaled in-cylinder conditions. Each element of the kernel is computed individually. The elements of the kernels used in this work are:

- square exponential (SE):

$$k_{SE}(\bar{s}_{ICC}, \bar{s}'_{ICC}) := \varphi_f^2 \exp\left(\tfrac{1}{2} r(\bar{s}_{ICC}, \bar{s}'_{ICC})^2\right) \quad (24)$$

  with the set of hyperparameters $\phi = \{\varphi_f, \Phi_l\}$;
- Matérn with $\nu = \frac{3}{2}$:

$$k_{\text{Matérn}}(\bar{s}_{ICC}, \bar{s}'_{ICC}) := \varphi_f^2 \left(1 + \sqrt{3} r(\bar{s}_{ICC}, \bar{s}'_{ICC})\right) \exp\left(-\sqrt{3} r(\bar{s}_{ICC}, \bar{s}'_{ICC})\right) \quad (25)$$

  with the set of hyperparameters $\phi = \{\varphi_f, \Phi_l\}$;
- Matérn with $\nu = \frac{5}{2}$:

$$k_{\text{Matérn}}(\bar{s}_{ICC}, \bar{s}'_{ICC}) := \varphi_f^2 \left(1 + \sqrt{5} r(\bar{s}_{ICC}, \bar{s}'_{ICC}) + 5 r(\bar{s}_{ICC}, \bar{s}'_{ICC})^2\right) \times \exp\left(-\sqrt{5} r(\bar{s}_{ICC}, \bar{s}'_{ICC})\right) \quad (26)$$

  with the set of hyperparameters $\phi = \{\varphi_f, \Phi_l\}$;
- rational quadratic (RQ):

$$k_{RQ}(\bar{s}_{ICC}, \bar{s}'_{ICC}) := \varphi_f \left(\tfrac{1}{2\varphi_\alpha} r(\bar{s}_{ICC}, \bar{s}'_{ICC})^2\right)^{\varphi_\alpha} \quad (27)$$

  with the set of hyperparameters $\phi = \{\varphi_f, \varphi_\alpha, \Phi_l\}$.

For each kernel, a distinction is made between with and without automatic relevance determination (ARD). In the case where ARD is not used, the hyperparameter Φ_l reduces to a scalar. In the case where ARD is used, the hyperparameter Φ_l is a diagonal matrix with unique elements on the diagonal. The hyperparameters are determined by maximising the marginal log-likelihood, as described in (13), using the training set.

For the studied combustion metrics, Tables 2 and 3 show the mean absolute error in the mean behaviour and in the standard deviation, respectively. For each combustion metric, the best result is in bold. In some cases, the difference between the best and second best option is negligible. The Matérn kernel with $\nu = \frac{3}{2}$ gives the best result for most of the combustion metrics in both mean behaviour and standard deviation for the data sets used. The resulting MAE of the mean-value behaviour shows a comparable or improved modelling error to those found in the literature, as illustrated in Table 4. Although the model accuracy of this work seems to be similar, the results have to be handled with care. It is difficult to give a fair comparison since most studies only give absolute errors and are unclear on the operating conditions.

Table 2. Mean absolute error in the mean behaviour of key combustion metrics for the validation set using different kernels with $n_{PC} = 8$. The best result for each combustion metric is in bold.

	Without ARD				With ARD			
	SE	**Matérn** $\nu = \frac{3}{2}$	**Matérn** $\nu = \frac{5}{2}$	**RQ**	**SE**	**Matérn** $\nu = \frac{3}{2}$	**Matérn** $\nu = \frac{5}{2}$	**RQ**
$IMEP_g$ [bar]	0.2255	**0.2061**	0.2088	0.2330	0.4161	0.2489	0.3006	0.2769
$\max(p(\theta))$ [bar]	2.4564	**1.6567**	1.8007	1.9383	2.5273	1.6653	2.0811	2.1632
$\max\left(\frac{dp}{d\theta}\right)$ [bar/CAD]	0.8269	0.7880	0.7962	0.7896	0.7546	**0.7515**	0.7987	0.7724
CA50 [CAD]	0.6121	0.5499	0.5591	0.5489	0.9507	0.5580	**0.5258**	0.5323
CA75-CA25 [CAD]	0.7178	0.6795	0.6884	0.6570	0.9371	0.5712	0.6364	**0.5554**
R_b [-]	0.1456	0.1407	0.1403	0.1325	0.2616	0.1239	0.1669	**0.1284**

Table 3. Mean absolute error in the standard deviation of key combustion metrics for the validation set using different kernels with $n_{PC} = 8$. The best result for each combustion metric is in bold.

	Without ARD				With ARD			
	SE	**Matérn** $\nu = \frac{3}{2}$	**Matérn** $\nu = \frac{5}{2}$	**RQ**	**SE**	**Matérn** $\nu = \frac{3}{2}$	**Matérn** $\nu = \frac{5}{2}$	**RQ**
$IMEP_g$ [bar]	0.4775	**0.3280**	0.3770	0.3685	0.4960	0.4426	0.4113	0.4210
$\max(p(\theta))$ [bar]	1.6716	**0.9952**	1.2239	1.1792	1.7561	1.4948	1.3761	1.4194
$\max\left(\frac{dp}{d\theta}\right)$ [bar/CAD]	0.1177	0.1183	0.1152	**0.1116**	0.1288	0.1461	0.1571	0.1466
CA50 [CAD]	0.2806	**0.2261**	0.2379	0.2448	0.2664	0.2276	0.2533	0.2329
CA75-CA25 [CAD]	0.6130	0.5248	0.5424	0.5327	0.5144	**0.4510**	0.4930	0.4740
R_b [-]	0.1340	**0.1296**	0.1340	0.1355	0.2616	0.1393	0.1518	0.1584

Table 4. Comparison of the mean behaviour MAE of key combustion metrics between this work and studies in the literature [11–13,18,20,24]. The best result for each metric is in bold.

	This work	**[11]**	**[12]**	**[13]**	**[18]**	**[20]**	**[24]**
$IMEP_g$ [bar]	0.21	-	-	0.43	0.22	0.033	**0.008**
$\max(p(\theta))$ [bar]	1.66	-	-	-	-	-	**0.20**
$\max\left(\frac{dp}{d\theta}\right)$ [bar/CAD]	0.79	-	-	-	6.4	**0.015**	0.71
CA50 [CAD]	0.6	0.3	0.36	1.0	0.22	0.7	**0.2**
CA75-CA25 [CAD]	0.7	**0.2**	-	-	2.4	-	-
R_b [-]	**0.14**	-	-	-	-	-	-

5. Validation of the Prediction Quality of the Combustion Model

The main goal of this work is to predict the in-cylinder pressure and cycle-to-cycle variation. In this section, the outcome of the model is compared to measurements using the validation data set. The hyperparameters shown in Table 5 are used. These choices for hyperparameters give the overall best prediction for the used data set, as discussed in Section 4.

Table 5. Selected hyperparameters and kernel used during the validation in Section 5.

Parameter	Value
n_{train}	75
n_{cyc}	50
n_{PC}	8
Kernel	Matérn with $\nu = \frac{3}{2}$

5.1. Overall Prediction Quality

First, the overall quality of the prediction is assessed. To evaluate the quality of the predicted in-cylinder pressure over a complete combustion cycle, the following measure is used for the mean behaviour:

$$e_p(s_{\text{ICC}}) = \frac{1}{n_{\text{CA}}} \sum_{i=1}^{n_{\text{CA}}} (\hat{p}(\theta_i, s_{\text{ICC}}) - p(\theta_i, s_{\text{ICC}})) \tag{28}$$

and for cycle-to-cycle variation:

$$e_{\sigma_p}(s_{\text{ICC}}) = \frac{1}{n_{\text{CA}}} \sum_{i=1}^{n_{\text{CA}}} (\hat{\sigma}_p(\theta_i, s_{\text{ICC}}) - \sigma_p(\theta_i, s_{\text{ICC}})). \tag{29}$$

To assess the prediction quality of the combustion metrics, the observed average, minimum, and maximum relative differences between the predicted and measured combustion metrics are determined for the mean behaviour and cycle-to-cycle variation. In all metrics, a positive value is related to predicting higher values compared to the measured values.

The results are summarised in Table 6. For the validation set, the data-based combustion model is capable of accurately predicting the mean in-cylinder pressure curve: absolute errors are smaller than 0.59 bar. The variance error is also small. Furthermore, this table shows that, except for $\max\left(\frac{dp}{d\theta}\right)$, the mean behaviours have a good prediction quality (with a mean relative error up to −3.9%). According to the minimum and maximum relative difference, both over- and under-prediction are observed. Only the mean behaviour of $\max\left(\frac{dp}{d\theta}\right)$ shows a bad prediction quality and the model always under-predicts these values. This is expected, since peak pressure rise rates are difficult to predict; see also Figure 4. The prediction quality of the cycle-to-cycle variations can be improved, since most of the time the amount of cycle-to-cycle variation is over-predicted. Again, the worst performance is observed in $\max\left(\frac{dp}{d\theta}\right)$.

Table 6. Prediction quality of the full in-cylinder pressure (absolute error) and of the related combustion metrics (relative error) for the validation set.

	Mean Behaviour			Cycle-to-Cycle Variation		
	Mean	Minimum	Maximum	Mean	Minimum	Maximum
e_p [bar]	0.051	−0.40	0.59			
e_{σ_p} [bar^2]				0.24	0.10	0.56
IMEP_g	2.3%	−15.9%	31.3%	65.6%	7.4%	89.0%
CA50	−3.9%	−39.5%	17.9%	21.5%	−40.3%	54.7%
CA75-CA25	−0.5%	−24.7%	12.4%	56.0%	37.5%	68.5%
R_b	−1.2%	−21.6%	16.6%	57.0%	18.8%	79.1%
$\max(p(\theta))$	0.4%	−5.9%	4.7%	37.0%	−3.4%	84.5%
$\max\left(\frac{dp}{d\theta}\right)$	−22.7%	−48.1%	−7.2%	−85.1%	−252.6%	39.5%

5.2. Variation in Start-of-Injection of Directly Injected Fuel

Figure 5 shows the modelled mean-value and cycle-to-cycle variations of important combustion parameters over a range of SOI_{DI} and the nominal conditions shown in Table 7. The results are in line with the results shown in Table 6. Except for the peak pressure rise rate, the mean value of the model is similar to that of the measurements. The modelled trend in the peak pressure rise rate seems to correspond to the measured values. The standard deviation of the model only matches with $\max(p(\theta))$. The trend in the standard deviation of the model of $\max\left(\frac{dp}{d\theta}\right)$ and R_b seems correct, but it is either too high or too low. The standard deviation of the model does not match the measurements for the $IMEP_g$, CA50, and CA75-CA25.

Table 7. Nominal operating conditions of the simulated model for the results shown in Figures 5 and 6. For reference, the ranges in experiments are indicated.

	Simulated	Measured
Q_{tot} [kJ]	2.3	2.2 to 2.4
BR [-]	0.8	0.75 to 0.85
SOI_{DI} [CADaTDC]	40	40
p_{im} [bar]	1.55	1.45 to 1.65
T_{im} [°C]	45	40 to 50
X_{EGR} [-]	0.2	0.1 to 0.3

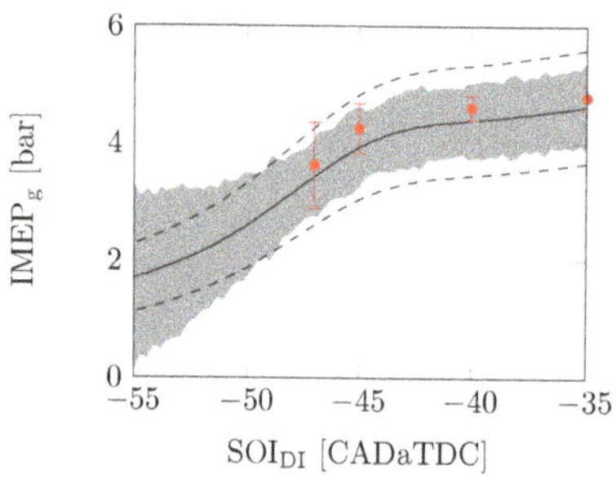

(**a**) Gross indicated mean effective pressure (solid) and 5% $cov(IMEP_g)$ (dashed)

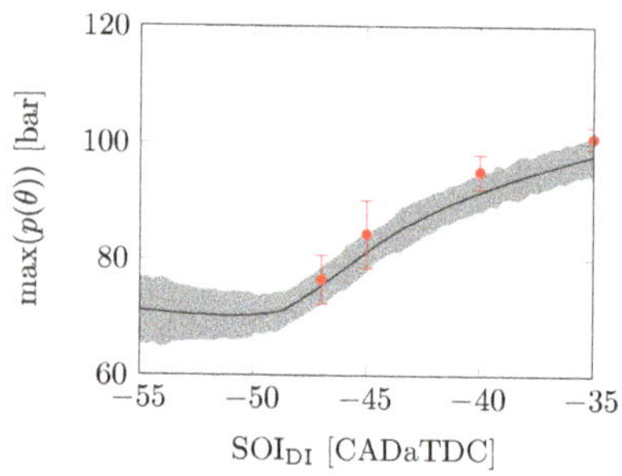

(**b**) In-cylinder peak pressure

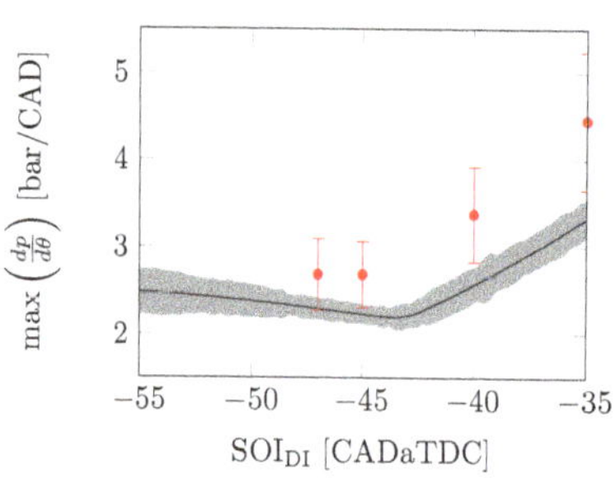

(**c**) In-cylinder peak pressure rise rate

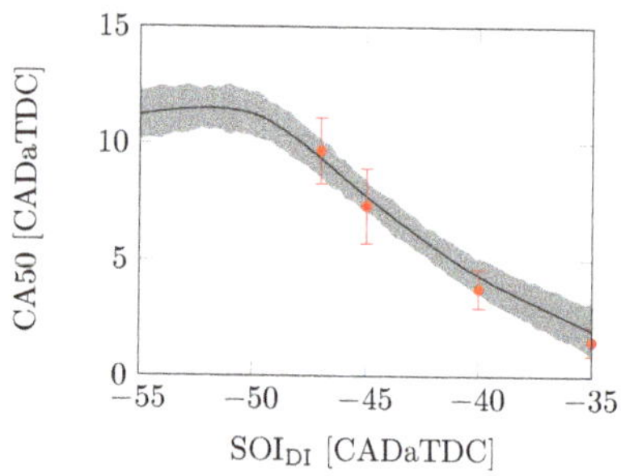

(**d**) Crank angle at 50% total heat release

Figure 5. *Cont.*

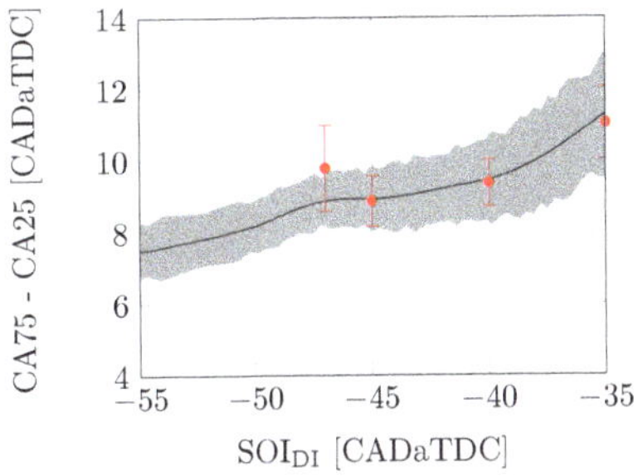

(e) Combustion duration

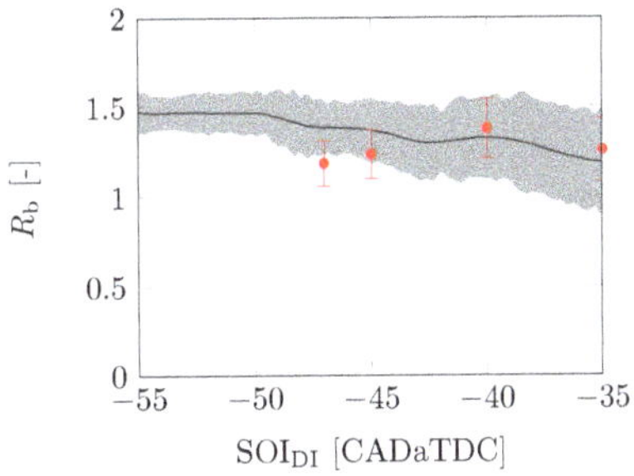

(f) Burn ratio

Figure 5. Average and cycle-to-cycle variations of important combustion metrics (black) and the measured distribution (red) for different values of SOI$_{DI}$ and the nominal conditions shown in Table 7 using the hyperparameters shown in Table 5.

5.3. Variation in Intake Manifold Temperature

Figure 6 shows the modelled mean-value and cycle-to-cycle variations of important combustion parameters over a range of T_{im} and the nominal conditions shown in Table 7. Again, the results are in line with the results shown in Table 6. Similarly to the sweep of SOI$_{DI}$, the mean value of the model is similar to that of the measurements except for the peak pressure rise rate. The modelled trend in the peak pressure rise rate seems to correspond the the measured values. The standard deviation of the model only matches with $\max(p(\theta))$ and CA50. The trend in the standard deviation of the model of CA75-CA25 seems correct, but it is too high. The standard deviation of the model does not match the measurements for the IMEP$_g$, $\max\left(\frac{dp}{d\theta}\right)$, and R_b.

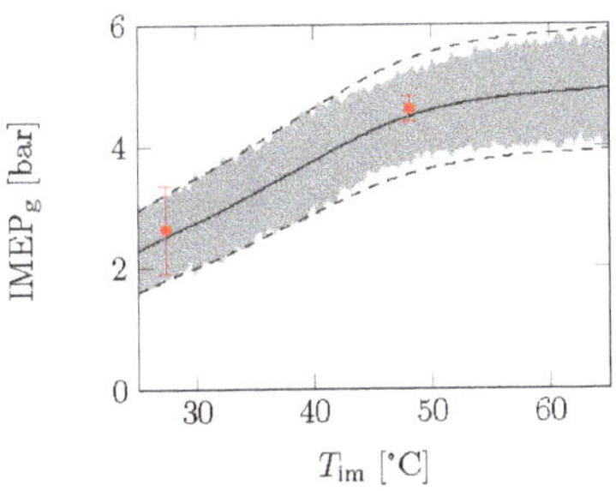

(a) Gross indicated mean effective pressure (solid) and 5% cov(IMEP$_g$) (dashed)

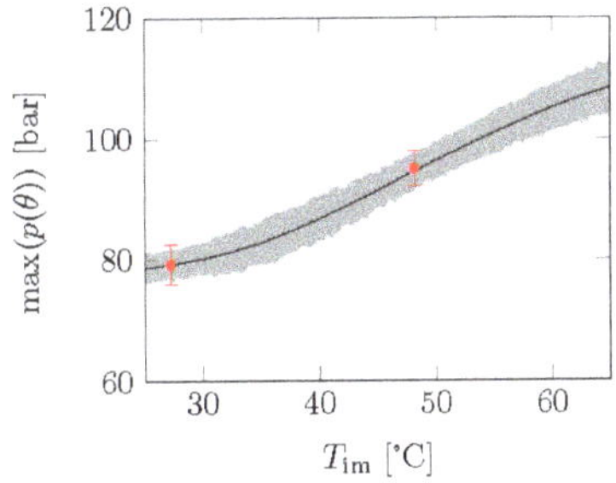

(b) In-cylinder peak pressure

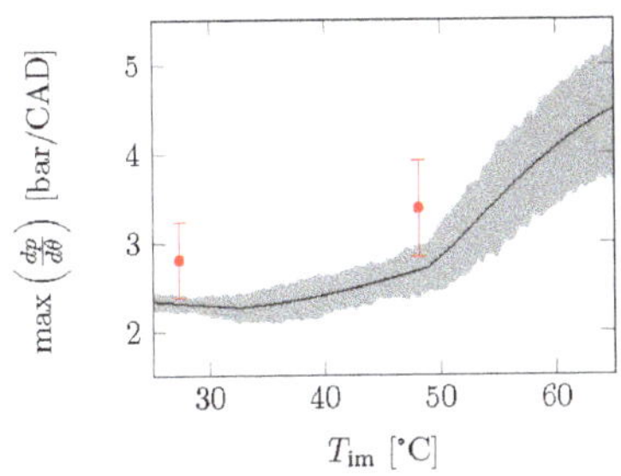

(c) In cylinder peak pressure rise rate

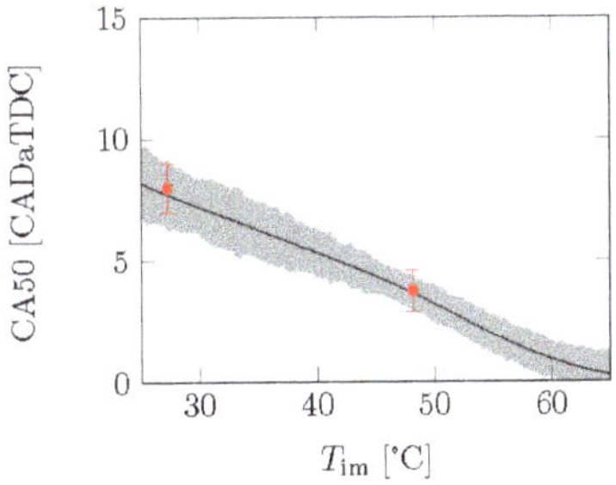

(d) Crank angle at 50% total heat release

Figure 6. *Cont.*

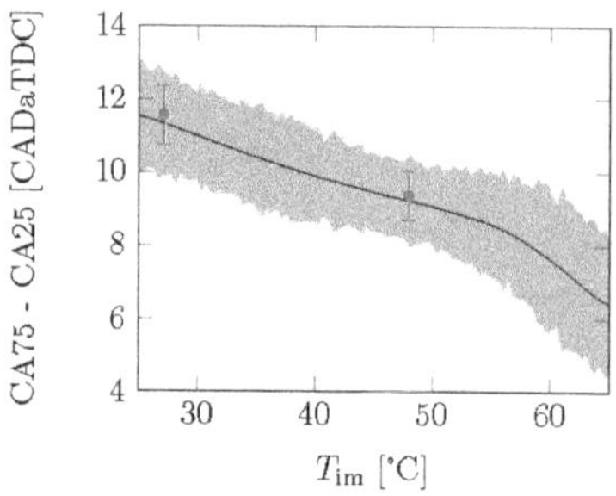

(e) Combustion duration

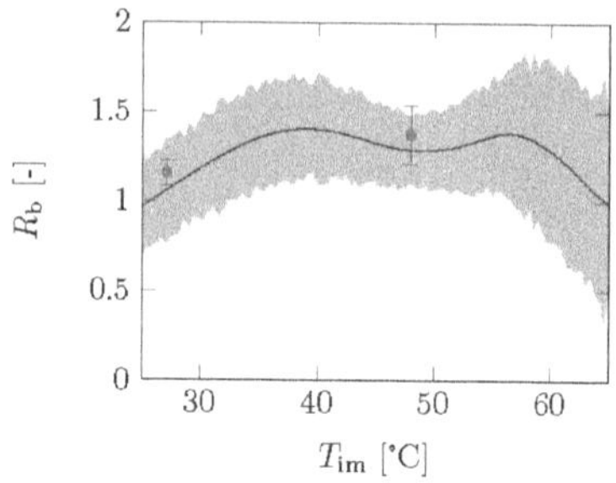

(f) Burn ratio

Figure 6. Average and cycle-to-cycle variations of important combustion metrics (black) and the measured distribution (red) for different values of T_{im} and the nominal conditions shown in Table 7 using the hyperparameters shown in Table 5.

5.4. Discussion on the GPR Modelling of Cycle-to-Cycle Variation

In both sweeps, the predicted standard deviations do not always match the measurements. In (8), $w_i(s_{\text{IVC}})$, and $w_j(s_{\text{IVC}})$ $\forall i, j \in \{1, 2, \ldots, n_{\text{PC}}\}$ are assumed to be independent to align with the available GPR literature; however, this independence is not necessarily the case. To evaluate the correlation between weights at a fixed s_{ICC}, the Pearson correlation matrix R is used. This is given by

$$[R(s_{\text{ICC}})]_{ab} = \frac{\sum_{k=1}^{n_{\text{cyc}}} (w_{a,k}(s_{\text{ICC}}) - \tilde{\mu}_{w_a}(s_{\text{ICC}}))(w_{b,k}(s_{\text{ICC}}) - \tilde{\mu}_{w_b}(s_{\text{ICC}}))}{\tilde{\sigma}_{w_a}(s_{\text{ICC}})\tilde{\sigma}_{w_b}(s_{\text{ICC}})}, \tag{30}$$

where $\tilde{\mu}_{w_i}(s_{\text{ICC}})$ and $\tilde{\sigma}_{w_i}(s_{\text{ICC}})$ are the mean and standard deviation of the measured weights at s_{ICC}, respectively. The values of R range from -1 to 1. When an element of R is zero, there is no correlation between the two variables. However, when an element is -1 or 1 there is full correlation between the two variables. The determinant of the R can be used as a measure for the amount of correlation, where $\det(R)$ ranges from 0 to 1. If $\det(R) = 1$ all variables are fully uncorrelated. However, if $\det(R) = 0$ at least two variables are fully correlated.

Figure 7 shows the distribution of the weights for 50 consecutive cycles of the first five PCs running at a constant $s^*_{\text{ICC}} \in \mathcal{S}^*$ with the least amount of coupling according to the determinant of the Pearson correlation matrix. In Figure 7, the weights have been scaled as

$$\tilde{w}_i(s_{\text{ICC}}) = \frac{w_i(s_{\text{ICC}}) - \tilde{\mu}_{w_i}(s_{\text{ICC}})}{\tilde{\sigma}_{w_i}(s_{\text{ICC}})} \tag{31}$$

to emphasise the coupling. The corresponding symmetric Pearson correlation matrix is given by

$$R = \begin{bmatrix} 1 & 0.6762 & -0.3265 & -0.2830 & 0.0240 \\ & 1 & -0.5037 & 0.0088 & 0.2896 \\ & & 1 & 0.1368 & -0.3906 \\ & & & 1 & -0.2157 \\ & & & & 1 \end{bmatrix} \tag{32}$$

with $\det(R) = 0.23$. This shows that the distributions between some of the weights are significantly correlated, as is also illustrated in Figure 7. Therefore, it is no surprise that the quality of the prediction of the cycle-to-cycle variation deviates from the proposed model. This emphasises the importance of developing GPR methods that include the correlation between the outputs.

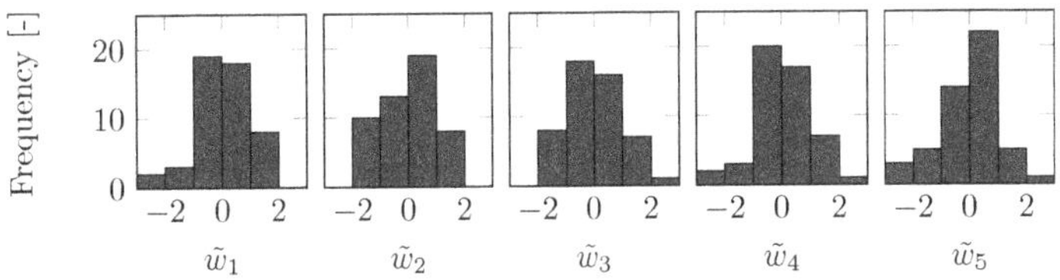

(**a**) Histogram of the individual PC weights

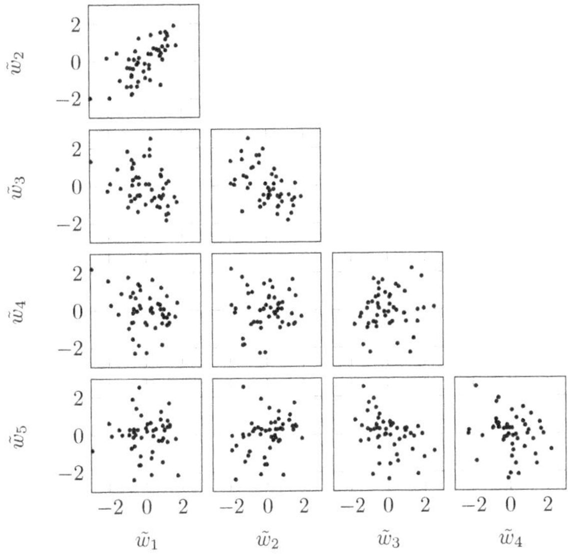

(**b**) Joint distribution between the PC weights

Figure 7. Distribution of the weights for $n_{\mathrm{cyc}} = 50$ cycles for the first five PCs for a constant $s^*_{\mathrm{ICC}} \in \mathcal{S}^*$ with the least amount of coupling according to the Pearson correlation matrix.

6. Conclusions

In this study, a data-based model for the in-cylinder pressure and the corresponding cycle-to-cycle variations is proposed. This model combines a PCD of the in-cylinder pressure and GPR to map in-cylinder conditions and account for cyclic variations.

The proposed data-based modelling approach is successfully applied to an experimental RCCI engine setup. The assumption that the model can be split into a general principal component part and operating-condition-dependent weights is confirmed. A detailed analysis of the hyperparameters for the PCD and GPR is performed. It is found that, for the used data set, more than eight PCs do not further improve the accuracy of the decomposition based on important combustion metrics. For the GPR, the Matérn kernel with $\nu = \frac{3}{2}$ and without ARD gives the best results. The average prediction error of the mean in-cylinder pressure over a complete combustion cycle is 0.051 bar and the corresponding mean cycle-to-cycle variation is 0.24 bar^2. The prediction quality of the mean behaviour of the evaluated combustion metrics has a relative inaccuracy ranging from −3.9% to 2.3%. The prediction error of the cycle-to-cycle variation of the evaluated combustion metrics ranges from 21.5% to 65.5%. The peak pressure rise rate is traditionally hard to predict; in the proposed model it has an inaccuracy of −22.7% in mean behaviour and −85.1% in cycle-to-cycle variation.

In the presented approach, the correlation between $w_i(s_{\mathrm{IVC}})$ and $w_j(s_{\mathrm{IVC}})$ has been neglected for ease of implementation. To improve the accuracy of the cycle-to-cycle variations, this correlation should be added. However, there are very few approaches that extend the GPR framework to including correlation between model outputs known in the literature.

In conclusion, the mean-value performance of our model is comparable or shows improvements compared to models found in the literature. This shows that, even when neglecting correlation, the model performs well. The model can be used for in-cylinder pressure shaping as proposed in Vlaswinkel and Willems [10]. Furthermore, it can be used in model-based optimisation approaches that take into account cycle-to-cycle variations and safety criteria. When combined with the PCD-based emission model of Henningsson et al. [25], the model provides a base for optimisation approaches with emission constraints.

Author Contributions: Conceptualisation, M.V. and F.W.; data curation, M.V.; formal analysis, M.V.; funding acquisition, F.W.; investigation, M.V.; methodology, M.V. and F.W.; project administration, F.W.; resources, F.W.; software, M.V.; supervision, F.W.; validation, M.V. and F.W.; visualisation, M.V.; writing—original draft, M.V.; writing—review and editing, M.V. and F.W. All authors have read and agreed to the published version of the manuscript.

Funding: The research presented in this study is financially supported by the Dutch Technology Foundation (STW) under project number 14927.

Data Availability Statement: The datasets presented in this article are not readily available because of agreements made for project funding. Requests to access the datasets should be directed to m.g.vlaswinkel@tue.nl.

Acknowledgments: The authors would like to thank Marnix Hage, Michel Cuijpers and Bart van Pinxten from the Zero Emission Lab at the Eindhoven University of Technology for their support during experimentation. The authors would like to thank Lucy Pao for proofreading this manuscript.

Conflicts of Interest: The authors declare no conflicts of interest.

References

1. Leach, F.; Kalghatgi, G.; Stone, R.; Miles, P. The Scope for Improving the Efficiency and Environmental Impact of Internal Combustion Engines. *Transp. Eng.* **2020**, *1*, 100005. [CrossRef]
2. Duarte Souza Alvarenga Santos, N.; Rückert Roso, V.; Teixeira Malaquias, A.C.; Coelho Baêta, J.G. Internal Combustion Engines and Biofuels: Examining Why This Robust Combination Should Not Be Ignored for Future Sustainable Transportation. *Renew. Sustain. Energy Rev.* **2021**, *148*, 111292. [CrossRef]
3. Benajes, J.; García, A.; Monsalve-Serrano, J.; Guzmán-Mendoza, M. A Review on Low Carbon Fuels for Road Vehicles: The Good, the Bad and the Energy Potential for the Transport Sector. *Fuel* **2024**, *361*, 130647. [CrossRef]
4. Dempsey, A.B.; Walker, N.R.; Gingrich, E.; Reitz, R.D. Comparison of Low Temperature Combustion Strategies for Advanced Compression Ignition Engines with a Focus on Controllability. *Combust. Sci. Technol.* **2014**, *186*, 210–241. [CrossRef]
5. Kokjohn, S.L.; Hanson, R.M.; Splitter, D.A.; Reitz, R.D. Fuel Reactivity Controlled Compression Ignition (RCCI): A Pathway to Controlled High-Efficiency Clean Combustion. *Int. J. Engine Res.* **2011**, *12*, 209–226. [CrossRef]
6. Reitz, R.D.; Duraisamy, G. Review of High Efficiency and Clean Reactivity Controlled Compression Ignition (RCCI) Combustion in Internal Combustion Engines. *Prog. Energy Combust. Sci.* **2015**, *46*, 12–71. [CrossRef]
7. Li, J.; Yang, W.; Zhou, D. Review on the Management of RCCI Engines. *Renew. Sustain. Energy Rev.* **2017**, *69*, 65–79. [CrossRef]
8. Paykani, A.; Garcia, A.; Shahbakhti, M.; Rahnama, P.; Reitz, R.D. Reactivity Controlled Compression Ignition Engine: Pathways towards Commercial Viability. *Appl. Energy* **2021**, *282*, 116174. [CrossRef]
9. Willems, F. Is Cylinder Pressure-Based Control Required to Meet Future HD Legislation? *IFAC-PapersOnLine* **2018**, *51*, 111–118. [CrossRef]
10. Vlaswinkel, M.; Willems, F. Cylinder Pressure Feedback Control for Ideal Thermodynamic Cycle Tracking: Towards Self-learning Engines. *IFAC-PapersOnLine* **2023**, *56*, 8260–8265. [CrossRef]
11. Khodadadi Sadabadi, K.; Shahbakhti, M.; Bharath, A.N.; Reitz, R.D. Modeling of Combustion Phasing of a Reactivity-Controlled Compression Ignition Engine for Control Applications. *Int. J. Engine Res.* **2016**, *17*, 421–435. [CrossRef]
12. Guardiola, C.; Pla, B.; Bares, P.; Barbier, A. A Combustion Phasing Control-Oriented Model Applied to an RCCI Engine. *IFAC-PapersOnLine* **2018**, *51*, 119–124. [CrossRef]
13. Raut, A.; Irdmousa, B.K.; Shahbakhti, M. Dynamic Modeling and Model Predictive Control of an RCCI Engine. *Control. Eng. Pract.* **2018**, *81*, 129–144. [CrossRef]
14. Sui, W.; González, J.P.; Hall, C.M. Combustion Phasing Modelling of Dual Fuel Engines. *IFAC-PapersOnLine* **2018**, *51*, 319–324. [CrossRef]

15. Irdmousa, B.K.; Rizvi, S.Z.; Veini, J.M.; Nabert, J.D.; Shahbakhti, M. Data-Driven Modeling and Predictive Control of Combustion Phasing for RCCI Engines. In Proceedings of the 2019 American Control Conference (ACC), Philadelphia, PA, USA, 10–12 July 2019; pp. 1617–1622. [CrossRef]
16. Kakoee, A.; Bakhshan, Y.; Barbier, A.; Bares, P.; Guardiola, C. Modeling Combustion Timing in an RCCI Engine by Means of a Control Oriented Model. *Control Eng. Pract.* **2020**, *97*, 104321. [CrossRef]
17. Bekdemir, C.; Baert, R.; Willems, F.; Somers, B. Towards Control-Oriented Modeling of Natural Gas-Diesel RCCI Combustion. In Proceedings of the SAE 2015 World Congress & Exhibition, London, UK, 21 April 2015. [CrossRef]
18. Klos, D.; Kokjohn, S.L. Investigation of the Sources of Combustion Instability in Low-Temperature Combustion Engines Using Response Surface Models. *Int. J. Engine Res.* **2015**, *16*, 419–440. [CrossRef]
19. Xia, L.; de Jager, B.; Donkers, T.; Willems, F. Robust Constrained Optimization for RCCI Engines Using Nested Penalized Particle Swarm. *Control. Eng. Pract.* **2020**, *99*, 104411. [CrossRef]
20. Basina, L.N.A.; Irdmousa, B.K.; Velni, J.M.; Borhan, H.; Naber, J.D.; Shahbakhti, M. Data-Driven Modeling and Predictive Control of Maximum Pressure Rise Rate in RCCI Engines. In Proceedings of the 2020 IEEE Conference on Control Technology and Applications (CCTA), Montreal, QC, Canada, 24–26 August 2020; pp. 94–99. [CrossRef]
21. Verhaegh, J.; Kupper, F.; Willems, F. Data-Driven Air-Fuel Path Control Design for Robust RCCI Engine Operation. *Energies* **2022**, *15*, 2018. [CrossRef]
22. Pan, W.; Korkmaz, M.; Beeckmann, J.; Pitsch, H. Unsupervised Learning and Nonlinear Identification for In-Cylinder Pressure Prediction of Diesel Combustion Rate Shaping Process. In Proceedings of the 13th IFAC Workshop on Adaptive and Learning Control Systems ALCOS, Winchester, UK, 4–6 December 2019; Volume 52, pp. 199–203. [CrossRef]
23. Vlaswinkel, M.; De Jager, B.; Willems, F. Data-Based In-Cylinder Pressure Model Including Cyclic Variations of an RCCI Engine. *IFAC-PapersOnLine* **2022**, *55*, 13–18. [CrossRef]
24. Mishra, C.; Subbarao, P. Design, Development and Testing a Hybrid Control Model for RCCI Engine Using Double Wiebe Function and Random Forest Machine Learning. *Control Eng. Pract.* **2021**, *113*, 104857. [CrossRef]
25. Henningsson, M.; Tunestål, P.; Johansson, R. A Virtual Sensor for Predicting Diesel Engine Emissions from Cylinder Pressure Data. *IFAC Proc. Vol.* **2012**, *45*, 424–431. [CrossRef]
26. Panzani, G.; Ostman, F.; Onder, C.H. Engine Knock Margin Estimation Using In-Cylinder Pressure Measurements. *IEEE/ASME Trans. Mechatronics* **2017**, *22*, 301–311. [CrossRef]
27. Panzani, G.; Pozzato, G.; Savaresi, S.M.; Rösgren, J.; Onder, C.H. Engine Knock Detection: An Eigenpressure Approach. *IFAC-PapersOnLine* **2019**, *52*, 267–272. [CrossRef]
28. Rasmussen, C.E.; Williams, C.K.I. *Gaussian Processes for Machine Learning*; The MIT Press: Cambridge, MA, USA, 2005. [CrossRef]
29. Eriksson, L.; Thomasson, A. Cylinder State Estimation from Measured Cylinder Pressure Traces—A Survey * This Project Was Financed by the VINNOVA Industry Excellence Center LINK-SIC. *IFAC-PapersOnLine* **2017**, *50*, 11029–11039. [CrossRef]
30. Wilhelmsson, C.; Tunestål, P.; Johansson, B. *Model Based Engine Control Using ASICs: A Virtual Heat Release Sensor*; Institut Francais du Petrole: Rueil-Malmaison, France, 2006.

energies

Article

Evaluating Combustion Ignition, Burnout, Stability, and Intensity of Coal–Biomass Blends Within a Drop Tube Furnace Through Modelling

Garikai T. Marangwanda [1,2,*] and Daniel M. Madyira [1]

[1] Department of Mechanical Engineering Science, University of Johannesburg, Johannesburg 2026, South Africa; dmadyira@uj.ac.za
[2] Centre of Applied Research and Innovation in the Built Environment, University of Johannesburg, Johannesburg 2026, South Africa
* Correspondence: gmarangwanda@uj.ac.za; Tel.: +27-62-025-9829

Abstract: This study focused on evaluating the combustion ignition, burnout, stability, and intensity of Hwange coal and Pinus sawdust blends within a drop tube furnace (DTF) through modelling. The cocombustion of coal with biomass is gaining attention as a strategy to improve fuel efficiency and reduce emissions. Hwange coal, a key energy source in Zimbabwe, produces significant emissions, while Pinus sawdust offers a renewable alternative with favourable combustion properties. Optimising cocombustion performance is highly dependent on understanding various mass- and energy-conservation-related parameters in detail, hence the motivation of this study. The fuels of interest were blended through increasing the Pinus sawdust mass percentages up to 30%. A DTF that is 2 m long and 0.07 m in diameter was modelled and validated successfully using particle residence time and temperature profiles. An increase in blending resulted in an increase in combustion intensity, as made apparent by the heat of reaction profiles, which were also shown to be dependent on the kinetic rate of the reaction between CO and O_2 to form CO_2. The burnout rate profiles demonstrated that as blending increased, heat was released more abruptly over a short distance; hence, combustion became less stable. The burnout rate profiles were shown to be dependent on the kinetic rate of reaction between char and O_2 to form CO. The effect of DTF wall temperatures (1273, 1473, and 1673 K) was also studied, with the results showing that at a low temperature, the reaction zone was delayed to a distance of 0.8 m from the injection point, as compared to 0.4 m at 1673 K. In summary, this study demonstrated that combustion ignition, burnout, and intensity increased with the blending ratio of Pinus sawdust, whilst combustion stability decreased.

Keywords: ignition; burnout; combustion stability; combustion intensity; drop tube furnace

1. Introduction

The cocombustion of coal with biomass waste has been under investigation for a long time with the main aim of reducing emissions traditionally obtained from coal combustion on its own [1]. Varied experimental techniques ranging from pilot-scale to actual power plant combustion scenarios have been studied over the years with success [2,3]. With the continued developments in technology, the modelling of cocombustion has also taken a prominent role as engineers seek to optimise the combustion process by extracting maximum energy and maximum waste incineration at reduced emissions and minimum cost [4,5]. Some researchers then saw it fit to characterise the change in combustion parameters from a technical point of view, thus studying the influence of cocombustion on burnout, ignition, stability, and intensity [6]. Being able to predict the influence of cocombustion on various combustion parameters is of practical importance to furnace operators in industry. Good and predictable ignition properties imply that combustion reactions commence within the expected time range, thus avoiding problems caused by premature or delayed ignition [7]. Consequently, the energy absorbed from the furnace to offset ignition is accurately predicted, leaving furnace operators with useful optimum operating parameters.

Burnout can be described as the ability of a fuel sample to extinguish its combustible material, which includes the volatile matter and char constituents. With respect to solid fuel combustion, most researchers monitor burnout in the same manner by measuring the char content only at the furnace exit in comparison to the initial fixed carbon content [7]. Benim et al. [2] during their studies with pre-dried lignite and torrefied biomass and Gao et al. [8] during their studies with Colombian coal and woodchip directly or indirectly showed improvement in the overall burnout as blending increased when they monitored the char content in the products. It is important not to overlook the homogenous gaseous combustion products when analysing burnout since carbon monoxide and other unburnt gaseous constituents also influence the overall burnout percentage [9]. Some heavy volatile components released during the devolatilisation process of solid fuels require relatively long residence times to attain complete combustion [10]. As far as the researchers are concerned, no CFD studies have been reported that have managed to monitor both carbon in ash and carbon monoxide in flue gases concurrently as a function of burnout percentage.

The application of mass conservation towards cocombustion entails looking at the quantity and composition as well as the rate of production or depletion of combustion species. By extension, the validation of a cocombustion model is usually based on measuring the species molar fraction along and across the furnace. The application of energy conservation towards cocombustion modelling entails monitoring temperature-related parameters. The basic parameters that are mainly used for cocombustion model validation are temperature and wall heat flux [11]. Of equal importance, but less alluded to during the combustion modelling of boilers, are flame stability and intensity. Combustion stability, a measure relatively linked to ignition and homogenous combustion, is generally used to monitor the predictability of a combustion process [6]. When fuel blending is being monitored, it is imperative that stability is maintained to avoid any flashback or liftoff flames or, in extreme cases, flame extinction [12]. Ma et al. [13] monitored the heat flux profiles that are exhibited by blended samples, which can be used to evaluate combustion stability and intensity. To be more accurate, it makes sense to monitor combustion stability through plotting rate of heat release profiles, as residence time is also incorporated into the analysis. It is expected that homogeneous volatile combustion is less stable compared to heterogeneous char combustion, which means that highly blended samples are less stable when comparing the activation energy required in each case [14,15].

Interestingly, the lesser the stability, the higher the combustion intensity, which is closely linked to the heat released, as demonstrated by Marangwanda et al. [6] during their studies with coal and sawdust. As far as experimental procedures are concerned, the use of image processing techniques will likely give a complete understanding of flame behaviour, which can help with the evaluation of combustion intensity. Unfortunately, most of these experimental procedures require specialised equipment [16,17].

Table 1 summarises different modelling studies that have been carried out and how they reported on ignition, burnout, intensity, and stability. The ability to capture almost all of these combustion parameters by CFD makes it an indispensable and affordable tool with regard to cocombustion optimisation. This study addresses the gap existent with regard to combining most of these combustion parameters whilst optimising the cocombustion process.

Table 1. Summary of methods used to report on combustion parameters.

Description	Fuels	Method of Reporting on			
		Ignition	Burnout	Intensity	Stability
Combustion in a 150 kW reactor [18]	Coal and wheat straw	Temperatures	Char percentage	-	-
Lab-scale bubbling fluidised bed combustor [19]	Coal, lignite, spruce wood, wheat straw, and hazelnut shell	-	Unburnt carbon	-	-
Cocombustion in a 660 MW tangentially fired boiler [4]	Coal and sewage sludge	-	-	Heat flux	-
Experimental study on combustion with nitrogen [20]	Coal, poplar wood, and corn stalks	Temperature Radiation spectrum	-	-	Image processing
Cocombustion in a 500 MW boiler [21]	Coal slurry	Temperature	Species molar fractions	-	-

2. Experimental Methods

2.1. Fuel Blend Characterisation

The physical and chemical fuel specifications of the fuel blends of interest have been detailed in a prior publication by the same authors [22]. Hwange coal, a bituminous thermal-grade coal, is widely used in Zimbabwe's power generation due to its high availability and established infrastructure for extraction and utilisation. However, its combustion leads to significant greenhouse gas emissions, making it necessary to explore alternative co-firing strategies. Pinus sawdust, on the other hand, is an abundant waste biomass in Zimbabwe, primarily sourced from sawmill operations in the Eastern Highlands. Its high volatile matter and lower sulphur content make it a suitable candidate for cocombustion with coal, contributing to emission reduction and improved combustion characteristics [6,22]. Using the fuel characterisation results, the volatile composition for each blend was determined. It was assumed that each volatile had a lumped form represented by $C_aH_bO_cN_dS_e$ regardless of the various constituents of volatiles such as CH_4, CO, etc. Using proximate, ultimate, and calorific experimental data, the lumped volatile composition for each fuel blend was derived through analytical methods [23], as demonstrated by the equations in Table 2.

Table 2. Volatile composition determination.

Constituent	%Weight	%Mol
C	$\frac{C_{daf} - FC_{daf}}{VM_{daf}}$	$\frac{1}{12} \times \left[\frac{C_{daf} - FC_{daf}}{VM_{daf}}\right] \times Mr_{volatile} = a$
H	$\frac{H_{daf}}{VM_{daf}}$	$\frac{1}{1} \times \left[\frac{H_{daf}}{VM_{daf}}\right] \times Mr_{volatile} = b$
O	$\frac{O_{daf}}{VM_{daf}}$	$\frac{1}{16} \times \left[\frac{O_{daf}}{VM_{daf}}\right] \times Mr_{volatile} = c$
N	$\frac{N_{daf}}{VM_{daf}}$	$\frac{1}{14} \times \left[\frac{N_{daf}}{VM_{daf}}\right] \times Mr_{volatile} = d$
S	$\frac{S_{daf}}{VM_{daf}}$	$\frac{1}{32} \times \left[\frac{S_{daf}}{VM_{daf}}\right] \times Mr_{volatile} = e$

Here, VM and FC represent volatile matter and fixed carbon, whilst the subscript daf denotes "dry ash free". C, H, N, S, and O represent carbon, hydrogen, nitrogen, sulphur, and oxygen. To determine the enthalpy of formation for the volatile ($H_{f,vol}$), Equation (3) was employed after rearranging Equations (1) and (2).

$$H_{f,fuel} = FC_{daf} \times H_{f,char} + VM_{daf} \times H_{f,vol} \quad (1)$$

$$H_{f,char}: \ C + O_2 = CO_2 \ (-32.8 MJ/kg); H_{f,char} = LCV_{coal} \quad (2)$$

$$H_{f,char} = LCV_{coal} = \frac{HCV_{coal} - h_{latent,H20} \times MC_{ar}}{1 - MC_{ar} - Ash_{ar}} - \frac{H_{ar} \times Mr(H_2O)}{2 \times Mr(H)} \times h_{latent,H20} \quad (3)$$

This method offered a more comprehensive approach to representing fuel properties, avoiding the assumption that the fuel sample consists solely of methane or butane, as suggested by some authors [24,25]. The proximate and volatile compositions of the blends that were used are presented in Table 3.

Table 3. Chemical compositions of fuel blends.

Fuel Blend	Proximate Analysis on a Dry Basis	Volatile Molar Composition	Volatile Molar Mass	Enthalpy of Formation for Volatile ($H_{f,vol}$)
100HC	53.97FC, 23.10VM, 22.93Ash	$C_{0.292}H_{2.200}O_{0.618}N_{0.086}S_{0.014}$	17.24	-5.904×10^7
90HC10PS	48.21FC, 29.91VM, 21.88Ash	$C_{0.535}H_{2.102}O_{0.556}N_{0.0646}S_{0.0111}$	18.68	-7.862×10^7
80HC20PS	46.35FC, 31.82VM, 21.83Ash	$C_{0.559}H_{2.36}O_{0.618}N_{0.0585}S_{0.011}$	20.11	-9.819×10^7
70HC30PS	46,02FC, 33.74VM, 20.24Ash	$C_{0.579}H_{2.615}O_{0.683}N_{0.052}S_{0.011}$	21.55	-1.178×10^8
100PS	15.62FC, 80.68VM, 3.70Ash	$C_{1.107}H_{2.37}O_{0.99}N_{0.01}$	31.63	-2.548×10^8

The molar ratio of carbon showed an increase with blending from 0.292 up to 0.579 for a 30% blending ratio, as summarised in Table 3. Similar trends were also experienced with hydrogen and oxygen though to a lesser magnitude. Nitrogen in the volatile decreased with blending from 0.086 to 0.052 as blending increased, whilst the sulphur content decreased marginally with blending. This is supported by other authors indirectly through the assertion that biomass contains higher volatile matter than coal. This translates to a gradual decrease in energy content since the H/C ratio of blended volatiles is lower than that of the unblended sample [26]. Maisyarah et al. [27] also suggested that a lower H/C ratio tended to result in a higher mass loss for the overall sample at the end of the combustion process. As such, a higher mass loss was expected for the blended fuel samples. The kinetic parameters of the fuel blends as obtained from thermogravimetric analysis are summarised in Table 4, representing the homogenous (volatile) and heterogenous (char) combustion stages. Detailed information about kinetic parameter determination is available in a previous paper by the same authors [22].

Table 4. Activation energy and pre-exponential factor values for the fuel blends heated in air.

Fuel Blend	Stage	E_a (kJ/mol)	A (s^{-1})
100HC	Volatile combustion	92.98	5.84×10^5
	Char combustion	52.90	1.16×10^3
90HC10PS	Volatile combustion	107.89	5.05×10^6
	Char combustion	68.99	2.78×10^4
80HC20PS	Volatile combustion	104.95	2.94×10^8
	Char combustion	90.52	9.53×10^5
70HC30PS	Volatile combustion	106.05	2.72×10^9
	Char combustion	103.85	6.12×10^7

2.2. Drop Tube Furnace Experimental Setup

The drop tube furnace (DTF) experimental setup consisted of an alumina tube serving as the furnace, with a total height of 2032 mm and an internal diameter of 70 mm. A detailed description of the setup and procedure is available in the work of Marangwanda and Madyira [28]. The furnace had an alumina silicate refractory lining, with heating tubes systematically positioned around the alumina tube. Fuel blends were introduced through a 2 mm cooled injector at 373 K, while a 6 mm cooled sampling probe at the reactor exit facilitated collection. Within the fuel feeding system, 15 g of fuel blends was placed in a test tube, which was connected to the primary carrier gas flowing at an average rate of 2.5 NL/min (within a 1.5–3.5 NL/min range). An electromechanical vibrator enabled controlled fuel feeding, ensuring fluidisation by the primary carrier gas, which transported the fuel into the outgoing pipe leading to the furnace. The water-cooled sampling probe maintained a suction flow rate of 6–15 NL/min, while the secondary gas was preheated to 1273 K and supplied at 10–20 NL/min, depending on the required furnace residence time. The collecting probe was positioned at variable distances 520 mm, 920 mm, and 1320 mm along the furnace centreline at pre-determined experimental conditions. A bag filter was located at the outlet such that char and ash could be collected for subsequent analysis.

During devolatilisation, the system operated at a single set temperature of 1273 K, whereas during char combustion, three furnace temperature settings (1273 K, 1473 K, and 1673 K) were investigated. The injector was modelled to introduce fuel in a normal direction to the inlet, making an assumption that no swirl effects existed. The experimental procedure was conducted in two distinct stages. The first stage was performed under a nitrogen atmosphere, simulating the devolatilisation process, while the second stage took place in an oxygen-enriched environment, representing char combustion. During the first stage, only solid char was collected and analysed, as the released volatiles were carried away with the exhaust gases. Given that devolatilisation requires short residence times, a 520 mm probing position was deemed adequate for data collection. Following char collection and characterisation, the second stage involved reintroducing the char into the DTF under an oxygen-enriched atmosphere (3% O_2, 97% N_2), leading to the final production of ash. The use of a two-stage approach was necessitated by DTF operating temperature limitations and combustion atmosphere constraints. A controlled 3% O_2 concentration ensured the development of gradual combustion profiles suitable for this study. Higher O_2 concentrations tend to accelerate combustion reactions, potentially leading to intensified reaction zones or even explosive conditions, making precise measurement difficult with the available instrumentation.

These observations were experimentally demonstrated by Zou et al. [29] whilst studying the combustion of pulverised coal under various atmospheres ranging from 21% to 50% O_2.

Within a DTF, the fuel is exposed to an isothermal heating rate within the range of 10^4 and 10^5 K/s, which is quite high when compared to any TGA analysis. Liu et al. [30] studied the cocombustion kinetic parameters of biomass and plastic at high heating rates within the range of 10^4 K/s. Their findings demonstrated that in as much as combustion parameters vary by large factors when the heating rate increases, kinetic parameters such as the activation energy and pre-exponential factor succumb to the compensation effect at high heating rates. The heating rate increased with activation energy at first; it then decreased and eventually remained at the same value for heating rates of 750 and 1000 K/min. However, Czajka et al. [31], whilst extrapolating the kinetic parameters of

South African coal during the pyrolysis process, suggested using corrections, as given in Equations (4) and (5).

$$\frac{k_{heat1}}{k_{heat2}} = \left(\frac{A_{heat1}}{A_{heat2}}\right)^n = \left(\frac{\beta_{heat1}}{\beta_{heat2}}\right)^n \quad (4)$$

$$\frac{A_{heat1}}{\beta_{heat1}} = \vartheta_1.\exp(\vartheta_2.E) \quad (5)$$

where k denotes the kinetic rate, β is the heating rate, n is a factor dependent on the amount of weight released, heat1 and heat2 refer to parameters at the 1st and 2nd heating rates, and ϑ is a parameter which is closely related to the compensation effect [32]. Through inference from their findings, the authors determined the corresponding kinetic parameters for DTF simulation. As noted, during the first stage of the experiment, which was conducted under a nitrogen atmosphere, the released volatiles were entrained in the carrier gas, while the solid char was collected using a bag filter at the outlet. In the second stage, which was under an oxygen-enriched atmosphere, the char underwent complete combustion, leaving behind ash, which was collected at the outlet as well. The properties of the particles that were not directly measurable were estimated indirectly using experimental data, which included the use of fluid velocity and particle size distribution.

3. Numerical Methods

ANSYS FLUENT 2021 R1 was used to implement the cocombustion model that had been developed which is based on the conservation equation, as given in Equation (6) [33].

$$\frac{\partial}{\partial t}(\rho\varphi) + \frac{\partial}{\partial x_i}(\rho U_i \varphi) = \frac{\partial}{\partial x_i}\left(\Gamma \frac{\partial \varphi}{\partial x_i}\right) + S_\varphi \quad (6)$$

where the following variables are used:

ρ: represents the density;

φ: variable (mass, specific enthalpy, or species mass fraction);

U_i: velocity (u, v, w);

Γ: variable diffusion coefficient;

S_φ: variable source or sink.

3.1. Furnace Geometry and Meshing

The boundaries drawn and meshed were simplified to only capture the important physics that influence the combustion process.

Since the experimental DTF had a cylindrical shape, only 30° of the DTF was modelled due to the following:

- There was a need to reduce the computational power required to run the setup;
- The geometry was cylindrical about the furnace axis;
- A 3D analysis would be able to capture the axial and radial variation in parameters with more precision as compared to a 2D analysis;
- The inlets (fuel inlet and secondary carrier gas inlet) and outlet boundaries were all normal to the symmetrical planes that were defined;
- The expected flow was going to be repeated periodically about the axis since the fuel and secondary carrier gas inlet flows were distributed evenly and normal to their corresponding boundaries.

A structured mesh was employed for the furnace meshing; thus, hexa-meshing was performed. During meshing, emphasis was placed on achieving acceptable determinant (>0.2), angle (>18°), and warpage (<10) values for all the meshes that were generated so as to reduce any errors associated with bad meshing in the later stages. All the structured meshes had some wall inflation to capture the boundary layer and a bit of refinement towards the axis to anticipate the reaction zone position. Illustrations of the meshing employed within this study are given in Figures 1 and 2.

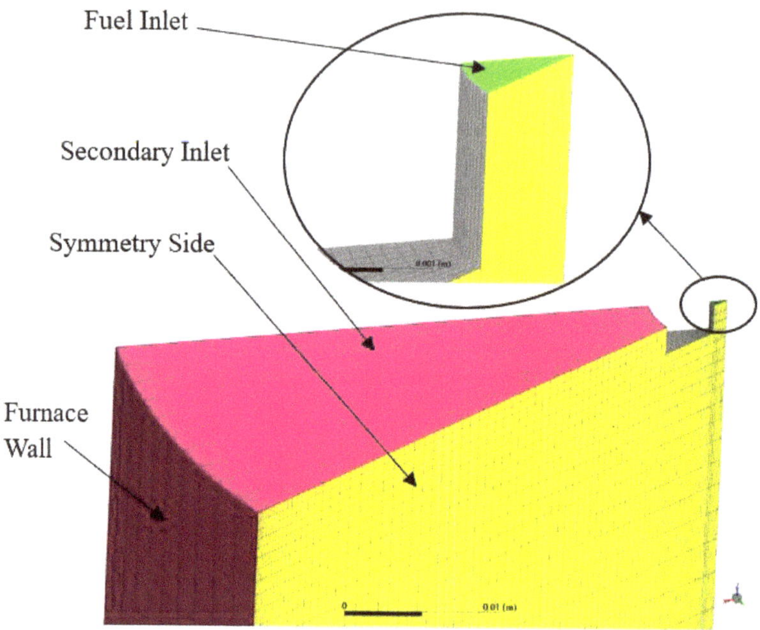

Figure 1. DTF mesh showing inlets.

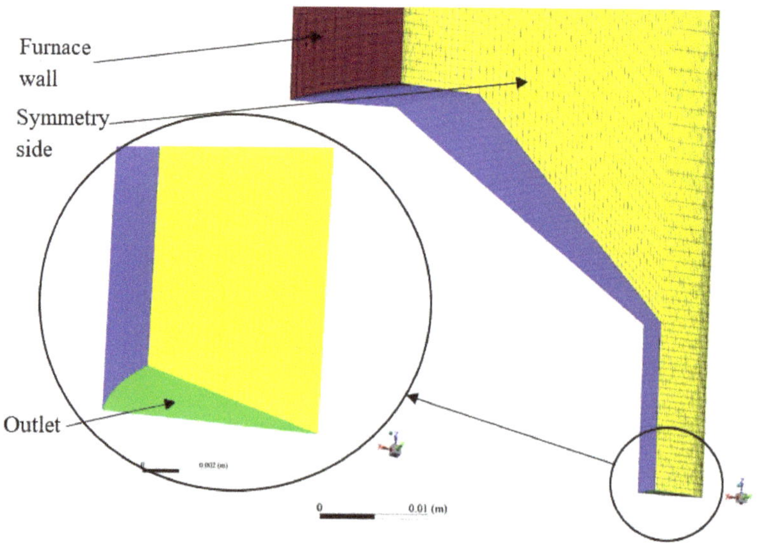

Figure 2. DTF mesh showing outlet.

3.2. Cocombustion Model Setup

The submodels that were used to capture combustion were compiled within ANSYS FLUENT 2021 R1. The main strength of ANSYS FLUENT 2021 R1 with regard to combustion modelling lies within its ability to handle user-defined functions (UDFs). UDFs allowed the researchers to include customised submodels other than those built in by default within ANSYS FLUENT 2021 R1. Due to their versatility and conciseness, C^{++} codes were written to act as the UDFs for incorporation into the ANSYS FLUENT 2021 R1 overall combustion model. The UDFs were used to customise mainly the furnace freeboard temperature profile, particle drag properties of the coal and biomass species as they travelled through the furnace, and specific heat capacities of the various participating gaseous species such as CO_2, N_2, O_2, and H_2O. Table 5 summarises the boundary conditions that were employed by the researchers.

Table 5. Combustion model boundary conditions.

Physics	Model
Turbulence	RNG k-epsilon, scalable wall function
Radiation	Discrete ordinate model, P1 model weighted sum of grey gases model (WSGGM)
Particle distribution	40 continuous-phase iterations per DPM iteration Rosin–Rammler diameter distribution
Inlets	Fuel velocity inlet as 1.85 m/s For a 30° modelled section, total fuel mass flow rate translated to 2.08×10^{-6} kg/s, primary carrier gas mass flow rate was 4.85×10^{-7} kg/s, and secondary carrier gas was 2.43×10^{-5} kg/s
Chemical rection	Species transport option, eddy dissipation concept

The sawdust particles utilised a shape factor of 0.83 since a cylindrical shape was assumed due to the fibrous nature of biomass. In as much as the pulverised Hwange coal particles were irregular in shape, they were assumed to be spherical because when small coal particles heat up, they tend to soften up and adopt a spherical shape [26].

Since the first stage was carried out under a nitrogen-enriched atmosphere, resembling the devolatilisation process, the main chemical reactions that were expected were related to the release of volatiles. As already investigated under the TGA experiments, devolatilisation was deduced to follow the single-rate kinetic devolatilisation submodel, as given in Table 4. The second stage was carried out under a 97% nitrogen and 3% oxygen atmosphere, resembling the heterogenous combustion of char. No homogenous combustion of volatiles was expected because all the volatiles were carried along with the carrier gas in the first stage and, thus, were not present during the second stage. The heterogenous surface reaction mechanisms are also known as multiple surface reaction mechanisms because the reactions are assumed to take place within the boundary layer and bulk flow simultaneously. The activation energy and pre-exponential factors during heterogenous combustion are given in Table 4, as determined by the researchers using TGA experiments. The gasification reactions of char with CO_2 or H_2O were overlooked since the reaction temperatures within the reactor were not high enough to warrant their influence. Furthermore, the concentrations of CO_2 and H_2O were low as compared to N_2 [34].

The furnace was calibrated using a different coal sample to verify the effectiveness of the thermal measurements. In as much as the furnace wall was set to have a certain temperature, the actual furnace wall temperature was slightly different from the setting, as shown in Figure 3. A known fuel sample made of South African coal was used to calibrate the furnace and validate the cocombustion model. Table 6 summarises the chemical properties of the South African coal used for furnace calibration and validation.

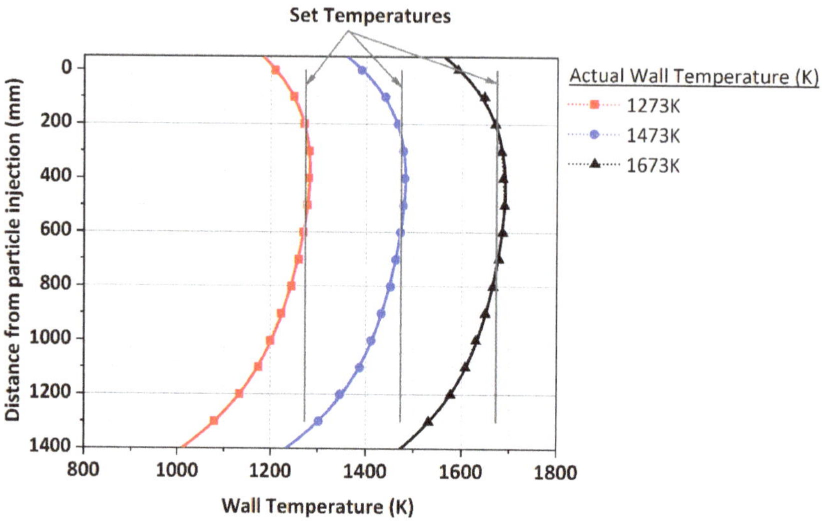

Figure 3. Furnace wall temperature calibration.

Table 6. South African coal chemical analysis.

Ash	VM	FC	MC	C	H	N	S	Carbonates	O	HHV (MJ/kg)
28.9	20.4	47.9	2.8	54.75	2.41	1.30	1.47	3.96	4.41	21.00

3.3. *Drop Tube Furnace Model Sensitivity Analysis*

The grid convergence index (GCI), as propounded by Roache [35], looks at the effect of mesh density on the variation in performance parameters. This index is adapted for fluid-based analysis through its ability to analyse boundary layer effects by the use of near-wall inflation monitoring. Only the GCI analysis was conducted, as it was deemed enough to analyse the combustion parameters under consideration. More focus was placed on the near-wall inflation parameters, hence the GCI. The performance parameters were those variables of interest to the user during the modelling practice. As such, mesh independence was achieved when the index approached unity. In this case, the particle final residence time and particle peak temperature were used to evaluate the GCI. As illustrated in Figures 1 and 2, the shortest edge had a distance of 1.109 mm, which was used to guide the initial seed element size for the course mesh.

Refinement was then carried out depending on the required emphasis, as presented. The 399,200-cell mesh (mesh 1), 159,755-cell mesh (mesh 2), and 60,208-cell mesh (mesh 3) were evaluated to give a refinement ratio (r) of around 2.576. Equations (7) to (10) were thus employed to evaluate the GCI based on the parameters of interest, as presented in Table 7.

$$p = \ln\left|\frac{f_3 - f_2}{f_2 - f_1}\right| / \ln(r) \quad (7)$$

$$GCI_{12} = \frac{F_s \times \left|\frac{f_2 - f_1}{f_1}\right|}{r^p - 1} \quad (8)$$

$$GCI_{23} = \frac{F_s \times \left|\frac{f_3 - f_2}{f_2}\right|}{r^p - 1} \quad (9)$$

$$\frac{GCI_{23}}{r^p \times GCI_{12}} \approx 1 \quad (10)$$

where the following variables are used:

p: order of convergence;

f: performance parameter;

r: refinement ratio = 2.576;

GCI: grid convergence index;

F_s: factor of safety (in this case, 2 levels of refinement = 1.25).

As illustrated in Table 7, the asymptotic range values are both within the range of unity, highlighting how the solution is mesh-independent for the three meshes under consideration. Potgieter et al. [36] made a similar conclusion with regard to the grid convergence index during their heat transfer studies, which further supports the robust nature of this approach.

Table 7. Mesh sensitivity based on grid convergence index.

Mesh Cells	Mesh Label	Performance Parameter	p	GCI_1	GCI_2	Asymptotic Range Value
		[Particle residence time (s)]				
399,200	1	1.2991				
159,755	2	1.2407	0.4986	0.0932	0.1565	1.0471
60,208	3	1.3343				
		[Particle peak temperature (K)]				
399,200	1	1006.14				
159,755	2	1029.03	1.1840	0.0137	0.0413	0.9778
60,208	3	958.85				

Figure 4 shows a trace of the average particle residence time as it passes through the furnace for the South African fuel sample under a 1273 K furnace temperature with a 520 mm probing position setting. All CFD meshes demonstrated a good initial estimation of particle residence time up to around 250 mm from the injection point. Eventually, the course mesh (60,208 cells) produced the least deviation from the experimental values by staying closer to the 95% confidence band up to the furnace exit.

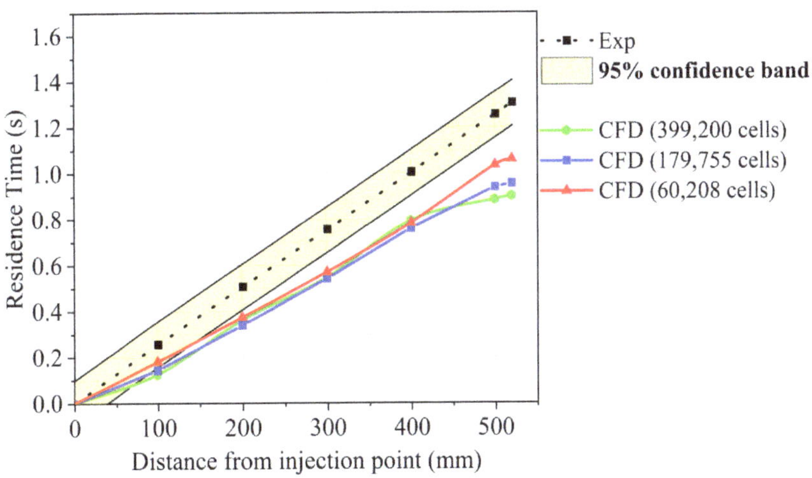

Figure 4. Mesh sensitivity analysis.

In conclusion, mass conservation suggested the use of a medium-density mesh, whilst analysis based on energy conservation suggested the use of a course-density mesh. Correlation with the experimental parameters suggested the use of the course mesh; thus, all subsequent analysis employed a mesh density similar to the course mesh.

3.4. Cocombustion Model Validation

The char produced during stage 1 was then passed through the DTF under an oxidative atmosphere of 0.3% O_2 and 97% N_2. As expected, this triggered oxidation reactions as far as the char particle is concerned; hence, the experimental results related to the oxidation of SA coal char were subsequently used to validate the model further with respect to the particle residence time and particle temperature. The stage 2 setup allowed for extra probing positions at 520 mm, 920 mm, and 1320 mm from the fuel injection point. The stage 2 setup also allowed the researcher to investigate the effect of the heating rate as the SA char particle moved through the DTF by having different furnace wall temperatures of 1273, 1473, and 1673 K. Data obtained on the variation in the experimental and predicted residence time with respect to the position when the char from the SA coal sample was combusted within the DTF are presented in Table 8 and Figure 5.

Table 8. DTF stage 2 experimental and predicted residence times of SA coal sample.

Position	Designation	Residence Time at Various Furnace Wall Temperatures (s)		
		1273 K	1473 K	1673 K
520 mm	SA coal (Exp) SA coal (CFD)	1.3000 1.5217	1.1000 1.2755	1.0000 1.0826
920 mm	SA coal (Exp) SA coal (CFD)	2.2000 3.3290	1.9000 2.9557	1.7000 2.5335
1320 mm	SA coal (Exp) SA coal (CFD)	3.2000 4.7551	2.8000 3.3035	2.5000 2.9774
	RSME	1.1168	0.6828	0.5566

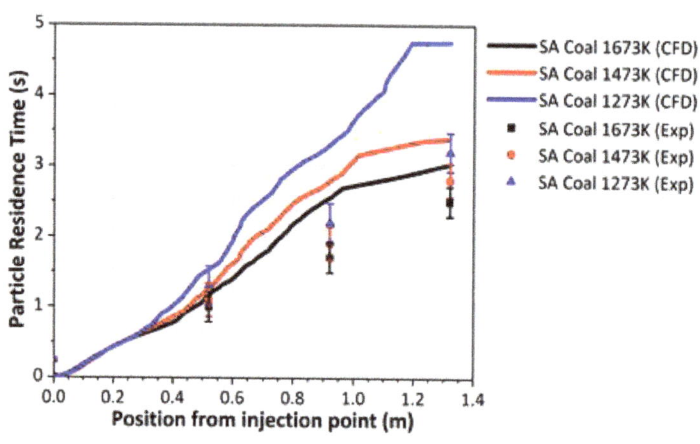

Figure 5. DTF stage 2 plot of experimental and predicted residence times of SA coal fuel.

The predicted values produced a similar trend as compared to experimental values, though an overprediction was experienced at all probing positions. A furnace temperature of 1273 K allowed the combusting particles to have a predicted residence time of 1.523 s when probed at 520 mm from the injection point. As the particle moved further within the furnace, the residence time increased to 4.755 s when probed at 1320 mm from the injection point. When compared to experimental residence times, a deviation of 17.05% was experienced at a 520 mm probing position. A deviation of 48.60% was experienced at a 1320 mm probing position. Averaging these deviations in the predicted particle residence times against the experimental residence times resulted in a root mean square error (RMSE) of 1.117 for a furnace temperature of 1273 K.

An increase in the DTF heating rate through an increase in the furnace temperature resulted in a decrease in the deviation between the experimental and predicted particle residence times. A furnace temperature of 1673 K resulted in a predicted particle residence time of 1.083 s at a probing position of 520 mm compared to an experimental value of 1.000 s, which represents a deviation of 8.30%. At a probing position of 1320 mm, the predicted particle residence time was 2.977 s compared to an experimental value of 2.500 s, representing a deviation of 19.08%. Similarly, averaging the deviations between the experimental and predicted values at a furnace temperature of 1673 K showed an RMSE value of 0.557.

The overprediction can be attributed to various factors related to the determination of the experimental particle residence time. The experimental particle residence time was determined indirectly through experimental measurements of fluid velocity and temperature at various positions. The evaluation of the experimental particle residence time assumed a constant particle diameter and density, though in reality, combustion causes a loss in mass for the char particle. This, in turn, gave a semblance of a heavy particle, whilst in reality, it is lighter in weight. Consequently, the terminal velocity was skewed to appear as if the particle was moving fast, which is far from reality (DTF used downward firing; thus, particles moved in the same direction as gravitational force).

Secondly, the drag model that was employed during CFD prediction is based on Haider and Levenspiel [37], whilst that used in the experimental determination is anchored on a constant derived from Stokes' Law. This allowed the prediction model to capture the drag induced by an irregular-shaped particle, unlike the approach used in the experimental determination. Lastly, the deduction used for the experimental particle residence time assumes that all particles travel in a one-dimensional direction without any recirculation. This, in turn, overlooks the effect of the recirculation and three-dimensional movement of particles, as utilised by the CFD model. Thus, the overprediction of particle residence time is justified and within the acceptable range for the predicted and experimental values.

As the fuel particle travelled through the furnace, the experimental and predicted values of particle temperature were also evaluated for validation purposes. Table 9 and Figure 6 summarise the experimental and predicted results. As the combusting particles travelled through the DTF, comparison between the experimental and predicted particle temperatures demonstrated a reduction in deviation. At a furnace wall temperature of 1673 K, the combusting particle attained an experimental temperature of 1676.80 K compared to a predicted value of 1289.18 K when probed at a distance of 520 mm from the injection point. This represented a negative deviation of 23.11%. When probed at 1320 mm from the injection point, a positive deviation of 4.79% was experienced between the experimental particle temperature of 1511.19 K and the predicted value of 1583.59 K. Monitoring these deviations at various probing positions showed that when the DTF is under a high heating rate, the RSME attains its lowest value of 0.1422. This is mirrored by the RSME values for the particle residence time, which also show better prediction as the furnace wall temperature increases.

Table 9. DTF stage 2 experimental and predicted particle temperatures of SA coal sample.

Position	Designation	Particle Temperature (K)		
		1273 K	1473 K	1673 K
520 mm	SA coal (Exp)	1248.00	1459.82	1676.80
	SA coal (CFD)	939.61	1048.27	1289.18
920 mm	SA coal (Exp)	1191.80	1411.48	1633.88
	SA coal (CFD)	997.21	1145.57	1519.20
1320 mm	SA coal (Exp)	1042.74	1276.33	1511.19
	SA coal (CFD)	970.60	1140.55	1583.59
	RSME	0.1759	0.2052	0.1422

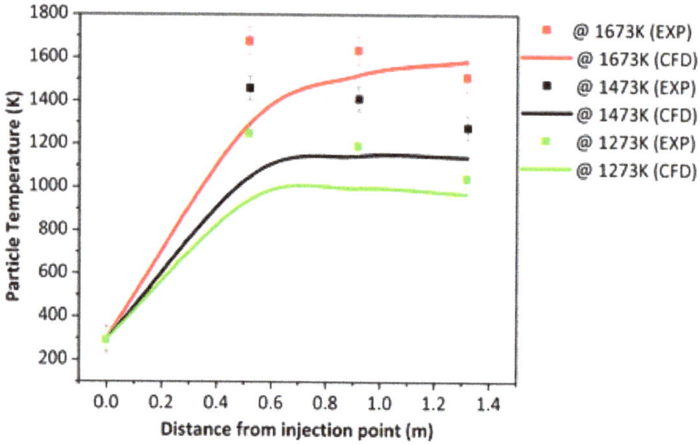

Figure 6. DTF stage 2 plot of experimental and predicted particle temperatures of SA coal sample.

The deviation between the experimental and predicted particle temperature values can be attributed to the manner in which the experimental particle temperatures were determined. The experimental particle temperatures were not determined directly but rather through experimental furnace wall temperatures and equations related to the theory of char combustion. One of the steps used to determine the experimental particle temperature involved the use of a single-step mechanism to represent the oxidation of char. The single-step mechanism stipulates that char oxidises to CO_2 without any formation of intermediate species. This notion contradicts what other researchers such as Zhou et al. [38] and Graeser et al. [39], who investigated coal char under different operating conditions, have suggested. As such, the cocombustion model employed the two-step mechanism which also catered for particle diffusion, conductivity, variable specific heat capacity, and the change in char porosity during the combustion process. In summary, the cocombustion model did manage to predict the char combustion of SA coal at different furnace temperatures with success. Measuring combustion parameters during drying, devolatilisation, char combustion, and volatile combustion is very difficult and largely dependent on sophisticated instruments since the residence time is very small. In as much as Table 9 and Figure 6 demonstrate deviations between the experimental and CFD temperature parameters, the trends correspond to each other. As articulated by Zhang et al. [40], at high heating rates, such as those encountered within a DTF, the validation of combustion models using experimental values tends to be centred around verifying if the trends correspond rather than obtaining accurate values.

4. Results and Discussion

Due to the release of heat during combustion, parameters related to the conservation of energy were monitored in detail for the different fuel blends using the developed cocombustion model (100HC, 90HC10PS, 80HC20PS, and 70HC30PS). These parameters included the particle temperature, sensible heat of the reaction, absolute heat of the reaction, and ignition, as well as the burnout properties. The heat released during char oxidation was modelled as a two-step reaction by assuming the formation of CO within the boundary layer and its subsequent oxidation to CO_2 in the bulk flow. Since coal char had a higher combustible fraction than sawdust char, as demonstrated in a publication by the same authors [22], more heat was expected from the unblended sample. Figure 7 shows the particle temperature of the 100HC case as well as the 70HC30PS case when combusted in the DTF with a wall temperature of 1473 K. Blending did not result in any significant variation in the particle temperatures along the axis or radially. One of the main characteristics that influences solid fuel combustion is surface area. The particle sizes that were employed were relatively small to warrant influence due to size difference. Mandø et al. [41] suggested that pulverised samples adopt a quasi-similar spherical shape during combustion unlike large particles, which was demonstrated in this study. Blending was performed on a mass basis; thus, the fuel flow rates that were employed were constant regardless of fuel blending.

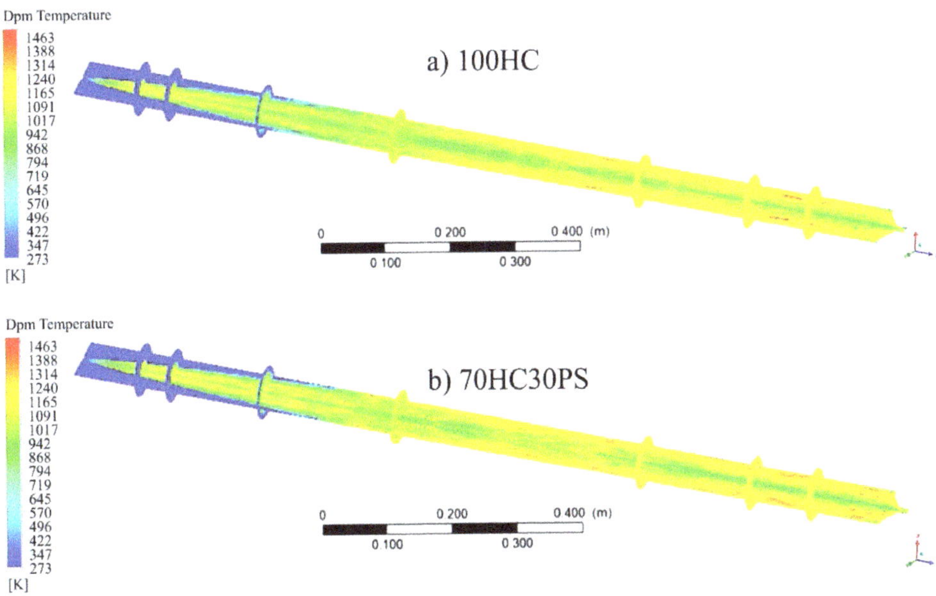

Figure 7. DTF stage 2 particle temperature on axial and radial planes for (**a**) 100HC and (**b**) 70HC30PS fuel samples at 1473 K DTF wall temperature.

With the small flow rates that were employed in this study, furnace wall temperatures rather than combustion reactions were able to influence the particle temperature, as demonstrated in Figure 8. The figure demonstrates the influence of having a DTF wall temperature of 1673 K and 1273 K on the combustion of a 70HC30PS fuel sample. As the flow developed in an almost similar fashion, the particles under a high DTF wall temperature were able to attain higher values along the axial direction. Correlating with Figure 9, which shows species molar fractions of CO_2 at various DTF wall temperatures, it is evident that particle temperature plays an important role in aiding the formation of CO_2, as higher temperatures promote the formation of CO_2.

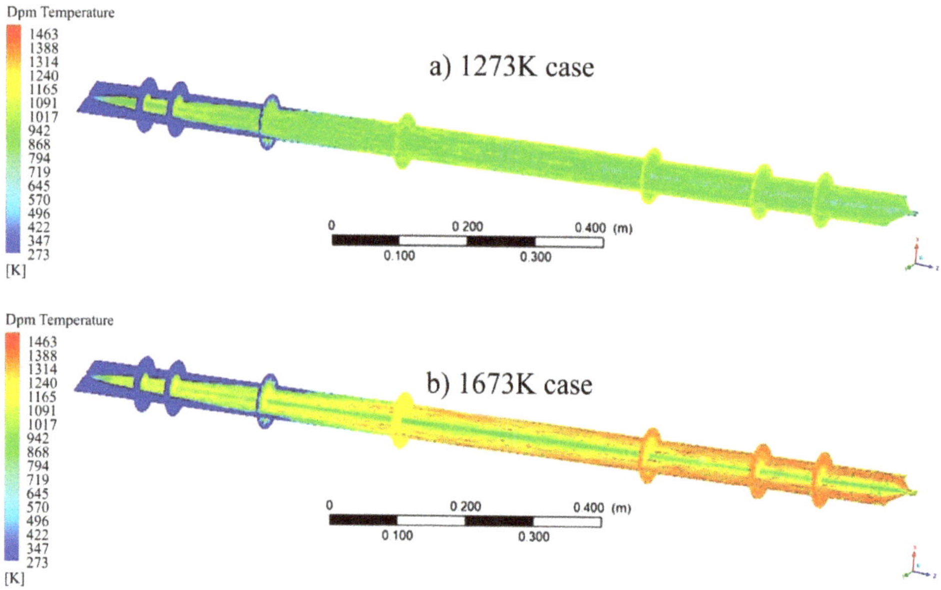

Figure 8. DTF stage 2 particle temperature against axial particle position for a 70HC30PS fuel sample at (**a**) 1273K and (**b**) 1673 K DTF wall temperatures.

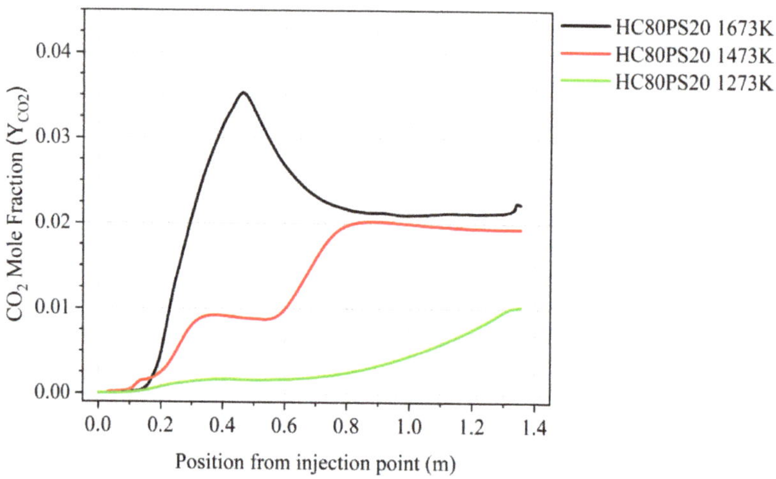

Figure 9. DTF stage 2 axial variation in CO_2 mole fraction for an 80HC20PS fuel sample at different DTF wall temperatures.

With respect to the heat released, Figure 10 shows heat of reaction contours along the axis and radially for the 70HC30PS case as a function of furnace temperature. Since the oxidation of char is treated as a two-step reaction, the contours of the kinetic reaction rates were also superimposed on the figures for better analysis. The setting up of the cocombustion model had to resemble the experimental conditions, which required an artificial atmosphere containing 97% N_2 and 3% O_2 by volume; Equations (11) and (12) were employed after assuming a 100% excess O_2 scenario. This was deduced through calculating the average experimental O_2-to-fuel ratio for the various fuel blends.

$$C + 0.5\,[O_2 + 32.333\,N_2] \rightarrow -CO + 16.167\,N_2 \tag{11}$$

$$CO + 1.0\,[O_2 + 32.333\,N_2] \rightarrow CO_2 + 32.333\,N_2 + 0.5\,O_2 \tag{12}$$

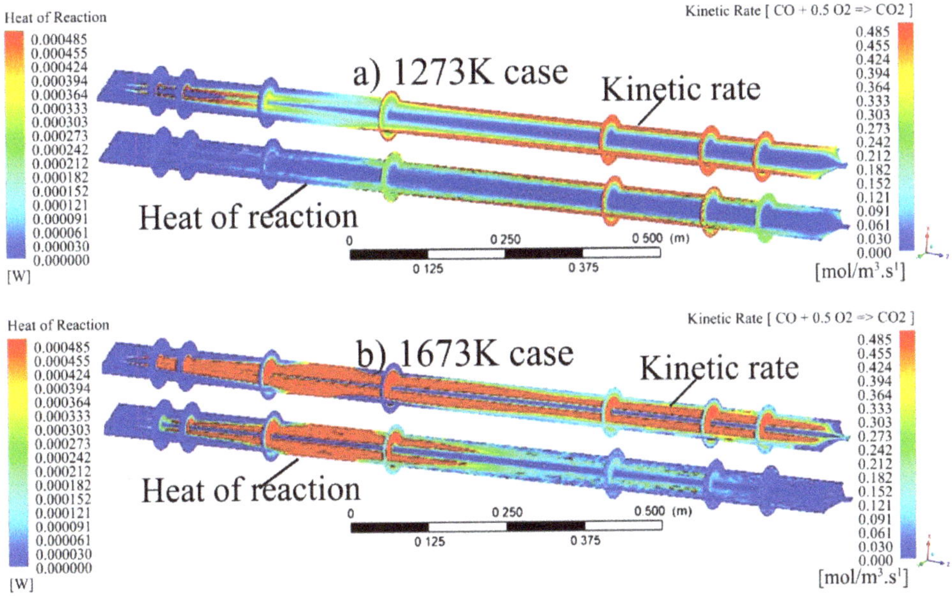

Figure 10. DTF stage 2 heat of reaction contours and kinetic reaction rate contours for a 70HC30PS fuel sample at (**a**) 1273 K and (**b**) 1673 K DTF wall temperatures.

The high N_2 content in the products of combustion due to the artificial atmosphere resulted in a reduced heat of reaction for all the cases. This can be attributed to the fact that N_2 has a higher specific heat capacity than O_2. This allows N_2 to absorb more heat than O_2 just to raise the temperature by a unit temperature [42]. By correlating with Figure 11, which shows discrete-phase particle burnout superimposed with contours of the Equation (11) kinetic rate of reaction, it becomes quite evident that most of the heat was released during the oxidation of CO. The initial heat released during char oxidation corresponded to the particle burnout profiles. Low DTF wall temperatures reduced the heat of the reaction, as shown in Figure 10. As expected, low DTF wall temperatures hindered the oxidation of CO to form CO_2, thus delaying the reaction zone to an average location of 0.8 m from the injection point as compared to the initial 0.4 m obtained at high furnace temperatures. The nature of the flow development resulted in a reaction zone that tended to be concentrated towards the centreline for high temperatures and near walls for low temperatures. The sensible heat released during combustion increased with temperature, as shown in Figure 12. As already discussed, high furnace temperatures increased the subsequent kinetic reaction rate associated with char oxidation and the heat of the reaction.

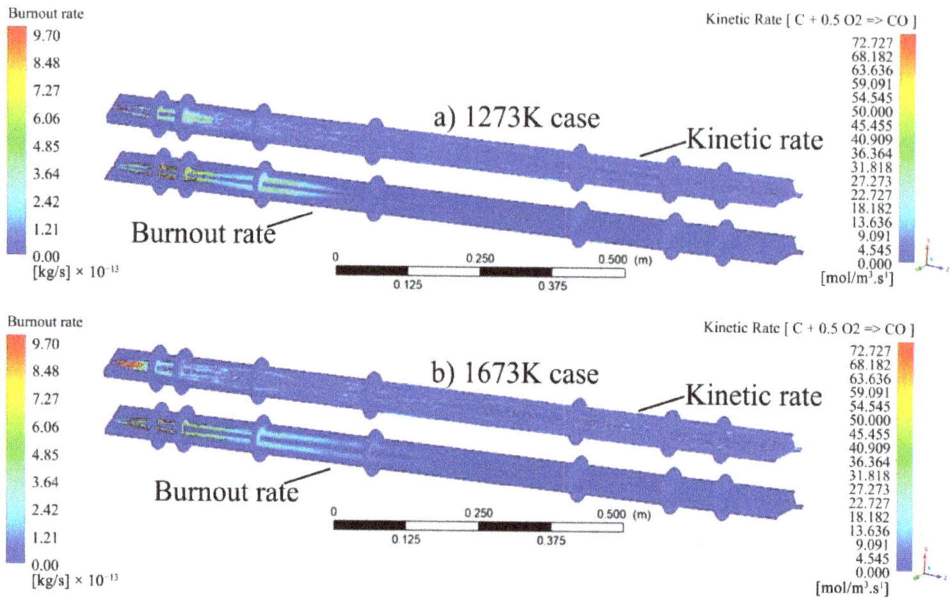

Figure 11. DTF stage 2 particle burnout contours and kinetic reaction rate contours for a 70HC30PS fuel sample at (**a**) 1273 K and (**b**) 1673 K DTF wall temperatures.

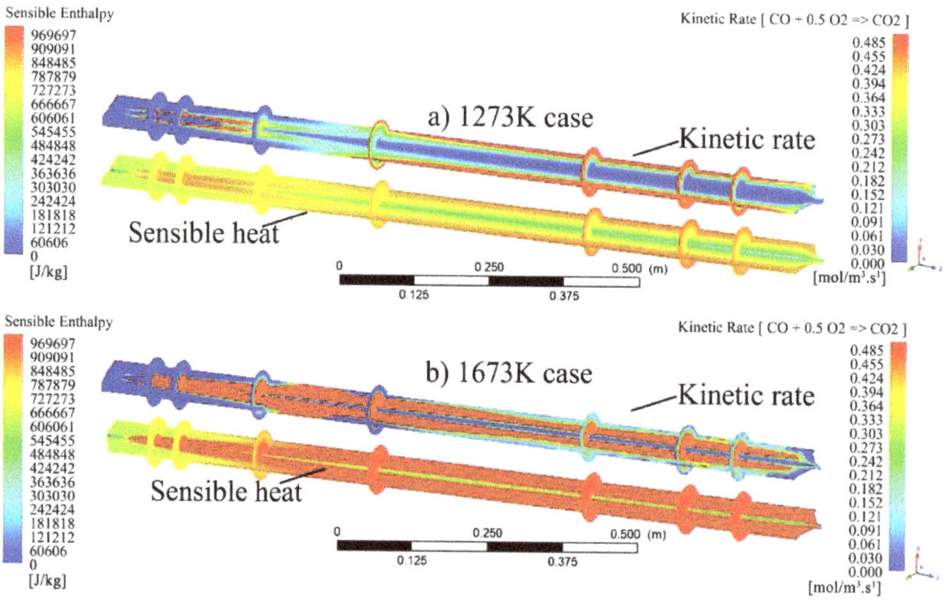

Figure 12. DTF stage 2 sensible heat contours and kinetic reaction rate contours for a 70HC30PS fuel sample at (**a**) 1273 K and (**b**) 1673 K DTF wall temperatures.

The effect of blending on the heat of the reaction is represented in Figure 13 with an illustration of axial and radial contours. Blending affected the heat of the reaction by delaying the reaction zone as well as increasing the intensity of combustion within the zone. The intensity of combustion is an index (Ψ_1) that can be investigated through thermogravimetric experiments as well. Thermogravimetric experiments demonstrated an increase with respect to the combustion intensity index from 17.0470 × 10^{-8} to 38.9440 × 10^{-8} as blending increased from 90HC10PS to 70HC30PS for a heating rate of 20 K/min [6]. Figure 13 supports these findings by highlighting where these high-combustion-intensity zones are located within the furnace with respect to the injection point.

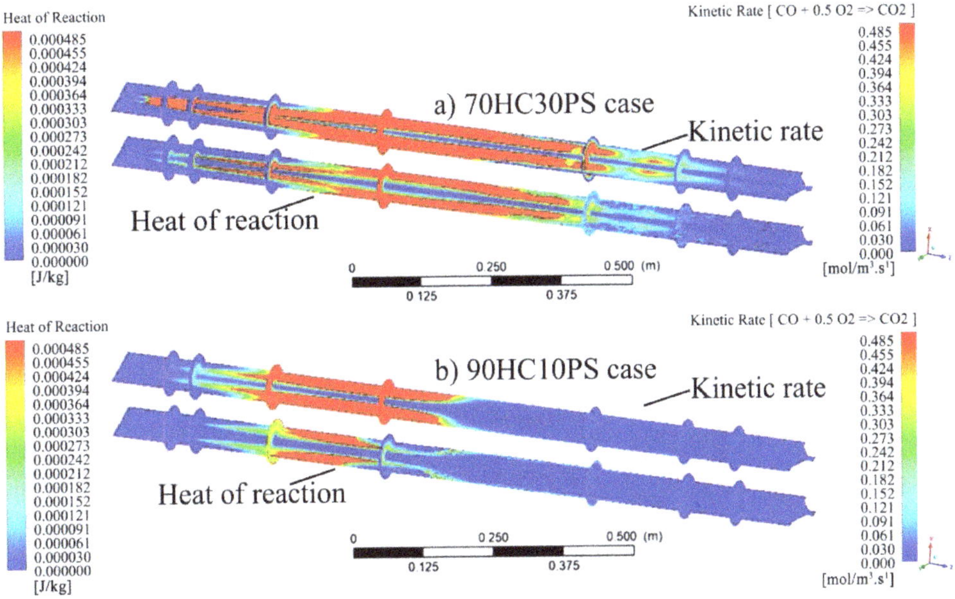

Figure 13. DTF stage 2 heat of reaction contours and kinetic reaction rate contours for (**a**) 70HC30PS and (**b**) 90HC30PS at 1473 K.

The 70HC30PS fuel blend contours show a gradual release of heat which is directly linked to the gradual burnout of the char particles. In contrast, the 90HC10PS fuel sample shows a shorter intense reaction zone. Even though a higher combustible matter fraction was associated with the 90HC10PS fuel, the rapid release of CO, hence burnout, was experienced due to its lower porosity as seen by the early onset of the CO oxidation. A low porosity is associated with high-density particles as with the coal char particle, hence less surface area available for combustion reactions. This assertion is supported by other researchers such as Tang et al. [43] during their studies with demineralised coal, Di Blasi [44] during his pyrolysis studies of wood, and Sadhukhan et al. [45], who focused on large coal particles. As demonstrated in Figure 14, blending increased particle burnout maximum values marginally, though the reaction zone for all the cases was located within the same region. The burnout contours showed dependence on the kinetic reaction rate of char with O_2 to form CO.

Specifically, HC exhibits a lower porosity compared to PS, and increasing the proportion of PS in the blend augments the fuel's porosity. This enhancement facilitates improved air permeability and oxygen diffusion during combustion, promoting more efficient combustion reactions and leading to an increased heat release rate. This aligns with the experimental data obtained by Marangwanda et al. [6] which demonstrate that a blend containing 30% PS by weight results in a 15% increase in combustion efficiency compared to pure HC. Further supporting this, Vyas et al. [46] emphasised that increased porosity in biomass fuels enhances internal surface area, facilitating better oxygen diffusion and more complete combustion.

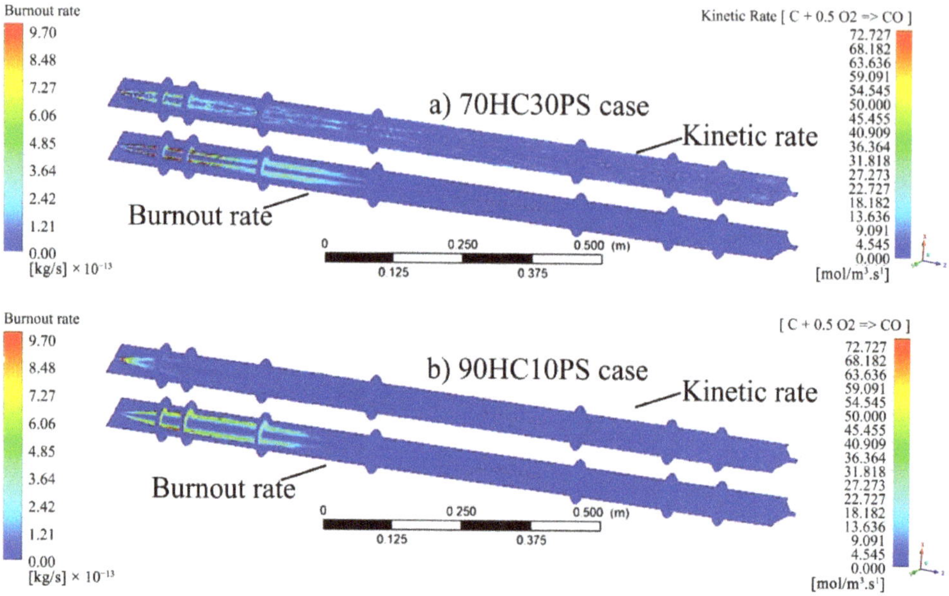

Figure 14. DTF stage 2 DPM burnout contours and kinetic reaction rate contours for (**a**) 70HC30PS and (**b**) 90HC10PS at 1473 K.

5. Conclusions

The cocombustion model was successfully employed to predict the combustion behaviour of coal and biomass fuel blends. In summary, the findings of this study are as follows:

- The variation in the particle residence time and temperature within the DTF was used to validate the cocombustion model. The predicted values produced a similar trend as compared to experimental values, though an overprediction was experienced with an average root mean square error (RMSE) of 1.117 at a 1273 K DTF wall temperature and 0.557 at a 1673 K DTF wall temperature. This overprediction was attributed to various factors related to the experimental procedure; hence, further research with regard to the characterisation of the char and volatiles produced by devolatilisation was suggested.
- Increasing the Pinus sawdust blending ratio resulted in more volatiles being released, as mirrored by the proximate composition of the fuel blends. As such, the volatile composition on the fuel blends showed that the molar ratio of carbon increased with blending (0% to 30% sawdust) from 0.292 up to 0.579. The hydrogen and oxygen molar ratios also increased with Pinus sawdust blending, though to a lesser magnitude. The nitrogen molar ratio decreased from 0.086 to 0.052 as blending with Pinus sawdust increased, whilst the sulphur molar ratio decreased marginally.
- The cocombustion model was able to bring synergy between various submodels of interest which tend to be overlooked in certain instances. The eddy dissipation concept submodel captured the combustion mechanisms successfully; the weighted sum of grey gases model captured the radiation from the combustion products successfully.
- The visualisation of various profiles highlighted the co-dependency of certain combustion parameters on others. The discrete-phase particle burnout profiles were shown to be dependent on the oxidation of CO to form CO_2 kinetic rate of reaction. It was also made evident that low DTF wall temperatures hindered the oxidation of

CO to form CO_2, thus delaying the reaction zone to an average location of 0.8 m from the injection point as compared to the initial 0.4 m obtained at high DTF wall temperatures.

- Blending affected the heat of the reaction by promoting the onset of the reaction zone as well as increasing the combustion intensity within the reaction zone. The gradual release of heat was shown to be directly linked to the gradual burnout of the char particle. In conclusion, the reaction zone was modelled successfully to highlight the important combustion parameters.

Author Contributions: G.T.M. was responsible for conceptualisation, formal analysis, investigation, methodology, and writing of this manuscript. D.M.M. was responsible for funding acquisition and resources, as well as reviewing and editing of this manuscript. All authors have read and agreed to the published version of the manuscript.

Funding: This research received no external funding.

Data Availability Statement: The raw data supporting the conclusions of this article will be made available by the authors on request.

Acknowledgments: Special thanks to the University of Johannesburg, South Africa, for supporting the authors during their research.

Conflicts of Interest: The authors declare no conflicts of interest.

Nomenclature

The following units, symbols, and abbreviations were used in this manuscript:

Units	
E_a	Activation energy (kJ/mol)
T	Static temperature (K)
A	Pre-exponential factor (s^{-1})
U_i	Velocity (m/s)
Greek Symbols	
ρ	Density
φ	Variable (mass, specific enthalpy, or species mass fraction)
Γ	Variable diffusion coefficient
$S_φ$	Variable source or sink
Abbreviations	
HC	Bituminous coal
PS	Pinus sawdust
TGA	Thermogravimetric analysis
CFD	Computational fluid dynamics
FC	Fixed carbon
VM	Volatile matter
DTF	Drop tube furnace
EDC	Eddy dissipation concept
HHV	Higher heat value
GCI	Grid convergence index
Subscripts	
daf	Dry ash free
vol	Volatile

References

1. Zhou, C.; Gao, F.; Yu, Y.; Zhang, W.; Liu, G. Effects of Gaseous Agents on Trace Element Emission Behavior during Co-Combustion of Coal with Biomass. *Energy Fuels* **2020**, *34*, 3843–3849. [CrossRef]

2. Benim, A.C.; Deniz Canal, C.; Boke, Y.E. Computational Investigation of Oxy-Combustion of Pulverized Coal and Biomass in a Swirl Burner. *Energy* **2022**, *238*, 121852. [CrossRef]
3. Cardoso, J.S.; Silva, V.; Chavando, J.A.M.; Eusébio, D.; Hall, M.J. Numerical Modelling of the Coal Phase-out through Ammonia and Biomass Co-Firing in a Pilot-Scale Fluidized Bed Reactor. *Fuel Commun.* **2022**, *10*, 100055. [CrossRef]
4. Wei, D.; An, D.; Wang, T.; Zhang, H.; Guo, Y.; Sun, B. Influence of Fuel Distribution on Co-Combustion of Sludge and Coal in a 660 MW Tangentially Fired Boiler. *Appl. Therm. Eng.* **2023**, *227*, 120344. [CrossRef]
5. Lyu, Q.; Wang, R.; Du, Y.; Liu, Y. Numerical Study on Coal/Ammonia Co-Firing in a 600 MW Utility Boiler. *Int. J. Hydrogen Energy* **2023**, *48*, 17293–17310. [CrossRef]
6. Marangwanda, G.T.; Madyira, D.M.; Ndungu, P.G.; Chihobo, C.H. Combustion Characterisation of Bituminous Coal and Pinus Sawdust Blends by Use of Thermo-Gravimetric Analysis. *Energies* **2021**, *14*, 7547. [CrossRef]
7. Jin, W.; Si, F.; Kheirkhah, S.; Yu, C.; Li, H.; Wang, Y. Numerical Study on the Effects of Primary Air Ratio on Ultra-Low-Load Combustion Characteristics of a 1050 MW Coal-Fired Boiler Considering High-Temperature Corrosion. *Appl. Therm. Eng.* **2023**, *221*, 119811. [CrossRef]
8. Gao, H.; Runstedtler, A.; Majeski, A.; Boisvert, P.; Campbell, D. Optimizing a Woodchip and Coal Co-Firing Retrofit for a Power Utility Boiler Using CFD. *Biomass Bioenergy* **2016**, *88*, 35–42. [CrossRef]
9. Gu, T.; Ma, W.; Guo, Z.; Berning, T.; Yin, C. Stable and Clean Co-Combustion of Municipal Sewage Sludge with Solid Wastes in a Grate Boiler: A Modeling-Based Feasibility Study. *Fuel* **2022**, *328*, 125237. [CrossRef]
10. Chen, B.; Liu, B.; Shi, Z. Combustion Characteristics and Combustion Kinetics of Dry Distillation Coal and Pine Tar. *Int. J. Aerosp. Eng.* **2020**, *2020*, 8888556. [CrossRef]
11. Choi, C.R.; Kim, C.N. Numerical Investigation on the Flow, Combustion and NO_x Emission Characteristics in a 500 MW_e Tangentially Fired Pulverized-Coal Boiler. *Fuel* **2009**, *88*, 1720–1731. [CrossRef]
12. Mohanna, H. Combustion of Pulverized Biomass: Impact of Fuel Preparation and Flow Conditions. Ph.D. Thesis, Normandie Université, Montpellier, France, 2021.
13. Ma, W.; Zhou, H.; Zhang, J.; Zhang, K.; Liu, D.; Zhou, C.; Cen, K. Behavior of Slagging Deposits during Coal and Biomass Co-Combustion in a 300 KW Down-Fired Furnace. *Energy Fuels* **2018**, *32*, 4399–4409. [CrossRef]
14. Fang, Y.; Qiao, L.; Wang, C.; Luo, X.; Xiong, T.; Tian, R.; Tang, Z.; Sun, H. Study on the Influence of the Activation Energy on the Simulation of Char Combustion. *IOP Conf. Ser. Earth Environ. Sci.* **2019**, *300*, 042062. [CrossRef]
15. Xinjie, L.; Shihong, Z.; Xincheng, W.; Jinai, S.; Xiong, Z.; Xianhua, W.; Haiping, Y.; Hanping, C. Co-Combustion of Wheat Straw and Camphor Wood with Coal Slime: Thermal Behaviour, Kinetics, and Gaseous Pollutant Emission Characteristics. *Energy* **2021**, *234*, 121292. [CrossRef]
16. Wang, G.; Guiberti, T.F.; Cardona, S.; Jimenez, C.A.; Roberts, W.L. Effects of Residence Time on the NOx Emissions of Premixed Ammonia-Methane-Air Swirling Flames at Elevated Pressure. *Proc. Combust. Inst.* **2022**, *39*, 4277–4288. [CrossRef]
17. Sarroza, A.C.; Bennet, T.D.; Eastwick, C.; Liu, H. Characterising Pulverised Fuel Ignition in a Visual Drop Tube Furnace by Use of a High-Speed Imaging Technique. *Fuel Process. Technol.* **2017**, *157*, 1–11. [CrossRef]
18. Yin, C.; Rosendahl, L.A.; Kær, S.K. Towards a Better Understanding of Biomass Suspension Co-Firing Impacts via Investigating a Coal Flame and a Biomass Flame in a Swirl-Stabilized Burner Flow Reactor under Same Conditions. *Fuel Process. Technol.* **2012**, *98*, 65–73. [CrossRef]
19. Demirbaş, A. Biomass Co-Firing for Boilers Associated with Environmental Impacts. *Energy Sources* **2005**, *27*, 1385–1396. [CrossRef]
20. Xia, Y.; Zhang, J.; Tang, C.; Pan, W. Research and Application of Online Monitoring of Coal and Biomass Co-Combustion and Biomass Combustion Characteristics Based on Combustion Flame. *J. Energy Inst.* **2023**, *108*, 101191. [CrossRef]
21. Kuznetsov, V.A.; Minakov, A.V.; Bozheeva, D.M.; Dekterev, A.A. International Journal of Greenhouse Gas Control Oxy-Fuel Combustion of Pulverized Coal in an Industrial Boiler with a Tangentially Fired Furnace. *Int. J. Greenh. Gas Control* **2023**, *124*, 103861. [CrossRef]
22. Marangwanda, G.T.; Madyira, D.M.; Chihobo, C.H. Determination of Cocombustion Kinetic Parameters for Bituminous Coal and Pinus Sawdust Blends. *ACS Omega* **2022**, *7*, 32108–32118. [CrossRef] [PubMed]
23. ANSYS. *ANSYS FLUENT User's Guide*; ANSYS: Canonsburg, PA, USA, 2021; pp. 1–170.
24. Yin, C.; Rosendahl, L.A.; Condra, T.J.; Kær, S.K. Use of Numerical Modeling in Design for Co-Firing Biomass in Wall-Fired Burners. *Chem. Eng. Sci.* **2004**, *59*, 3281–3292. [CrossRef]
25. Al-Abbas, A.H.; Naser, J. Effect of Chemical Reaction Mechanisms and NOx Modeling on Air-Fired and Oxy-Fuel Combustion of Lignite in a 100-KW Furnace. *Energy Fuels* **2012**, *26*, 3329–3348. [CrossRef]
26. Marangwanda, G.T.; Madyira, D.M.; Babarinde, T.O. Coal Combustion Models: An Overview. *J. Phys. Conf. Ser.* **2019**, *1378*, 032070. [CrossRef]
27. Maisyarah, A.; Shiun, J.; Nasir, F.; Hashim, H. Ultimate and Proximate Analysis of Malaysia Pineapple Biomass from MD2 Cultivar for Biofuel Application. *Chem. Eng. Trans.* **2018**, *63*, 127–132. [CrossRef]

28. Marangwanda, G.T.; Madyira, D.M. Experimental Investigation on the Effect of Blending Bituminous Coal with Pinus Sawdust on Combustion Performance Parameters. *Heliyon* **2024**, *10*, e27287. [CrossRef]
29. Zou, C.; Cai, L.; Wu, D.; Liu, Y.; Liu, S.; Zheng, C. Ignition Behaviors of Pulverized Coal Particles in O_2/N_2 and O_2/H_2O Mixtures in a Drop Tube Furnace Using Flame Monitoring Techniques. *Proc. Combust. Inst.* **2015**, *35*, 3629–3636. [CrossRef]
30. Liu, M.; Han, B.; Bai, J.; Ru, J.; Wang, X.; Xing, L.; Li, H.; Lv, S.; Gao, A.; Wang, Y.; et al. Investigation on the Synergistic Effects and Thermokinetic Analyses during Co-Combustion of Corn Stalk and Polyethylene Plastic: Effect of Heating Rate and Placement Method. *Fuel* **2025**, *385*, 134032. [CrossRef]
31. Czajka, K.M.; Modliński, N.; Kisiela-Czajka, A.M.; Naidoo, R.; Peta, S.; Nyangwa, B. Volatile Matter Release from Coal at Different Heating Rates –Experimental Study and Kinetic Modelling. *J. Anal. Appl. Pyrolysis* **2019**, *139*, 282–290. [CrossRef]
32. Mularski, J.; Lue, L.; Li, J. Development of a Numerical Method for the Rapid Prediction of Ignition Performance of Biomass Particles. *Fuel* **2023**, *348*, 128520. [CrossRef]
33. Versteeg, H.K.; Malalasekera, W. *An Introduction to Computational Fluid Dynamics: The Finite Volume Method Approach*; Prentice Hall: Hoboken, NJ, USA, 1996.
34. de Souza-Santos, M.L. *Solid Fuels Combustion and Gasification*; CRC Press: Boca Raton, FL, USA, 2010; ISBN 9781420047509.
35. Roache, P.J. Perspective: A Method for Uniform Reporting of Grid Refinement Studies. *J. Fluids Eng.* **1994**, *116*, 405–413. [CrossRef]
36. Potgieter, M.S.W.; Bester, C.R.; Bhamjee, M. Experimental and CFD Investigation of a Hybrid Solar Air Heater. *Sol. Energy* **2020**, *195*, 413–428. [CrossRef]
37. Haider, A.; Levenspiel, O. Drag Coefficient and Terminal Velocity of Spherical and Nonspherical Particles. *Powder Technol.* **1989**, *58*, 63–70. [CrossRef]
38. Zhou, Z.; Chen, L.; Guo, L.; Qian, B.; Wang, Z.; Cen, K. Computational Modeling of Oxy-Coal Combustion with Intrinsic Heterogeneous Char Reaction Models. *Fuel Process. Technol.* **2017**, *161*, 169–181. [CrossRef]
39. Graeser, P.; Schiemann, M. Investigations on the Emissivity of Burning Coal Char Particles: Influence of Particle Temperature and Composition of Reaction Atmosphere. *Fuel* **2020**, *263*, 116714. [CrossRef]
40. Zhang, J.; Zheng, S.; Chen, C.; Wang, X.; ur Rahman, Z.; Tan, H. Kinetic Model Study on Biomass Pyrolysis and CFD Application by Using Pseudo-Bio-CPD Model. *Fuel* **2021**, *293*, 120266. [CrossRef]
41. Mandø, M.; Rosendahl, L.A.; Yin, C.; Sørensen, H. Pulverized Straw Combustion in a Low-NOx Multifuel Burner: Modeling the Transition from Coal to Straw. *Fuel* **2010**, *89*, 3051–3062. [CrossRef]
42. Turns, S. *An Introduction to Combustion: Concepts and Applications*, 2nd ed.; McGraw-Hill Series in Mechanical Engineering; McGraw-Hill: New York, NY, USA, 1996; ISBN 9780077418946.
43. Tang, L.; Xiao, J.; Mao, Q.; Zhang, Z.; Yao, Z.; Zhu, X.; Ye, S.; Zhong, Q. Thermogravimetric Analysis of the Combustion Characteristics and Combustion Kinetics of Coals Subjected to Different Chemical Demineralization Processes. *ACS Omega* **2022**, *7*, 13998–14008. [CrossRef]
44. Di Blasi, C. Modeling Chemical and Physical Processes of Wood and Biomass Pyrolysis. *Prog. Energy Combust. Sci.* **2008**, *34*, 47–90. [CrossRef]
45. Sadhukhan, A.K.; Gupta, P.; Saha, R.K. Modeling and Experimental Investigations on the Pyrolysis of Large Coal Particles. *Energy Fuels* **2011**, *25*, 5573–5583. [CrossRef]
46. Vyas, A.; Chellappa, T.; Goldfarb, J.L. Porosity Development and Reactivity Changes of Coal–Biomass Blends during Co-Pyrolysis at Various Temperatures. *J. Anal. Appl. Pyrolysis* **2017**, *124*, 79–88. [CrossRef]

energies

MDPI

Article

TG-MS Analysis of the Effect of Variations in Coal Particle Size on Combustion Characteristics and Kinetic Parameters

Jinyang Zhang [1,2,3], Guoliang Song [1,2,3,*], Weijian Song [1,3] and Hongliang Ding [1,3]

1 Institute of Engineering Thermophysics, Chinese Academy of Sciences, Beijing 100190, China; zhangjinyang@iet.cn (J.Z.); songweijian@iet.cn (W.S.); dinghongliang@iet.cn (H.D.)
2 University of Chinese Academy of Sciences, Beijing 100049, China
3 State Key Laboratory of Coal Conversion, Institute of Engineering Thermophysics, Chinese Academy of Sciences, Beijing 100190, China
* Correspondence: songgl@iet.cn

Abstract: This study investigated the influence of particle size on combustion performance using equivalent characteristic spectrum analysis (ECSA) on a TG-MS platform. The experiments were conducted at heating rates of 10 °C/min and 20 °C/min for three granular coal types with particle sizes of 1 mm, 4 mm, and 8 mm. The results showed that the ignition temperature, burnout temperature, and burnout time generally increased with particle size, while the combustion characteristic index for the 8 mm particles was 28.81% lower than that for 1 mm particles. The particle size effects were more pronounced at lower heating rates. Combustion kinetics revealed that the pre-combustion endothermic stage had a significant impact on the ignition temperature, followed by the volatilization stage. For Shenmu bituminous coal (SBC), a 1 kJ/mol reduction in apparent activation energy during the endothermic stage increased the ignition temperature by 13.02 °C (10 °C/min) or 17.11 °C (20 °C/min). Similar trends were observed for Datong bituminous coal (DBC) and Jincheng anthracite coal (JAC). A gas product analysis indicated that the peak release temperatures rose with particle size, and particle size variations affected the maximum release rates and combustion stage duration. Smaller particles generally released less NO during combustion.

Keywords: gaseous coal; combustion characteristic; kinetic analysis; gaseous release; particle size variation; apparent activation energy

1. Introduction

With the continuous and increasing penetration rate of renewable energy generation, building a power system that is highly adaptable to the growing proportion of new energy sources is essential for achieving carbon neutrality [1–3]. This approach also constitutes one of the most practical and effective strategies for promoting the transformation of society towards high-quality development [4]. However, the inherent instability of renewable energy inevitably imposes significant peak load pressures on the power grid, thus requiring the transformation of thermal generator sets from providing basic power generation guarantees to offering system regulation and supporting grid security. Due to its unique circulation loop and heat storage capabilities regarding bed materials, the circulating fluidized bed (CFB) boiler is bound to play an increasingly significant role in the development of this modern power system [5].

Against the aforementioned research background, numerous scholars have employed diverse methods to investigate the rapid peaking of the CFB, achieving certain progress.

Arias B et al. [6,7] investigated an oxygen-enriched CFB boiler unit equipped with a TES system and two silos for solid/bed material storage at different temperatures. This design led to a 50% reduction in fuel consumption during 12-h flexible operations. Hu X et al. [8] demonstrated that modifying the system configuration improved the operational flexibility of the unit. Gao M et al. [9] presented an advanced energy balance control system, increasing the maximum variable load rate of sub-critical 350 MW CFB units to approximately 2.7% per minute. Hou Y et al. [10] explored the challenges related to control optimization in supercritical CFB units and developed an advanced fast-response coordinated control system. In Europe, only a few studies have been conducted; however, successful operations with a load change rate of 4% per minute were reported at the Lagisza and Turow power plants in Poland [11]. Further investigations revealed that CFB boilers exhibited a relatively sluggish variable load rate, mainly owing to their unique combustion characteristics. This slow load variation was predominantly influenced by three types of inertia and pollutant emissions control, namely, flow inertia, reaction inertia, and heat transfer inertia [12]. The combustion reaction rate of the fuel was a crucial factor in determining the variable load rate of CFB boilers [13]. When the combustion reaction rate was sufficiently rapid, adjusting the fuel input could effectively regulate the combustion rate, thereby enabling quicker load adjustments [14]. In response to this requirement, the Institute of Engineering Physics of the Chinese Academy of Sciences devised a high-temperature preheating modification technology for fuel. Specifically, at a low air equivalence ratio (0.3 to 0.5), the fuel preheating modification device converted solid fuel into high-temperature gas–solid mixed fuel. The high-temperature gas–solid fuel was composed of high-calorific-value gas and high-physical-sensible-heat semi-coke [15,16]. The combustible components within the gas comprised CO, H_2, and CH_4. Compared with the raw fuel, the high-temperature semi-coke obtained through preheating and modification had smaller particle sizes and a more elaborate pore structure [17,18]. These characteristics conspicuously mitigated the reaction resistance exhibited by the fuel to the CFB boiler and made a substantial contribution to the rapid and efficient combustion within the CFB furnace. Zhu S et al. [19] explored the variable load characteristics of a fuel preheating modification device on a 1 MW pilot test platform through the implementation of the aforementioned modification technology. The findings demonstrated that the device was capable of achieving a variable load rate of 3.33% per minute, thereby facilitating the rapid load adjustment of CFB boilers. Furthermore, Tang Z et al. [20] conducted an investigation into the 0.1 MW CFB test platform to study its combustion and emissions characteristics under different loads, as well as the variations in temperature and pollutant emissions during load adjustments. The results indicated that, at lower loads, the temperature distribution along the height of the furnace decreased.

In conclusion, as the proportion of renewable energy in the power system rises, traditional coal-fired thermal power plants face greater challenges in flexibility transformations. The preheating modification technology has been used in CFB boilers, showing practicality and potential. However, the current research mainly focuses on pulverized coal size, while industrial CFB boilers use granular coal of 0–10 mm. There is a lack of research on granular coal. Thus, the existing research is insufficient for various fuel sizes. The study of granular coal combustion characteristics is highly valuable.

2. Experiments

2.1. Fuel Properties

Due to their distinct volatile content, SBC, DBC, and JAC were selected as the research samples in this experiment. The study of its combustion reaction mechanism could be of great significance in realizing the industrial application of granulated coal. As comparative samples, SBC was sourced from Shenmu, Shaanxi Province; Datong DBC was sourced from Datong, Shanxi Province; and JAC was sourced from Jincheng, Shanxi Province. All samples were dried at 105 °C for 24 h, followed by crushing and sieving to obtain granular coal fuel with particle sizes of approximately 1 mm, 4 mm, and 8 mm. The particle sizes of the obtained granular coal are illustrated in Figure 1, while the properties of SBC, DBC, and JAC are detailed in Table 1.

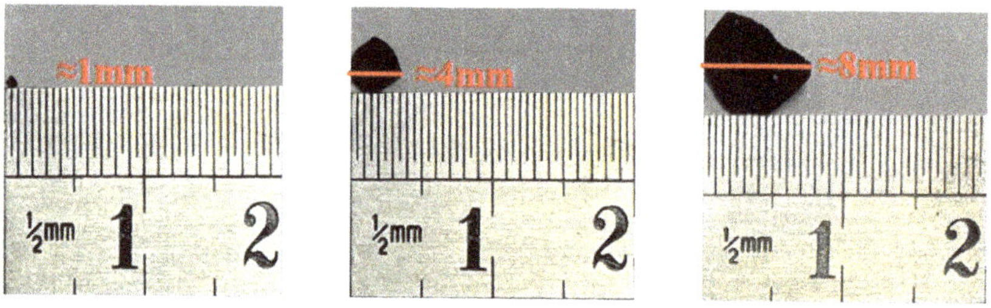

(**a**) 1 mm (**b**) 4 mm (**c**) 8 mm

Figure 1. Coal particles of different sizes.

Table 1. The properties of SBC, DBC, and JAC.

Items	SBC	DBC	JAC
Ultimate analysis (wt. %, air dry)			
Carbon (C_{ad})	73.60	45.99	54.93
Hydrogen (H_{ad})	4.30	2.94	2.28
Oxygen (O_{ad})	11.43	8.98	0.82
Nitrogen (N_{ad})	0.94	0.90	3.65
Sulfur (S_{ad})	0.32	0.36	1.09
Proximate analysis (wt. %, air dry)			
Moisture (M_{ad})	4.93	1.56	0.80
Volatile matter (V_{ad})	32.59	22.59	7.68
Fixed carbon (FC_{ad})	58.00	36.58	55.09
Ash (A_{ad})	4.48	39.27	36.43
Low heating value (MJ/kg)	28.49	17.22	20.49

Note: corner mark 'ad' stands for air-dried samples.

2.2. Experimental Methods

TG-MS Experiment: Thermogravimetric analysis coupled with mass spectrometry (TG-MS) is a widely recognized and essential technique for investigating combustion characteristics and the gaseous products generated during the combustion process. Previous research has validated the feasibility of using TG-MS to study the combustion behavior of granular coal. The TG-MS system monitored the ionic current intensity corresponding to specific mass-to-charge ratios, allowing for the quantification of gas product concentrations. This study focused on the *TG* testing and kinetic analysis of granular coal combustion characteristics. The structure of the TG-MS system is illustrated in Figure 2. The detailed operational procedures were as follows: During the preparation phase, approximately 50 ± 0.5 mg of the sample was placed in an Al_2O_3 crucible. Gas cylinders supplied gases that were mixed according to specified ratios. Upon introduction into the *TG* system, the flow rates were maintained at 20 mL/min for O_2 and 80 mL/min for Ar. The samples were initially held at room temperature before being heated from ambient conditions to 1200 °C at a predetermined heating rate over a period of 30 min. The gases introduced into the *TG* system include CO_2, O_2, and Ar, all supplied by cylinders with purities exceeding 99.999%.

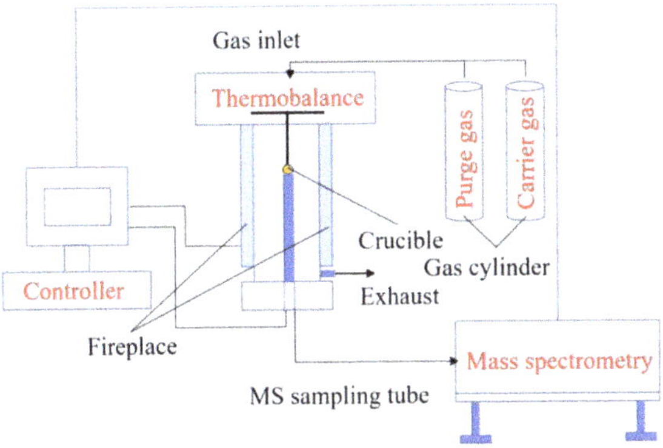

Figure 2. Diagram of the TG-MS system.

In the TG-MS experiment, the mass flow of each gas was analyzed using the ECSA method. ECSA can effectively separate the mass spectrum of gases released during TG-MS into distinct component spectra, even in cases where characteristic peaks overlap [21]. By employing the equivalent characteristic spectrum, the flow rate, concentration, and distribution of each component gas can be precisely determined, thereby enabling the accurate calculation of the pressure of all gases. This study mitigated the mass resolution effect in ECSA mass spectrometry, the diffusion effect in *TG*, and the temperature-dependent ion flow effect [22,23]. By integrating the escaping gas flow rates, the temporal variation in total gas emissions can be quantified. Detailed experimental conditions are provided in Table 2.

Table 2. TG-MS experimental conditions.

Sample	Case	Particle Size (mm)	Heating Rate (°C/min)
SBC	SBC-1 SBC-2	1	10 20
	SBC-3 SBC-4	4	10 20
	SBC-5 SBC-6	8	10 20
DBC	DBC-1 DBC-2	1	10 20
	DBC-3 DBC-4	4	10 20
	DBC-5 DBC-6	8	10 20
JAC	JAC-1 JAC-2	1	10 20
	JAC-3 JAC-4	4	10 20
	JAC-5 JAC-6	8	10 20

3. Results and Discussion

3.1. Effect of Particle Size Change on the Combustion Performance Parameters of Granular Coal

The combustion performance of the fuel is assessed based on its ignition difficulty and degree of burnout. The ignition temperature (T_i) can be determined using the tangent method on the *TG* curve [24,25], while the burnout temperature (T_h) is defined as the temperature at which 98% of the fuel's weight has been lost [26]. These characteristic parameters from the *TG* curve provide an intuitive reflection of the combustion performance of pulverized coal. Specific parameters including the ignition temperature, burnout temperature, and burnout time are shown in Table 3.

Table 3. T_i, T_h, and t_h of coal particles of varying sizes under different heating rates.

Samples	Case	T_i (°C)	T_h (°C)	t_h (min)
SBC	SBC-1	384.521	571.435	55.811
	SBC-2	403.545	597.557	57.535
	SBC-3	433.835	627.175	61.217
	SBC-4	398.235	810.957	39.935
	SBC-5	441.914	804.235	39.557
	SBC-6	438.275	756.717	37.114
DBC	DBC-1	455.257	761.593	74.353
	DBC-2	475.736	770.137	75.214
	DBC-3	483.278	815.653	79.778
	DBC-4	452.635	848.517	41.553
	DBC-5	504.314	877.132	43.114
	DBC-6	499.185	898.514	44.252
JAC	JAC-1	433.914	702.367	68.314
	JAC-2	468.358	717.335	69.953
	JAC-3	497.221	750.114	73.414
	JAC-4	460.835	821.178	40.353
	JAC-5	501.614	846.153	41.614
	JAC-6	510.936	936.135	45.823

Note: t_h (min) stands for the burnout temperature.

When the heating rate was maintained at 10 °C/min, the T_i and T_h of the three types of coal particles demonstrated a mostly steady ascending tendency along with the increase in particle size, although the magnitudes varied. Notably, the T_i of JAC exhibited the most remarkable change, with an increase of 14.59%. When the heating rate was increased to 20 °C/min, the trends in T_i and T_h diverged for different coal species as the particle size increased. Specifically, when the particle size increased from 1 mm to 4 mm, the T_i rose by approximately 50 °C across all coal types, among which SBC showed the largest increase of 13.65%. Conversely, further increases in particle size from 4 mm to 8 mm led to negligible changes in T_i. Based on previous studies [27–29], it can be inferred that, under a given heating rate, the T_i stabilizes beyond a certain critical particle size, while higher heating rates lead to a lower critical particle size. However, due to the inherent differences among coal types, the trends in T_h remain inconsistent.

The aforementioned phenomenon can be ascribed to the non-uniform heating of granular coal within the crucible during the experimental heating procedure. A temperature gradient existed between the surface and internal areas of the granules [30], causing a delay in the volatilization of the internal components. This led to ignition delay [31,32], which was more prominent for larger particle sizes. The T_i and T_h of the fuel were related to the coal rank, both T_i and T_h increased significantly as the volatile matter content decreased across different types of coal.

To more comprehensively evaluate the combustion characteristics of the fuel, several key parameters are typically employed to reflect the combustion performance of coal. The specific calculation formulas are presented as Equations (1) to (3):

(1) Flammability parameter (F_{ith}): the F_{ith} quantifies the initial reactivity of pulverized coal. It is characterized by the trend from the ignition point to the peak combustion reaction rate on the *TG* curve. A higher F_{ith} signifies the enhanced ignition performance of the fuel. This parameter is determined using the following formula [33]:

$$F_{ith} = \frac{(d\alpha/dt)_{max}}{T_i^2} \tag{1}$$

where $(d\alpha/dt)$max denotes the peak weight loss rate, %/min; and T_i denotes the ignition temperature of the fuel, °C.

(2) Burnout characteristic parameter (C_b): the C_b reflects the combustion reactivity of coal during the later stages of combustion. This parameter is influenced by factors such as the heating rate and burnout time. A higher value of the burnout C_b indicates the superior burnout performance of pulverized coal in the later stages of combustion. The calculation formula is as follows [34]:

$$C_b = \frac{f_i \times f_h}{t_h} \tag{2}$$

where f_i represents the initial burnout rate, defined as the ratio of the combustion weight loss at the ignition point on the TG curve to the total combustible content in the coal powder, %; f_h denotes the late burnout rate, which is the difference between the final burnout rate of the combustible substances in the granular coal and the initial burnout rate, %; and t_h is the time corresponding to the burnout temperature, min.

(3) Combustion characteristic parameter (S): combustion characteristics encompass parameters such as the ignition and burnout properties of granular coal. These parameters are crucial for a comprehensive evaluation of the combustion performance of granular coal. A higher value of S indicates superior combustion characteristics. The expression is as follows [35]:

$$S = \frac{(d\alpha/dt)_{max} \times (d\alpha/dt)_{meam}}{T_i^2 \times T_h} \tag{3}$$

where $(d\alpha/dt)_{mean}$ represents the mean weight loss rate, %.

The combustion performance parameters of coal fuels with diverse particle sizes are delineated in Table 4, while the tendencies of each parameter influenced by changes in particle size are depicted in Figure 3. At a heating rate of 10 °C/min, the F_{ith} for different coal types diminished as the particle size increased. Nevertheless, the trends for the C_b and S exhibited inconsistency. Specifically, for SBC, the C_b increased while the S declined with the increase in the particle size. For DBC, the C_b initially decreased before experiencing an upswing, peaking at 8 mm, while the S registered a reduction of 17.16%. In the case of JAC, the C_b initially rose before falling, while the S demonstrated a steadily decreasing trend, with both indices reaching their lowest values at 8 mm. At a heating rate of 20 °C/min, the variation trends of the combustion parameters among different coal types became more distinct. For SBC, both the F_{ith} and S initially decreased before surging significantly, with an increase rate exceeding 20% relative to their respective minimum values. For DBC, as the particle size increased, the C_b rose marginally, while the F_{ith} and combustion characteristic indices declined. For JAC, the combustion performance parameters initially decreased before rebounding with the increase in particle size.

Table 4. Combustion performance parameters of coal particles of varying sizes under different heating rates.

Samples	Case	F_{ith} (10^{-6}·min^{-1}·°C^{-2})	C_b (10^{-3}·min^{-1})	S (10^{-8}·min^{-2}·°C^{-3})
SBC	SBC-1	38.895	2.232	11.654
	SBC-2	36.653	2.235	10.193
	SBC-3	33.205	2.802	8.265
	SBC-4	44.342	2.776	13.097
	SBC-5	41.171	3.983	12.384
	SBC-6	49.533	6.029	16.853
DBC	DBC-1	15.707	1.565	1.697
	DBC-2	15.338	1.463	1.404
	DBC-3	14.995	2.645	1.403
	DBC-4	22.034	2.523	3.817
	DBC-5	16.751	2.585	2.695
	DBC-6	14.642	2.743	2.242
JAC	JAC-1	16.245	1.901	2.157
	JAC-2	15.903	2.988	2.025
	JAC-3	12.381	1.593	1.434
	JAC-4	19.057	4.163	3.661
	JAC-5	13.573	2.643	2.013
	JAC-6	16.201	3.308	2.937

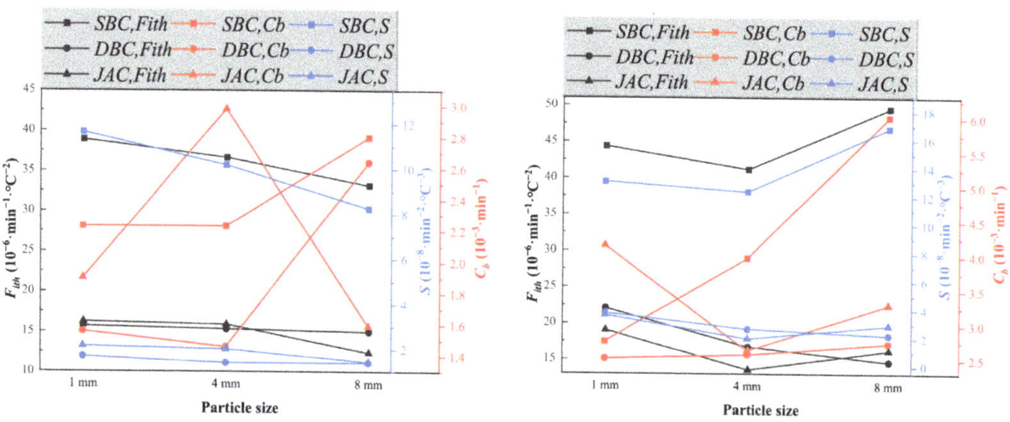

(**a**) 10 °C/min. (**b**) 20 °C/min.

Figure 3. Variation in combustion characteristic parameters with particle size.

In summary, diverse coal types manifested distinct combustion characteristics at varying particle sizes. Specifically, when the heating rate was 10 °C/min, the T_i of 8 mm coal particles in JAC was 14.59% higher than that of the 1 mm particles. Normally, as the particle size increased, the T_i, T_h, and t_h tended to ascend, resulting in a significantly lower S for the 8 mm particles compared to the 1 mm particles, with an average reduction of 28.81%. Additionally, it was observed that the changes in particle size had a more pronounced effect on the combustion performance parameters of granular coal at lower heating rates.

3.2. Effect of Changes in Particle Size on the Combustion Kinetics of Granular Coal

In this study, we used the Coats–Redfern integral expression within the single scan rate method to calculate the activation energy of granular coal fuel with varying particle sizes and coal types. The Coats–Redfern integral expression is as follows [36]:

$$\alpha = \frac{m_0 - m_t}{m_0 - m_f} \times 100\% \tag{4}$$

$$\ln\left[\frac{G(\alpha)}{T^2}\right] = \ln\left(\frac{AR}{\beta E_a}\right) - \frac{E_a}{RT} \tag{5}$$

where m_0 denotes the initial mass of pulverized coal (mg); m_t represents the instantaneous mass during the combustion process (mg); m_f signifies the residual mass after combustion (mg); α indicates the conversion rate of the combustion process (%); A is the pre-exponential factor (min^{-1}); E_a represents the apparent activation energy of the combustion reaction (kJ/mol); R is the ideal gas constant (8.314 J·mol^{-1}·K^{-1}); and $G(\alpha)$ symbolizes the integral of the mechanism function. For the combustion reaction of pulverized coal, $G(\alpha)$ is commonly expressed as $-\ln(1-\alpha)$.

The linear correlation between *1/T* and $\ln(\beta/T^2)$ was utilized to conduct a linear regression analysis of the aforementioned equation, from which the apparent activation energy E_a was derived based on the obtained slope. By examining the variations in E_a values, the reaction process was segmented and reflected in the *DTG* curve. According to changes in E_a, the entire coal combustion process from 200 °C to the end of burning was divided into five distinct stages [37–39] (the dehydration process below 200 °C was not considered to involve significant chemical reactions and thus was not analyzed in this study). (1) The endothermic phase (*EP*): during this phase, the temperature was insufficient to meet the ignition requirements of the coal particles. Consequently, instead of releasing heat, the coal absorbed heat, resulting in a negative E_a. For ease of comparison, E_a values are presented as absolute values below. (2) The volatiles release phase (*VRP*): as the temperature rose, volatile components within the coal begin to escape. Upon reaching adequate temperatures, these volatiles could ignite after their release. (3) The volatile oxidation phase (*VOP*): upon reaching the ignition temperature, volatile components ignited first, releasing a significant amount of heat, which subsequently initiated the ignition of coke. (4) The coke oxidation phase (*COP*): during this phase, the primary combustible components of the fuel underwent combustion, typically accounting for the longest duration of chemical reactions in the entire process. (5) The burnout phase (*BP*): as the combustible components diminished, the combustion rate decelerated, marking the transition into the burnout phase.

The division and changes in the E_a of SBC under varying particle sizes during the combustion stage are illustrated in Figure 4. At a heating rate of 10 °C/min, the duration of the *EP*, *VOP*, and *COP* intervals was progressively extended as the particle size increased. Notably, the most pronounced effect was observed in the *EP* interval, with an increase of 79.23%, followed by 55.38% for the *VOP* interval and 18.37% for the *COP* interval. This phenomenon could be attributed to the fact that, during the preparation phase preceding combustion, the E_a values for the *EP* and *VRP* intervals for a particle size of 1 mm were higher than those for larger particle sizes, thereby facilitating the rapid release of volatile fractions. Consequently, upon reaching a sufficient temperature, ignition occurred more swiftly, transitioning into the *VOP* interval and subsequently driving the combustion of the residual carbon structure. This clarified the observed increase in T_i and T_h with escalating particle size at a heating rate of 10 °C/min. When the heating rate was 20 °C/min, during the preparation stage, $E_{a,\,EP}$ and $Ea_{,\,VRP}$ initially rose, then dropped, and subsequently continued to increase along with the growth in particle size. The changing trend in the T_i was identical to that of $Ea_{,EP}$. It was hypothesized that the influence of $E_{a,EP}$ on the T_i was greater than that of $E_{a,VRP}$ during the preparation stage. Moreover, as the particle size increased, the maximum reaction rate also rose correspondingly. This is because, at a higher heating rate, the thermal stress disparity between the interior and exterior of large particles intensified with the increase in the particle size. Thus, larger particles were more prone to fracture, thereby exposing the internal structure of the particles to release volatiles [40,41]. They ignited rapidly upon reaching the T_i, generating additional heat to accelerate the combustion of coke, which is reflected as E_a as $E_{a,VRP8} > E_{a,VRP4} > E_{a,VRP1}$. This explains that the T_h of large particles was reduced with the increase in particle size after the reaction when the T_i was elevated.

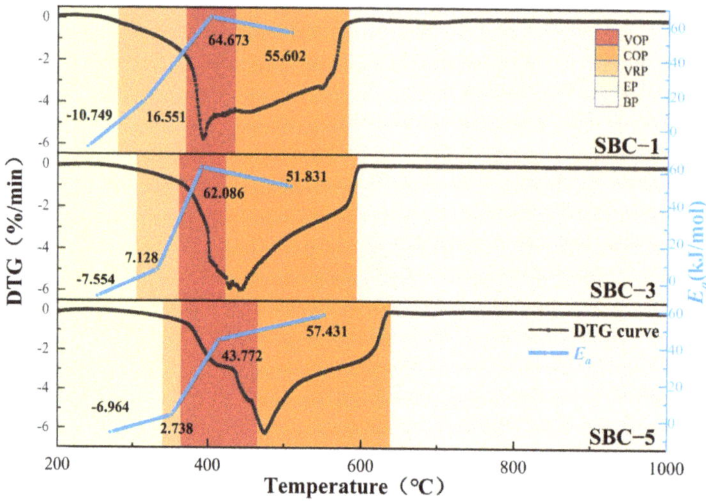

(**a**) 10 °C/min.

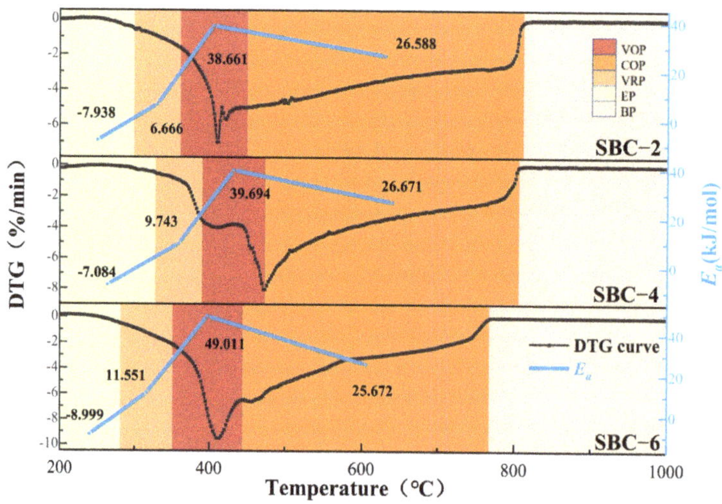

(**b**) 20 °C/min.

Figure 4. *DTG-E_a* curve of SBC as a function of particle sizes.

As depicted in Figure 5, the division of the combustion stage and the variations in the E_a of DBC under different particle sizes were examined. When the heating rate was 10 °C/min, the intervals of *EP* and *VRP* initially expanded and subsequently narrowed with the increase in particle size. In contrast, the intervals of *VOP* and *COP* presented the opposite tendency, initially narrowing and then widening. Concerning E_a, both *EP* and *VRP* were verified to be critical factors influencing the T_i; a higher $E_{a,VRP}$ indicated that the DBC particles were ignited more readily. Furthermore, the maximum reaction rate emerged at a particle size of 8 mm, which coincided with the peak values of $E_{a,VRP}$. It can be deduced that, under this heating rate, the maximum reaction rate of the DBC particles could be characterized by $E_{a,VRP}$. Higher $E_{a,VRP}$ values indicated greater energy demands for the reaction, resulting in more intense reactions. Despite the occurrence of the maximum rate at 8 mm, the consumption of combustible substances within the *VOP* interval and the carbonate removal reaction around 750 °C led to lower $E_{a,VRP}$ values in the *COP* interval, thereby prolonging the time needed for complete combustion. This explained the increase in the T_i and T_h of DBC particles as the particle size increased at a heating rate of 10 °C/min. When the heating rate was 20 °C/min, as the particle size increased, the *EP* interval initially broadened and then narrowed, the *VRP* interval expanded steadily, the *VOP* interval contracted, and the *COP* interval initially narrowed and subsequently widened. Among these changes, the *VRP* region presented the most significant amplitude of change, with the 4 mm and 8 mm regions increasing by 98% and 134%, respectively, compared to the 1 mm region, while the *VOP* interval decreased by 21% and 33%, respectively. The underlying principle behind this phenomenon was that smaller particles exhibit more uniform heat distribution, thereby substantially enhancing the heat absorption efficiency. This shows that, when the T_i of large-particle-size coal was higher, it was $E_{a,EP1} > E_{a,EP4} > E_{a,EP8}$, which is conducive to particle ignition. Additionally, the difference between the total width of the *VRP* interval and the *VOP* interval under various particle sizes was less than 10%. Under these circumstances, it could be suggested that the small particle size, which was more prone to ignition, was largely analogous to the poor coal particle fuel with a broader *VOP* interval and a larger $E_{a,VOP}$, while the overall combustion duration was shorter and the T_h was lower. This explains the fact that the T_i and T_h of the DBC particles increased with the increase in the particle size when the heating rate was 20 °C/min. Although the results were comparable when the heating rate was 10 °C/min and 20 °C/min, the variations in the width of the reaction intervals and E_a still differed.

As illustrated in Figure 6, we classified the combustion process stages of JAC and the variations in E_a for different particle sizes. When the heating rate was 10 °C/min, the *EP* interval gradually expanded along with the increase in particle size. Given the relatively low volatile content of JAC, it was determined that there was no obvious limit for the release of volatiles before combustion within the *VRP* interval and *VOP* interval. Hence, the *VRP* interval and *VOP* interval were combined and are discussed together (hereafter referred to as the *VP* interval), and $E_{a,VP}$ values were calculated. It was found that the *VP* interval gradually widened with the growth in particle size. When the particle size was 8 mm, the width of the *VP* interval increased by 22.86% compared to when the particle size was 1 mm, but $E_{a,VP}$ decreased by 9.07%. This indicates that, under this temperature condition, the ignition of the JAC particles was gradually delayed with the increase in particle size. In addition to the prolonged duration required in the endothermic stage, another crucial factor was that the reaction intensity of the *VP* interval was gradually weakened, and the length of the *COP* interval was influenced by the reaction intensity of the *VP* stage. The more intense the reaction in the *VP* interval, the fewer flammable substances remained in the *COP* interval; thus, the *COP* interval demonstrated a gradually narrowing trend compared to the *VP* interval. The numerical manifestation indicated that $E_{a,COP8}$ exceeded $E_{a,COP4}$, which in turn surpassed $E_{a,COP1}$. In contrast to $E_{a,VP}$, the *COP* interval of all particle sizes increased, with an increase of 95.39% for 8 mm. When the heating rate was 20 °C/min, the interval widths of *EP*, *VP*, and *COP* expanded concurrently with the increase in the particle size, while $E_{a,EP}$, and $E_{a,COP}$ gradually decreased, and $E_{a,VP}$ initially rose and then dropped. This suggests that the maximum reaction rate of coal combustion in the JAC was 4 mm, but the T_h at 4 mm was slightly higher than that at 1 mm due to the relatively weak reaction in the front and back intervals. The conclusion was similar to that of DBC. Even though the T_i and T_h of granular coal rose along with the increase in particle size at the heating rates of 10 °C/min and 20 °C/min, the E_a of each interval length did not change steadily.

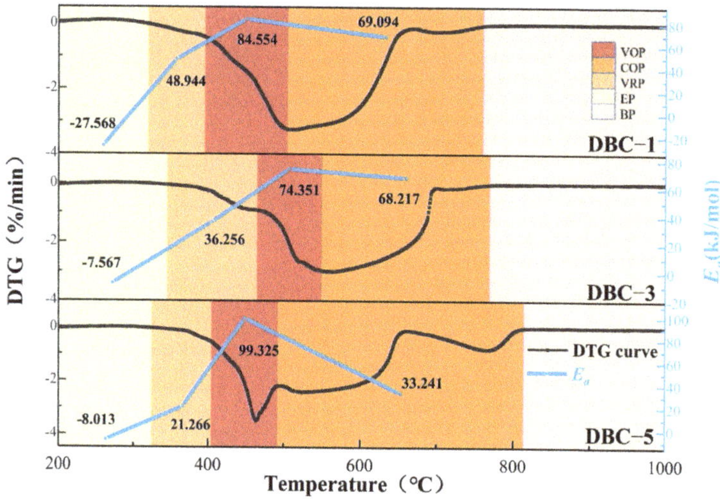

(**a**) 10 °C/min.

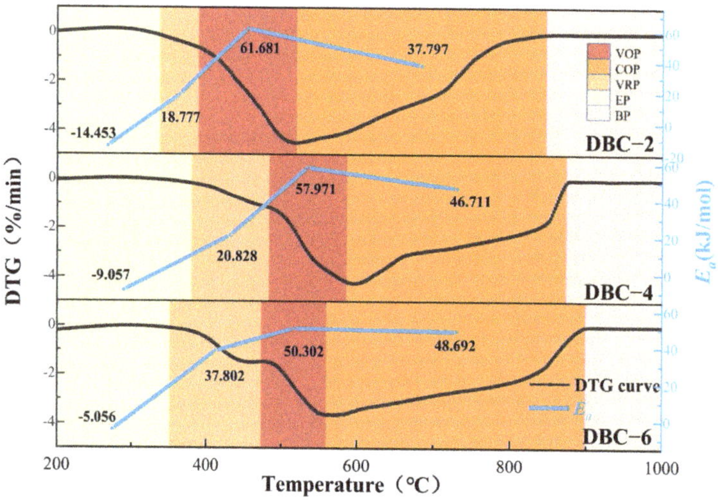

(**b**) 20 °C/min.

Figure 5. *DTG-E_a* curve of DBC as a function of particle sizes.

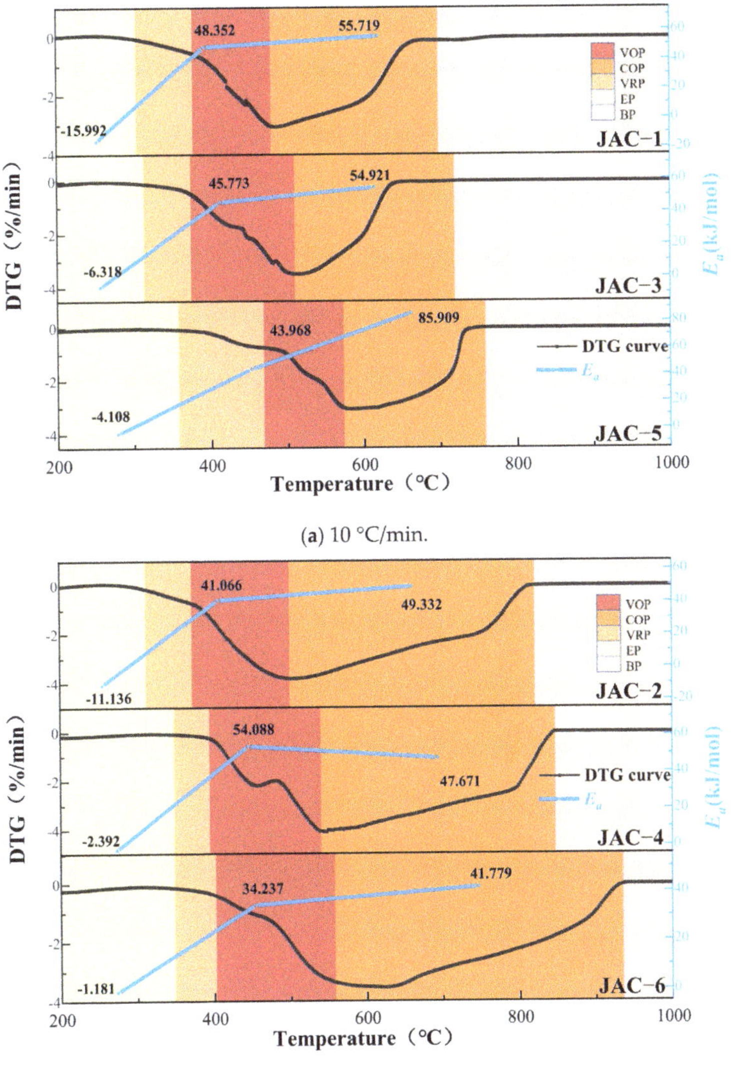

(a) 10 °C/min.

(b) 20 °C/min.

Figure 6. *DTG-E_a* curve of JAC as a function of particle size.

To sum up, the following two points can be delineated:

(1) The preheating stage prior to combustion had an impact on the ignition temperature of the coal particles. The endothermic stage was of the utmost importance, followed by the volatilization stage. Our calculations revealed that, for SBC, for every 1 kJ/mol reduction in the E_a during the endothermic stage, the T_i rose by 13.02 °C and 17.11 °C at heating rates of 10 and 20 °C/min, respectively. At the aforementioned heating rates, for DBC and JAC, the corresponding temperature alterations were 1.43 °C and 4.94 °C, and 5.32 °C and 5.03 °C, respectively.

(2) In this experiment, the T_i and T_h of the majority of coal particles increased with particle size, yet the variations in the reaction range and E_a were dissimilar. Owing to the lower volatile matter content and higher ash content, the regularity of the combustion process of coal particles in each region was more pronounced.

3.3. Effect of Changes in Particle Size on the Release Rates and Total Number of Particulate Gas Products

In the experiment, the programmed heating combustion temperature was significantly lower than the initial temperature for the formation of thermal NO_x, and the rapid NO_x could be disregarded. Therefore, NO_x originated from the nitrogen within the fuel. To further illuminate the impact of particle size variation on nitrogen-phase products during granular coal combustion, ECSA analysis was employed to calculate the release rate of related gas products, and the total releases of related gas products under each working condition were obtained through the temporal integration of the related gas products. For the same sample, the temperature range of NO production and that of CO_2 production were approximately identical, with the peak value of the release rate emerging in both instances, and the corresponding temperature of the peak value of the NO release rate and the CO_2 release rate being close. HCN and NH_3 were regarded as the main precursors for NO generation [42]. However, based on the experimental data, it was discovered that the release rate of CO_2, NO, and HCN was considerable, the signal strength and disturbance of NH_3 were of the same order of magnitude, and the amount of NH_3 generated was minuscule. Thus, the yield of NH_3 was neglected in the experiment, and no NO_2 generation was witnessed throughout the entire experiment. Hence, only HCN is preliminarily discussed in this paper, and its possible reactions are presented in Equations (6)–(9) [43–45]:

$$2\,HCN + 2.5\,O_2 \rightarrow 2\,NO + 2\,CO + H_2O \tag{6}$$

$$2\,HCN + 3\,NO \rightarrow 2.5\,N_2 + 2\,CO + H_2O \tag{7}$$

$$2\,C + O_2 \rightarrow 2\,CO \tag{8}$$

$$2\,CO + O_2 \rightarrow 2\,CO_2 \tag{9}$$

In the highly oxidizing combustion environment of the *TG* experiment, the reductive effect of CO on NO was negligible. Due to the relatively slow heterogeneous reaction kinetics between C and NO, NO could not be reduced to N_2 within the limited experimental period. Moreover, the background carrier gas contained no N_2, causing the observed N_2 release rate to remain consistently at the same level as the signal noise. Additionally, the volatile oxidation process during coal combustion was highly complex, involving numerous intermediate reaction products and NO reduction reactions. Therefore, this analysis mainly focuses on a simplified explanation of the temperature at the first peak for the generation of HCN, NO, CO_2, and CO, as well as the total amount of NO and HCN released in relation to particle size variations.

Figures 7 and 8 depict the release rate curve and total release amount curve of SBC under various particle sizes. When the heating rate was 10 °C/min, with the increase in particle size, the peak temperature of the release rates of the main gas products gradually rose, and the maximum release rates of NO and CO_2 gradually increased. The maximum release rates of HCN and CO initially rose and then decreased, but the length of the temperature interval varied. Among them, the temperature range corresponding to HCN gradually widened, and the temperature difference gradually expanded from 43 °C for 1 mm to 118 °C for 8 mm, an increase of approximately 175%, which indicates that, under this heating rate, fuel with a larger particle size is more likely to generate HCN. Additionally, when the particle size was 1 mm and 4 mm, the peak temperatures of each gas product were comparable. However, when the particle size was 8 mm, a platform with a temperature range of 440 °C to 460 °C emerged at the peak of CO, and the temperatures at both ends of the platform were respectively similar to the peak temperatures of NO and HCN. Meanwhile, the CO_2 release rate kept escalating while the CO release rate remained constant within this temperature range. According to the above Equation (6), HCN preferentially bound O_2 and continuously transformed into NO. The total amounts of HCN and NO concurrently increased along with the increase in the particle size. The total amount of HCN rose from 0.027 mg to 0.093 mg, indicating an increase of 244.44%, while the total amount of NO rose from 0.292 mg to 0.642 mg, with an increment of 119.86%. When the heating rate was 20 °C/min, as the particle size expanded, the peak temperature of the main-gas-phase product release rate initially rose and then declined, and the peak temperature of 8 mm emerged earlier than that of 1 mm. It was hypothesized that the potential cause could be that, under this heating rate, the larger particle size had a higher local temperature, causing a certain area to react initially. The release curve of HCN was unimodal, and the temperature difference between the peaks initially increased and then decreased with the particle size, reaching the minimum and maximum values at 1 mm and 4 mm, respectively, with the temperature difference between peaks being approximately 100 °C and 160 °C. The total amounts of HCN and NO also presented a pattern of initially rising and then falling, with the maximum values of HCN and NO increasing by 59.94% and 29.07%, respectively, compared to the minimum values. In conclusion, HCN, acting as a precursor of NO, had a considerable influence on the production of NO. As the total amount of HCN released from SBC increased, the total amount of NO released also rose. Therefore, the variation in particle size was capable of influencing the peak temperature of HCN release and the temperature disparity within the release interval of SBC particles, thereby exerting an impact on the total amount of NO release.

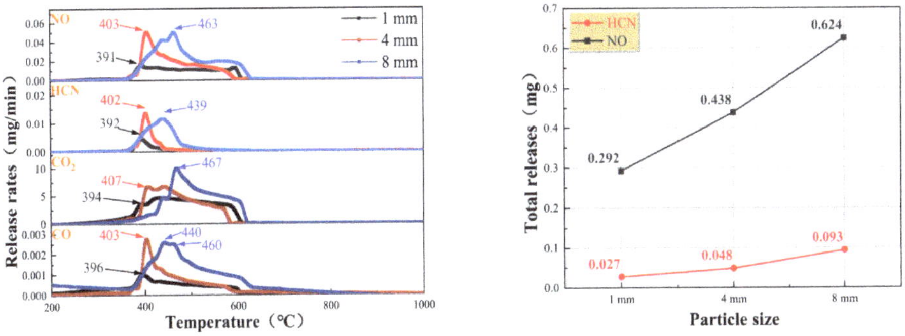

(**a**) Release rates of gas products. (**b**) Total released amount of HCN and NO.

Figure 7. The release rates of gas products and the total release amount of SBC at 10 °C/min.

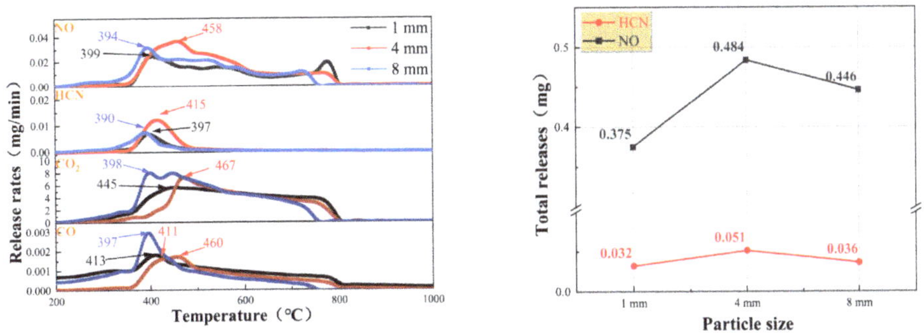

(**a**) Release rates of gas products. (**b**) Total released amount of HCN and NO.

Figure 8. Release rates of gas products and total release amount of SBC at 20 °C/min.

Figures 9 and 10 depict the release rate curves and the total release amounts of DBC under different particle sizes. At a heating rate of 10 °C/min, the first peak temperature and peak height of other gas products apart from CO_2 gradually rose with the increase in particle size, and the release rate of NO fluctuated repeatedly throughout the entire combustion process. The temperature difference in the peak interval of HCN initially increased and then decreased as the particle size expanded. When the particle size was 1 mm, HCN mainly transformed into NO at approximately 430 °C, and it then entered the plateau stage from 550 °C to 590 °C. When the particle size was 4 mm, the peaks of HCN, NO, and CO emerged successively within the range of 440 °C to 450 °C, and the release rate of CO_2 increased significantly at 476 °C, with the second peak temperature of NO and the first peak temperature of CO_2 being reached near 540 °C. When the particle size was 8 mm, the release rate of HCN reached its maximum at 451 °C, the release rates of CO and NO peaked at 455 °C, and, subsequently, the first peak of CO_2 occurred at 461 °C. With the increase in particle size, the total release amount of HCN initially increased and then decreased, with the maximum increase approximating 2.75 times. Nevertheless, the total release amount of NO rose as the particle size expanded, ranging from 0.352 mg to 0.535 mg, an increment of 51.99%. When the heating rate was 20 °C/min, with the growth in the particle size, the occurrence temperature of the first peak of the gas product fluctuated. Under the condition of a particle size of 1 mm, CO initially achieved the highest release rate at 421 °C; this was followed by HCN and NO reaching their peaks, respectively, at 429 °C and 435 °C. Ultimately, CO_2 attained the maximum release rate at 499 °C. When the particle size grew to 4 mm, the peak temperature of the HCN release rate was reached slightly earlier than that of CO, and an identical phenomenon transpired at 8 mm. The peak release rate of HCN initially decreased and then increased as the particle size expanded, and the temperature difference between the intervals gradually grew. The total amount of HCN released initially decreased and then increased, reaching a maximum value of 0.55 mg at 8 mm. Nevertheless, under the comprehensive influence, the release rate of NO in DBC initially increased and then decreased, attaining the minimum and maximum values at 1 mm and 4 mm, respectively, with the total release amounts being 0.441 mg and 0.544 mg. In conclusion, variations in particle size could affect the release rate and total release amount of HCN and NO; however, the specific regular variations produced more patterns, requiring additional in-depth research for clarification.

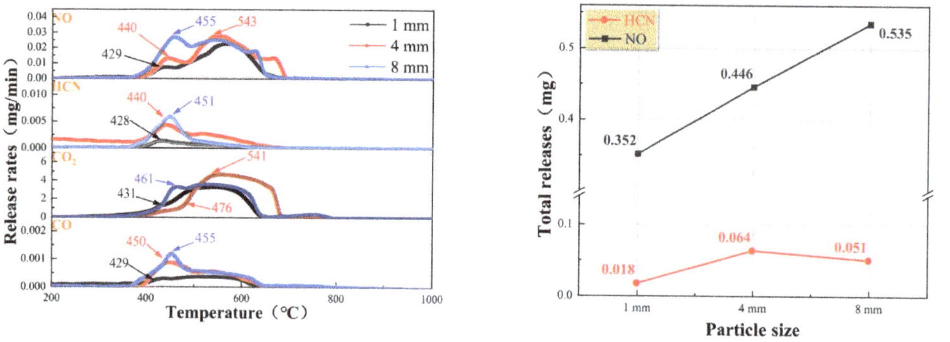

(**a**) Release rates of gas products. (**b**) Total released amounts of HCN and NO.

Figure 9. Release rates of gas products and total released amount of DBC at 10 °C/min.

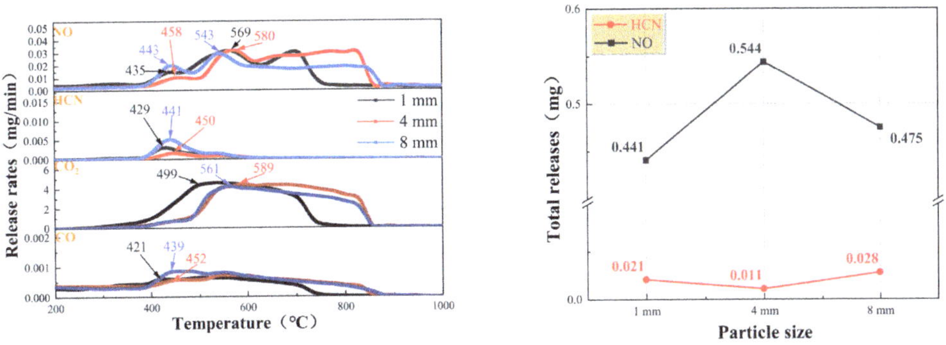

(**a**) Release rates of gas products. (**b**) Total released amounts of HCN and NO.

Figure 10. Release rates of gas products and total released amount of DBC at 20 °C/min.

Figures 11 and 12 present the curves of the release rates and total released amount of JAC under various particle sizes. When the heating rate was 10 °C/min, the temperature of the first peak and the peak height of the gas products gradually rose along with the increase in particle size, yet the height of the first peak varied. When the particle size was 1 mm, the release rates of HCN, NO, and CO reached their peaks successively within the range of 420 °C to 424 °C, and, subsequently, the release rate of CO_2 peaked at approximately 465 °C. When the particle size amounted to 4 mm, the first peak of the CO_2 release rate emerged within the range of 434 °C to 451 °C, and, subsequently, the maximum CO_2 release rate was attained at 494 °C. When the particle size was 8 mm, the first reaction peak occurred within the range of 442 °C to 458 °C, and, at 567 °C, the NO release rate reached its peak and then declined. Taking the CO_2 release rate as the basis for measuring the length of the platform, the calculation indicated that the length of the platform combustion stage initially decreased and subsequently increased with the increase in the particle size. This is in line with the principle that the total amount of NO released initially decrease and then rose, suggesting that the particle size would have an influence on the duration of the high-intensity combustion reaction and subsequently affect the total amount of gas products released [46]. Under the aforementioned experimental conditions, the total amount of HCN released initially rose and then declined, with a maximum increase of 44.44%, while the total release amount of NO showed the opposite effect, with a maximum of 0.453 mg at 1 mm and a minimum of 0.39 mg at 4 mm. When the heating rate was 20 °C/min, the sequence of the appearance of peaks underwent variations. Regarding the peak of CO, the sequence was 1 mm, 4 mm, and 8 mm, with corresponding temperatures of 426 °C, 434 °C, and 449 °C, respectively. Concerning HCN, the peak occurred in the sequence of 8 mm, 4 mm, and 1 mm, at 439 °C, 447 °C, and 458 °C, respectively. It was noted that the peak temperatures of HCN and CO gradually converged as the particle size increased. Meanwhile, the temperature difference between the maximum HCN release rate and the corresponding temperature range decreased with the growth in the particle size, resulting in a steady decrease in the total HCN release, with the maximum reduction reaching 74.61%. Therefore, although the main combustion stage at 8 mm was the longest, HCN, as the precursor, was the shortest, reducing the combination of HCN with O_2 to generate NO. Consequently, the total amount of NO released in the plant initially rose and then fell, with the minimum being 0.448 mg at 1 mm and the maximum being 0.598 mg at 4 mm.

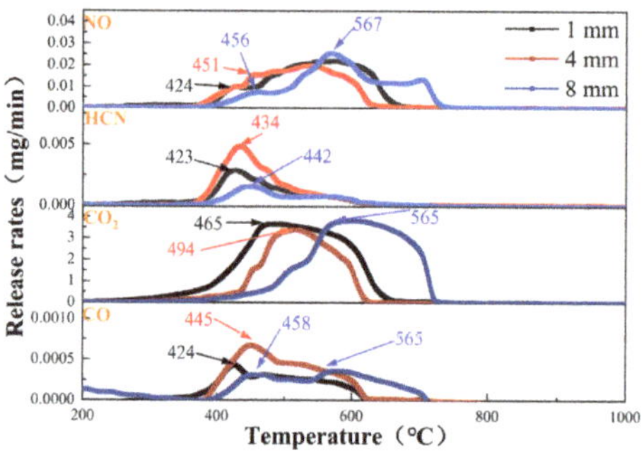

(**a**) Release rates of gas products.

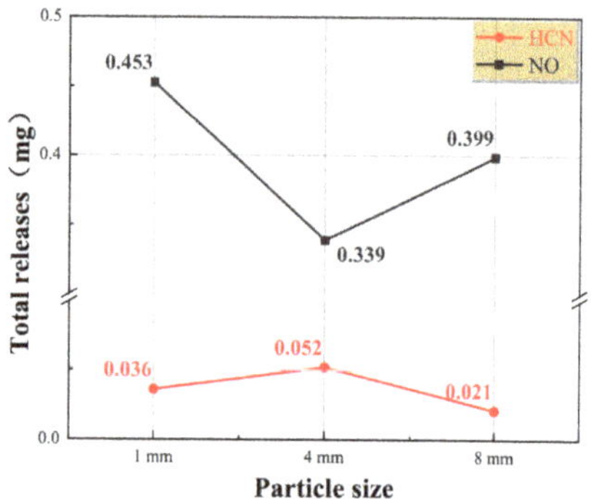

(**b**) Total released amounts of HCN and NO.

Figure 11. Release rates of gas products and the total released amount of JAC at10 °C/min.

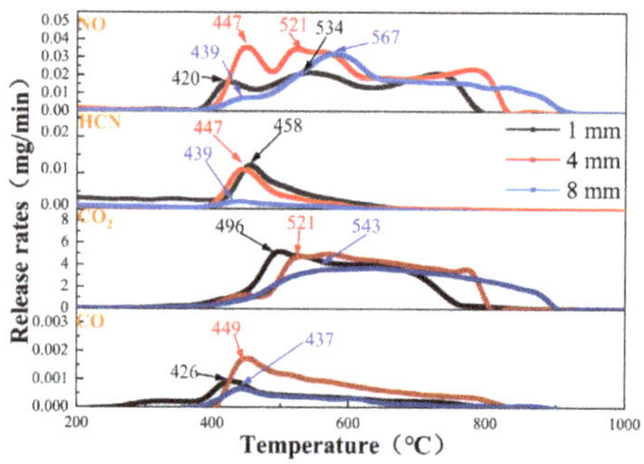

(**a**) Release rates of gas products.

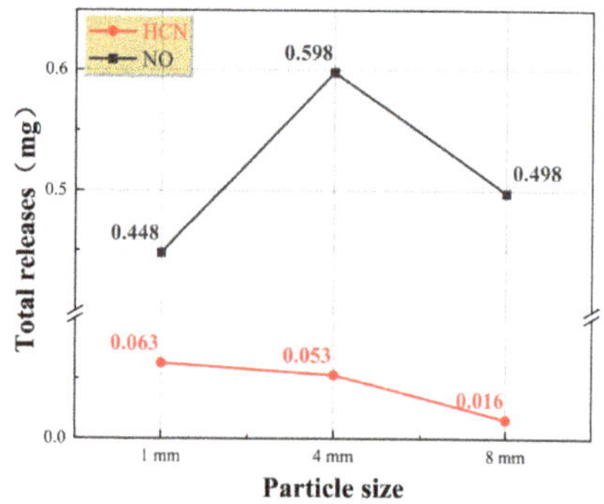

(**b**) Total released amounts of HCN and NO.

Figure 12. The release rates of gas products and total released amount of JAC at 20 °C/min.

In summary, this section expounded the influence of particle size on the gas products' release rates and the total released amount of various coal particles under different heating rates. The conclusions are as follows:

(1) The effect of particle size on the peak of each gas phase varied with the heating rate, and the temperature of the peak rose as the particle size increased.

(2) Changes in particle size not only change the maximum release rate of individual gas components but also prolonged or shortened the duration of the primary combustion phase for granular coal. By integrating the insights from previous studies, it becomes evident that the overall NO release stemmed from a confluence of multiple factors [47–49]. However, specific scenarios necessitate individual analysis. In most cases, smaller coal particles resulted in smaller amounts of NO being released during combustion, which is basically consistent with the research findings of Tang Z [50].

4. Conclusions

This study delved into and dissected the impacts of variations in granular coal particle size on the combustion performance parameters, combustion reaction kinetics, and the release rate and total quantity of gas phase products of coal particles from three perspectives. The principal conclusions were as follows:

(1) Regarding the combustion characteristic parameters, the different coal particle sizes exhibited differences in combustion performance. During the heating process in the experiment, as the granular coal was not heated completely uniformly in the crucible of the experimental instrument, a temperature disparity existed between the surface temperature and the internal temperature of the granular coal, leading to a reaction lag and ignition delay within the granular coal. This phenomenon was more pronounced for larger particle sizes. The most remarkable discovery was that, at a heating rate of 10 °C/min, the ignition temperature of the JAC with a particle size of 8 mm was 14.59% higher than that with a size of 1 mm. In most circumstances, as the particle size increased, the ignition temperature, burnout temperature, and burnout time rose. Consequently, the combustion characteristic index of the coal with an 8 mm particle size was conspicuously lower than the 1 mm sample, with an average reduction of 28.81%. Simultaneously, we found that the variation in particle size had a more conspicuous impact on the combustion performance parameters of granular coal at a lower heating rate.

(2) Regarding the kinetics of combustion reactions, the preparatory stage prior to combustion exerted a considerable influence on the ignition temperature of the granular coal. The endothermic stage had the most significant impact, followed by the volatilization stage. Our calculations showed that, for SBC, every 1 kJ/mol reduction in the E_a during the endothermic stage led to T_i increases of 13.02 °C and 17.11 °C at heating rates of 10 and 20 °C/min, respectively. For DBC and JAC at the same rates, the corresponding temperature changes were 1.43 °C and 4.94 °C, and 5.32 °C and 5.03 °C, respectively. Under the conditions of this experiment, the ignition temperature and burnout temperature of the majority of coal particles rose along with the increase in particle size, yet the reaction range and the corresponding apparent activation energy varied. Upon comparison, it was discovered that the length of the endothermic interval might be associated with the content of C in coal, and the length of the endothermic phase interval gradually rose with the reduction in C. Simultaneously, the length of the volatile oxidation phase interval also exhibited an inverse relationship with the volatile content.

(3) Concerning the release rates and total releases of gas products, particle size exerted diverse influences on the peak temperature of each gas phase at different heating rates. Generally speaking, with the increase in particle size, the peak temperature rose. Simultaneously, the variation in particle size affected the maximum release rate of each gas phase product and the duration of the main combustion stage of granular coal. Hence, the total amount of NO released was the result of comprehensive factors and demanded meticulous analysis. Nevertheless, in the majority of cases, the total amount of NO released during the combustion of granular coal with a small particle size was relatively low.

Author Contributions: Writing—original draft, J.Z.; Writing—review & editing, H.D.; Supervision, G.S. and W.S. All authors have read and agreed to the published version of the manuscript.

Funding: This work was supported by the "Key technology and demonstration of deep and flexible peaking for circulating fluidized bed boiler" project, Strategic Priority Research Program of the Chinese Academy of Sciences (XDA29010100).

Data Availability Statement: The original contributions presented in the study are included in the article, further inquiries can be directed to the corresponding author.

Conflicts of Interest: The authors declare no conflict of interest.

References

1. Wu, Z.; Zhou, M.; Wang, J.; Du, E.; Zhang, N.; Li, G. Profit-sharing mechanism for aggregation of wind farms and concentrating solar power. *IEEE Trans. Sustain. Energy* **2020**, *11*, 2606–2616. [CrossRef]
2. Kang, Z.; Yao, L. Key Scientific Issues and theoretical research framework for power systems with high proportion of renewable Energy. *Autom. Electr. Power Syst.* **2017**, *41*, 2–11.
3. Lu, Z.; Huang, H.; Shan, B.; Wang, Y.; Du, S.; Li, H. Morphological evolution model and power forecasting prospect of future electric power systems with high proportion of renewable energy. *Autom. Electr. Power Syst.* **2017**, *41*, 12–18.
4. Liu, X.; Zhang, X. Industrial agglomeration, technological innovation and carbon productivity: Evidence from China. *Resour. Conserv. Recycl.* **2021**, *166*, 105330. [CrossRef]
5. Li, S.; Yang, H.; Zhang, H. Measurements of solid concentration and particle velocity distributions near the wall of a cyclone. *Chem. Eng. J.* **2009**, *150*, 168–173. [CrossRef]
6. Arias, B. An analysis of the operation of a flexible oxy-fired CFB power plant integrated with a thermal energy storage system. *Int. J. Greenh. Gas Control* **2016**, *45*, 172–180. [CrossRef]
7. Arias, B.; Criado, Y.; Sanachez-Biezma, A.; Abanades, J.C. Oxy-fired fluidized bed combustors with a flexible power output suing circulating solids for energy storage. *Appl. Energy* **2014**, *132*, 127–136. [CrossRef]
8. Hu, X.; Deng, B.; Liu, H.; Yang, D.; Yang, H. Status and development trend of rapid load regulation technology for circulating fluidized bed boiler. *Clean Coal Technol.* **2023**, *29*, 11–23.
9. Gao, M.; Hong, F.; Liu, J.; Chen, F. Investigation on the energy conversion and load control of supercritical circulating fluidized bed boiler units. *J. Process Control* **2018**, *68*, 14–22. [CrossRef]
10. Hou, Y.; Wang, Z.; Wang, J. Optimization of coordinated control for rapid response to electric power in supercritical CFB unit. *Therm. Power Gener.* **2019**, *48*, 33–37.
11. Lockwood, T. *Techno-Economic Analysis of PC Versus CFB Combustion Technology*; IEA Clean Coal Centre: London, UK, 2013.
12. Wang, L.; Yang, D.; Shen, Z.; Mao, L.; Long, J. Thermal-hydraulic calculation and analysis of a 600 MW supercritical circulating fluidized bed boiler with annular furnace. *Appl. Therm. Eng.* **2016**, *95*, 42–52. [CrossRef]
13. Lyu, J.; Shang, M.; Ke, X.; Zhou, T.; Huang, Z.; Zhang, H.; Zhang, M.; Zhang, Y.; Wu, Y.; Yue, G. Powdered coal circulating fluidized bed combustion technology. *J. China Coal Soc.* **2023**, *48*, 430–437.
14. Lyu, J.; Tong, B.; Dong, J.; Wu, Y.; Pang, K. One-dimensional model of coal gangue combustion in circulating flu-idized bed boiler. *J. China Coal Soc.* **2016**, *41*, 2418–2425.
15. Liu, W.; Ouyang, Z.; Song, W.; Zhu, S.; Li, S. Experimental research on combustion characteristics and NO_x emission of three kinds of solid fuels preheated by a self-preheating burner. *Energy Fuels* **2019**, *33*, 8483–8490. [CrossRef]
16. Liu, W.; Ouyang, Z.; Cao, X.; Na, Y. Experimental research on flameless combustion with coal preheating technology. *Energy Fuels* **2018**, *32*, 7132–7141. [CrossRef]
17. Zhu, S.; Lyu, Q.; Zhu, J.; Liang, C. Experimental study on NO_x emissions of pulverized bituminous coal combustion preheated by a circulating fluidized bed. *J. Energy Inst.* **2019**, *92*, 247–256. [CrossRef]
18. Zhu, S.; Lyu, Q.; Zhu, J.; Li, J.; Man, C.; Liu, W. Preheating combustion characteristics of ultra-low volatile carbon-based fuel. *J. Therm. Sci.* **2019**, *28*, 772–779. [CrossRef]
19. Zhu, S.; Hui, J.; Lyu, Q.; Ouyang, Z.; Liu, J.; Zhu, J.; Zeng, X.; Zhang, X.; Ding, H.; Liu, Y.; et al. Experimental study on pulverized coal combustion preheated by a circulating fluidized bed: Preheating characteristics for peak shaving. *Fuel* **2022**, *324*, 124684. [CrossRef]
20. Tang, Z.; Song, G.; Yang, X.; Ji, Z. Research on Combustion and Emission Characteristics of Circulating Fluidized Bed During Load Changes. *J. Energy Inst.* **2022**, *105*, 334–341. [CrossRef]
21. Li, R.; Wei, K.; Huang, Q.; Xia, H. A novel method for precise measurement of unburnt carbon in boiler fly ash by ECSA based on TG-MS. *Fuel* **2020**, *264*, 116849. [CrossRef]
22. Xia, H.; Wei, K. Equivalent characteristic spectrum analysis in TG-MS system. *Thermochim. Acta.* **2015**, *602*, 15–21. [CrossRef]
23. Li, R.; Huang, Q.; Wei, K.; Xia, H. Quantitative analysis by thermogravimetry-mass spectrum analysis for reactions with evolved gases. *J. Vis. Exp.* **2018**, *140*, 58233.
24. Ying, Z.; Zheng, X.; Cui, G. Pressurized oxy-fuel combustion performance of pulverized coal for CO_2 capture. *Appl. Therm. Eng.* **2016**, *99*, 411–418. [CrossRef]
25. Wang, C.; Lei, M.; Yan, W.; Wang, S.; Jia, L. Combustion characteristics and ash formation of pulverized coal under pressurized oxy-fuel conditions. *Energy Fuels* **2011**, *25*, 4333–4344. [CrossRef]
26. Zheng, H.; Huang, Y.; Zeng, H.; Zhao, X.; Feng, Z.; Hou, R.; Bai, Z. Effects of alkali and alkaline metal on combustion characteristics of coal pyrolysis semi-coke. *Clean Coal Technol.* **2021**, *27*, 62–67.
27. Wang, Y.; Zhang, X.; Zhang, H.; Sasaki, K. Effects of temperature gradient and particle size on self-ignition temperature of low-rank coal excavated from inner Mongolia, China. *R. Soc. Open Sci.* **2019**, *6*, 190374. [CrossRef]

28. Luo, Z.; Yang, Y.; Ding, X.; Luo, C.; Zhang, F.; Zhang, M.; Shu, C.-M. Effects of particle size and oil-immersed concentration on the dust layer minimum ignition temperature and combustion characteristics of oil-immersed coal. *J. Therm. Anal. Calorim.* **2024**, *149*, 10297–10309. [CrossRef]
29. Mishra, D.; Azam, S. Experimental investigation on effects of particle size, dust concentration and dust-dispersion-air pressure on minimum ignition temperature and combustion process of coal dust clouds in a G-G furnace. *Fuel* **2018**, *227*, 424–433. [CrossRef]
30. Glushkov, D.; Kuznetsov, G.; Strizhak, P.; Sharypov, O. Ignition of a coal particle on a heated surface. *Solid Fuel Chem.* **2016**, *50*, 213–219. [CrossRef]
31. Petrukhin, N.; Grishin, N.; Sergeev, S. Ignition Delay Time an Important Fuel Property. *Chem. Technol. Fuels Oils* **2016**, *51*, 581–584. [CrossRef]
32. Salomatov, V.; Kuznetsov, G.; Syrodoy, S.; Gutareva, N. Ignition of coal-water fuel particles under the conditions of intense heat. *Appl. Therm. Eng.* **2016**, *106*, 561–569. [CrossRef]
33. Guo, F.; Zhong, Z. Optimization of the co-combustion of coal and composite biomass pellets. *J. Clean. Prod.* **2018**, *185*, 399–407. [CrossRef]
34. Zhou, Y.; Fan, D.; Duan, Q.; Fu, P. A new combustion index equation. *J. Chin. Soc. Power Eng.* **2007**, *3*, 381–384.
35. Jayaraman, K.; Kok, M.; Gokalp, I. Thermogravimetric and mass spectrometric (TG-MS) analysis and kinetics of coal-biomass blends. *Renew. Energy* **2017**, *101*, 293–300. [CrossRef]
36. Tong, W.; Liu, Q.; Ran, G.; Liu, L.; Ren, S.; Chen, L.; Jiang, L. Experiment and expectation: Co-combustion behavior of anthracite and biomass char. *Bioresour. Technol.* **2019**, *280*, 412–420. [CrossRef]
37. Pinchuk, V.; Sharabura, T.; Kuzmin, A.; Pinchuk, S. Engineering equations for determining coal-water fuel combustion stages. *J. Energy Inst.* **2020**, *93*, 1924–1933. [CrossRef]
38. Zhu, H.; Kai, S.; Zhang, Y.; Fang, S.; Wu, Y. The stage analysis and countermeasures of coal spontaneous combustion based on "five stages" division. *PLoS ONE* **2018**, *13*, e0202724. [CrossRef]
39. Li, Z.; Zhang, M.; Yang, Z.; Yu, J.; Liu, Y.; Wang, H. Division of coal spontaneous combustion stages and selection of indicator gases. *PLoS ONE* **2022**, *17*, e0267479. [CrossRef]
40. Zhang, H.; Cen, K.; Yan, J.; Ni, M. The fragmentation of coal particles during the coal combustion in a fluidized bed. *Fuel* **2002**, *81*, 1835–1840. [CrossRef]
41. Li, A.; Chi, Y. Thermal stress crushing theory of large carbon particles in fluidized bed combustion. *J. China Coal Soc.* **1998**, *23*, 4.
42. Zhang, S.; Hao, X.; Bo, F. Study on emission of nitrogen oxide in sintering process. *Res. Iron Steel* **2017**, *45*, 110–115.
43. Ouyang, Z.; Ding, H.; Liu, W.; Li, S.; Cao, X. Effect of the staged secondary air on NO_x emission of pulverized emi-coke flameless combustion with coal preheating technology. *Fuel* **2021**, *291*, 120137. [CrossRef]
44. Schafer, S.; Bonn, B. Hydrolysis of HCN as an important step in nitrogen oxide formation in fluidised combustion. Part l. Homogeneous reactions. *Fuel* **2000**, *79*, 1239–1246. [CrossRef]
45. Schafer, S.; Bonn, B. Hydrolysis of HCN as an important step in nitrogen oxide formation in fluidised combustion. Part II: Heterogeneous reactions involving limestone. *Fuel* **2002**, *81*, 1641–1646. [CrossRef]
46. Zhang, J.; Ye, L.; Ren, K.; Xu, R.; Teng, Z.; Zhu, J. Kinetic and experimental analysis of the effect of particle size on combustion performance of low-rank coals. *Fuel* **2023**, *349*, 128675. [CrossRef]
47. Yang, X.; Song, G.; Yang, Z.; Wang, C.; Ji, Z.; Zhang, X. Combustion and NO_x Emission Characteristics of Coal Slime Solid Waste at Different Feeding Positions. *J. Therm. Sci.* **2023**, *32*, 2351–2360. [CrossRef]
48. Ding, H.; Ouyang, Z.; Cao, X.; Zhu, S. Effects of the Secondary Air Excess Ratio on the Self-Preheating Combustion Characteristics and NO_x Emission of Semi-Coke. *J. Therm. Sci.* **2023**, *32*, 1263–1277. [CrossRef]
49. Song, G.; Xiao, Y.; Yang, Z.; Yang, X.; Lyu, Q. Experiment Study on the High-Temperature Thermal Treatment and Ultra-Low NO_x Control of Solid Waste Coal Slime in Circulating Fluidized Bed. *J. Therm. Sci.* **2022**, *31*, 2244–2251. [CrossRef]
50. Tang, Z.; Song, G.; Jiang, Y.; Yang, X.; Ji, Z.; Sun, L. Experimental Study on the Effect of Bed Material Amount and Fuel Particle Size on Load Change of Circulating Fluidized Bed. *J. Therm. Sci.* **2023**, *32*, 1758–1770. [CrossRef]

energies

MDPI

Article

Determining Pilot Ignition Delay in Dual-Fuel Medium-Speed Marine Engines Using Methanol or Hydrogen

Somayeh Parsa [1] and Sebastian Verhelst [1,2,*]

[1] Department of Electromechanical, Systems and Metal Engineering, Ghent University, Sint-Pietersnieuwstraat 41, 9000 Ghent, Belgium; somayeh.parsa@ugent.be
[2] Department of Energy Sciences, Lund University, Ole Römers väg 1, P.O. Box 118, SE-221 00 Lund, Sweden
* Correspondence: sebastian.verhelst@ugent.be

Abstract: Dual-fuel engines are a way of transitioning the marine sector to carbon-neutral fuels like hydrogen and methanol. For the development of these engines, accurate simulation of the combustion process is needed, for which calculating the pilot's ignition delay is essential. The present work investigates novel methodologies for calculating this. This involves the use of chemical kinetic schemes to compute the ignition delay for various operating conditions. Machine learning techniques are used to train models on these data sets. A neural network model is then implemented in a dual-fuel combustion model to calculate the ignition delay time and is compared using a lookup table or a correlation. The numerical results are compared with experimental data from a dual-fuel medium-speed marine engine operating with hydrogen or methanol, from which the method with best accuracy and fastest calculation is selected.

Keywords: ignition delay; chemical kinetics; ANN; machine learning; lookup table; dual fuel; multi-zone combustion model; methanol; hydrogen

1. Introduction

Maritime transport contributed nearly 3% of global anthropogenic CO_2 emissions in 2018 [1]. Within the EU, shipping industry accounted for an estimated 3 to 4% of total emissions, producing over 144 million tons of CO_2 in 2019 [2]. The maritime sector is expected to grow by between 25% and 180% by 2050 [2]. Projections indicate that that the carbon emissions caused by maritime transport could rise by up to 130% compared to 2008 levels by 2050. In response, the European Union has mandated reductions of 2% by 2025 and 80% by 2050 [1,3,4]. Therefore, urgent actions should be taken to meet the targets.

One of the promising ways to achieve these emission reduction targets is to transition to renewable fuels like hydrogen and methanol [5,6]. Gaseous hydrogen, as a clean energy carrier, is highly regarded as a promising fuel for the future, as it can be sourced from renewable source [7,8]. For marine applications, it is expected to be used where it outperforms the energy density of batteries, but where autonomy demands are still limited. If these demands become stricter, a denser energy carrier is needed, for which several hydrogen carriers are being considered. One of these is methanol [5,9,10]. Being liquid under ambient conditions, methanol facilitates easy handling and transportation; furthermore it is relatively straightforward to produce from biomass feedstocks and renewable electricity [11,12].

Hydrogen and methanol can serve as primary fuels in internal combustion engines (ICE). However, in compression ignition (CI) engines, the prime power plant for marine transportation, they cannot be used as such, due to their high autoignition temperature (reflected in their high octane numbers). A way around this is to use a pilot injection of a diesel-like fuel with low autoignition temperature (high cetane number), which then initiates combustion. Thus, methanol and hydrogen can be used in a dual-fuel mode. Hydrogen has a wide flammability range and fast flame speed, which makes it a promising solution for managing challenges associated with other gaseous fuels. Its application in dual-fuel diesel engines can improve thermal efficiency due to its faster flame velocity than other gaseous fuels [13]. Additionally, hydrogen does not contain structural carbon; it could thus reduce emissions of unburned hydrocarbons (UHC) and carbon monoxide (CO) in dual-fuel engines [7,14]. Methanol also demonstrates excellent engine performance with high efficiency and extremely low emissions compared to hydrocarbons such as petrol and diesel fuels [5,15,16], with its low flame temperatures leading to reduction in oxides of nitrogen (NO_x) and with no soot formation due to its carbon atom being bonded to an oxygen atom.

Various approaches are available for dual-fuel engines, primarily determined by the way hydrogen or methanol is injected into the engine. One of these methods is direct injection into the cylinder alongside a burning diesel jet [17,18], or they can be fumigated into the intake manifold at a single point or multiple points before reaching the intake valves [19,20]. In this study, port fuel injection (PFI) is focused on because it offers the most convenient solution for retrofitting engines. This is particularly important for the marine sector as the average age of vessels is over 20 years so one cannot solely rely on newbuilds to reduce the carbon emissions. The PFI method benefits from a low-pressure (and thus low cost) fuel circuit and requires very few engine modifications because methanol or hydrogen are introduced into the intake manifold [19,21]. This fumigation concept has been extensively studied and demonstrated over the past decade, but challenges persist. Hydrogen poses significant challenges with PFI, including issues like pre-ignition, knock, and backfiring due to its wide flammability range, low minimum ignition energy, and limited quenching distance. Furthermore, hydrogen displaces air in the intake, limiting the engine's power density [22]. For methanol, its high heat of vaporization can lead to potentially severe diesel knock under high load conditions [23]. The strong cooling caused by this high heat of evaporation also complicates ignition during cold starts, warming up, and low-load conditions [24,25]. Additionally, there is an increase in the levels of CO and THC [26] due to the fuel being premixed and thus entering combustion chamber crevices where a flame cannot propagate.

Conventional high-performance diesel engines normally have high valve overlap to ensure that residual gases are completely evacuated from the cylinder. Converting these engines to PFI dual-fuel operation introduces some challenges [27]. In methanol–diesel dual-fuel engines, some of the methanol–air mixture may escape into the exhaust during the valve overlap period, a phenomenon known as scavenging loss. This unburned fuel loss results in two main issues: first, it negatively impacts the engine's thermal efficiency; second, it leads to an increase in unburned methanol emissions. Zhenyu Sun et al. [27] found that valve timing affects methanol film formation and fuel scavenging losses. Therefore, it can significantly reduce thermal efficiency. They concluded that during the valve overlap period, unburned methanol escapes into the exhaust, resulting in reduced combustion efficiency. In hydrogen/diesel engines, hydrogen's low ignition energy and high diffusivity increase the risk of backfiring or early ignition during valve overlap. Studies show that high valve overlap makes this issue worse, but optimizing injection timing and valve overlap can help reduce the risk [28].

A comprehensive understanding of hydrogen/diesel and methanol/diesel co-combustion is vital to effectively addressing these challenges. One of the critical parameters for accurately simulating PFI dual-fuel combustion of hydrogen/diesel and methanol/diesel is the pilot ignition delay—the time between the start of pilot injection and the start of pilot combustion—as diesel is now injected into an air–fuel mixture with properties different from air. The results reported by Dierickx et al. [19], shown in Figure 1, illustrate how increasing the methanol energy fraction (MEF) significantly influences the ignition delay time of diesel in a PFI dual-fuel engine. As MEF increases, the start of combustion, shown by an increase in the pressure rise rate (marked with a red circle), occurs later in the cycle. This shift clearly reflects a longer ignition delay associated with higher levels of methanol substitution. If this time is incorrectly estimated, the whole combustion process will be calculated wrongly. Few studies have looked into how to calculate the ignition delay time in PFI dual-fuel engines operating on hydrogen or methanol under actual engine conditions.

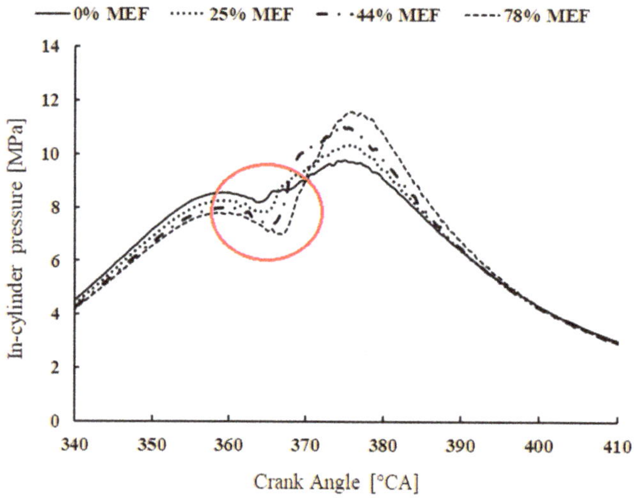

Figure 1. In-cylinder pressure for different methanol energy fractions (MEF) at 1500 rpm and BMEP of 1.23 MPa. Adapted from [19]. As MEF rises, the start of combustion, shown by the pressure rise rate increase (marked with a red circle), moves to a later point in the engine cycle.

In dual-fuel operation using hydrogen, there is no evaporative cooling, which influences the intake air temperature since hydrogen is usually introduced as a gas. Hydrogen's heat capacity at standard temperature (20 °C) and pressure (1 atm) is approximately 14 times greater than that of air, specifically 14.28 kJ/(kg K). This increased heat capacity of the mixture results in smaller rise in temperature while compressing [29,30]. Additionally, when hydrogen is added into the port, it displaces some of the intake air due to its low density, which reduces the volumetric efficiency compared to diesel operation and consequently lowers reduces the amount of oxygen in the intake mixture [28,31]. All these factors influence the temperature and the oxygen availability during diesel injection, affecting the ignition delay [30]. The chemical influence of premixed hydrogen or methanol on the ignition delay of a diesel pilot was investigated by Parsa et al. [32]. They concluded that premixed hydrogen did not notably impact the ignition delay of the pilot fuel, except when the percentage of hydrogen in the premixed fuel was significantly high. However, the ignition delay of the pilot fuel is greatly influenced by the presence of premixed methanol.

Introducing methanol into the incoming air in dual-fuel applications has been reported to prolong the ignition delay of the diesel pilot. This delay results from several elements. Firstly, methanol's high heat of vaporization [5,20] lowers the temperature of the intake air-fuel mix significantly, which reduces the temperature and pressure at the moment when diesel fuel starts being injected [19,33]. In addition, the higher heat capacity of the intake mixture also limits the rise in temperature as the mixture is compressed [33]. Additionally,

Yin et al. [34] found that at a methanol-to-air equivalence ratio of 0.1 and temperatures under 920 K, methanol slows down the initial chemical reaction rate of diesel autoignition. This effect has been linked to the temperature dependence of radical species conversion, particularly involving OH· and H_2O_2.

Few studies have analyzed the ignition delay of hydrogen/diesel and methanol/diesel dual-fuel engines under actual engine operation scenarios. Dhole et al. [35] conducted an experimental study on the combustion duration and ignition delay of a dual-fuel diesel engine using hydrogen, producer gas, and various mixtures of producer gas and hydrogen as secondary fuels with a diesel pilot. The experiments were performed on a 4-cylinder turbocharged and intercooled 62.5 kW generator set diesel engine at a constant speed of 1500 rpm. Their findings indicated that at low loads, replacing 30% (mass percentage) of the diesel fuel with hydrogen extended the combustion duration by 2.5 crank angle degrees (CA) and lengthened the ignition delay by 2° CA. However, at higher loads (80%) with 50% hydrogen substitution (mass percentage), both the ignition delay and combustion duration were reduced. The study also found that the ignition delay in dual-fuel engines is influenced not only by the type and concentration of gaseous fuels but also by the charge temperature, pressure, and oxygen content.

Some correlations have been proposed for estimating the ignition delay of methanol/diesel and hydrogen/diesel dual-fuel combustion. Dierickx et al. [36] proposed a correlation for the calculation of ignition delays in dual-fuel engines with hydrogen or methanol in a medium speed single cylinder engine. The newly developed methanol dual-fuel correlation incorporates the inhibition effect using the methanol-air equivalence ratio. Consequently, the ignition delay increases with a rising methanol-air equivalence ratio, resulting from an increased methanol content in the cylinder, besides temperature, pressure, and diesel equivalence ratio effects. In dual-fuel operation with hydrogen, this study discovered that the measured ignition delay was minimally affected by increasing the hydrogen energy fraction. However, they observed that ignition delay slightly decreased when they analyzed the temperature and pressure changes during the ignition delay and their impact on the various elements of the ignition delay correlations. Zong et al. [37] integrated a skeletal kinetic model developed by Xu et al. [38] for predicting the ignition delay of methanol/n-heptane into a diesel-methanol dual-fuel 3D CFD study. Also, Decan et al. [39] used tabulated ignition delays estimated by detailed chemistry schemes in the CFD simulation of a fumigated dual-fuel engine.

According to the literature review, an accurate estimation of the ignition delay in dual-fuel engines operating on hydrogen or methanol is necessary for any simulation of dual-fuel combustion, and it should be applicable as generally as possible. As discussed above, the ignition delay can be calculated in different ways: either through a correlation that typically incorporates some physics but is mostly phenomenological in nature, i.e., fitted to experimental data; or through chemical kinetic simulations. The latter are the best starting point, as they do not depend on engine-specific data. However, these require the selection of appropriate reaction mechanisms, that properly take the effects of both premixed fuel and pilot fuel into account. Also, as calculating the ignition delay time from chemical kinetics "on the fly", i.e., during engine combustion simulations, is computationally expensive, these mechanisms are typically used to generate a table beforehand, covering all conditions expected in engines. There are then various options on how to use this table during engine combustion simulations. It can be used to look up the relevant ignition delay time during engine combustion simulation, but with the recent improvements in machine learning, it could also be worthwhile to use machine learning to capture the table data in a mathematical formulation. Such an approach could potentially speed up the engine calculations as less time is needed for data retrieval.

To address these research gaps, the present study thus integrates ignition delay data from chemical kinetic mechanisms, calculated using Cantera over a wide range of operating conditions, into a dual-fuel multi-zone combustion model using two different methodologies: machine learning and look-up tables, and also compares these to the results when using the correlation from Dierickx et al. [36]. The accuracy of these methodologies is then evaluated against experimental data from a medium-speed marine engine. While traditional correlations require calibration based on engine specifications to yield accurate results, it is evaluated whether the methodologies proposed in this work allow using the calculated ignition delay times without the need for any pre-calibration. Additionally, as the data are derived from chemical kinetics, the chemical effect of the premixed fuel on the ignition delay of the pilot diesel can be further elaborated.

The main long-term objective of this work is to develop a more accurate 0D/1D dual-fuel combustion model (for which GT-Power V2023 serves as the simulation environment), serving as a virtual engine for future real engine development. Accurate prediction of ignition delay time is a critical step toward achieving full-cycle simulation accuracy. This study focuses specifically on improving the start of combustion by comparing ignition delay times predicted by different methodologies against experimental engine data under various operating conditions. While the start of combustion is crucial, accurately simulating it alone does not ensure improvements across the full combustion cycle. Other factors, such as laminar and turbulent burning velocities and flame surface area, also significantly impact the combustion process. However, this study focuses completely on establishing an accurate methodology for ignition delay time prediction, without attempting full combustion cycle evaluation, which will be tackled in follow-up work.

In the following sections, a brief explanation of the experimental setup used to collect the data and the study cases examined in this work will be provided. This will be followed by a detailed explanation of the numerical methodologies, and then the results and conclusions will be presented.

2. Experimental Setup

In the present study, data measured from the single-cylinder engine (SCE) shown in Figure 2 are utilized. The engine specifications can be found in Table 1. The SCE, developed in collaboration with Anglo Belgian Corporation, is located at the WTZ Roßlau laboratory. The experimental data was obtained from two different bore sizes: one with a bore size of 240 mm (SCE 1) and the other with a bore size of 256 mm (SCE 2).

Table 1. The engine specifications.

Engine Model Name	FM24
Cylinders	1
Compression ratio	12.1:1
Bore × stroke	240 × 290 mm (SCE1) 256 × 290 mm (SCE2)
Displacement volume	13.1 l (SCE1) and 14.9 l (SCE2)
Diesel injection system	Cam-driven Single Injection Pumps
Nominal power	200 kW (SCE1), 224 kW (SCE2)
Nominal speed	1000 rpm

In the single-cylinder engine, methanol was injected at a low pressure (below 1 MPa) into the intake port, whereas hydrogen injection was at a constant pressure, 0.05 MPa higher than the intake air pressure. The intake air temperature and pressure adjustment were based on the fuel and engine settings, as outlined in Table 2. An air intercooler positioned after the compressor, which controls the intake air pressure, regulates the intake air temperature. An intake air heater was employed to make final adjustments. The in-cylinder pressure sensor had a resolution of 1 crank angle degree. A pressure sensor was also used in the diesel high-pressure line of the pump-line-nozzle system. Both SCE1 and SCE2 are equipped with a single diesel injector that supplies diesel energy in both diesel and dual-fuel operations. The maximum diesel injection pressure is 100 MPa. Table 2 presents the different experimental cases whose data are used in this study. More details about the experimental setup and the cases can be found in the literature [36].

Figure 2. Picture of experimental setup. Methanol fuel supply system: 1: magnetic valve, 2: flexible hose pipe, 3: pressure sensor, 4: methanol injector, 5: surge tank [36].

Table 2. Experimental cases (MEF and HEF represent methanol energy fraction and hydrogen energy fraction, respectively [36]).

Campaign	Samples	Parameter Variations		
MEOH-SCE1	1	$P_{air} = 0.35$ MPa, $T_{air} = 324.15$ K,	MEF = 60%,	load = 75%
	2, 3	$P_{air} = 0.35$ MPa, $T_{air} = 348.15$ K,	MEF = 60, 70%,	load = 75%
	4, 5	$P_{air} = 0.35$ MPa, $T_{air} = 358.15$ K,	MEF = 70, 73%,	load = 75%
	6, 7	$P_{air} = 0.35$ MPa, $T_{air} = 324.15$ K,	MEF = 50, 60%,	load = 75%
	8, 9	$P_{air} = 3.5$ MPa, $T_{air} = 348.15$ K,	MEF = 50, 60%,	load = 75%
	10, 11	$P_{air} = 0.33$ MPa, $T_{air} = 324.15$ K,	MEF = 50, 60%,	load = 75%
	12, 13	$P_{air} = 0.35$ MPa, $T_{air} = 324.15$ K,	MEF = 50, 60%,	load = 75%
	14, 15	$P_{air} = 0.37$ MPa, $T_{air} = 324.15$ K,	MEF = 50, 60%,	load = 75%
MEOH-SCE2	1	$P_{air} = 0.37$ MPa, $T_{air} = 324.15$ K,	MEF = 46, 51%,	load = 75%
	2	$P_{air} = 0.36$ MPa, $T_{air} = 358.15$ K,	MEF = 50%,	load = 75%
	3	$P_{air} = 0.36$ MPa, $T_{air} = 348.15$ K,	MEF = 50%,	load = 75%
	4	$P_{air} = 0.31$ MPa, $T_{air} = 348.15$ K,	MEF = 45%,	load = 75%
	5	$P_{air} = 0.36$ MPa, $T_{air} = 348.15$ K,	MEF = 51%,	load = 75%
	6–8	$P_{air} = 0.19$ MPa, $T_{air} = 348.15$ K,	MEF = 50, 56 ,61%,	load = 75%
	9, 10	$P_{air} = 0.27$ MPa, $T_{air} = 348.15$ K,	MEF = 50, 56%,	load = 75%

Table 2. *Cont.*

Campaign	Samples	Parameter Variations		
H2-SCE1	1, 2	$P_{air} = 0.21$ MPa, $T_{air} = 309.15$ K,	HEF = 25, 40%,	load = 75%
	3, 4	$P_{air} = 0.21$ MPa, $T_{air} = 316.15$ K,	HEF = 25, 40%,	load = 75%
	5–8	$P_{air} = 0.2$ MPa, $T_{air} = 316.15$ K,	HEF = 25,35, 40,44%,	load = 75%
	9, 10	$P_{air} = 0.22$ MPa, $T_{air} = 316.15$ K,	HEF = 25, 40%,	load = 75%
	11–13	$P_{air} = 0.25$ MPa, $T_{air} = 316.15$ K,	HEF = 25, 40, 50%,	load = 75%
	14–16	$P_{air} =$ 0.26MPa, $T_{air} = 316.15$ K,	HEF = 25, 40, 50%,	load = 75%

3. Numerical Methodology

As mentioned previously, this study employs various methods to estimate the ignition delay time in dual-fuel marine engines. Initially, a constant volume batch reactor within the open-source Cantera code is used to calculate the ignition delay time of hydrogen/n-heptane and methanol/n-heptane mixtures under different operating conditions. Once the dataset is created, machine learning techniques including Artificial Neural Network (ANN) and Support Vector Regression (SVR) are applied in MATLAB R2023a to train the models. The trained neural network is then integrated into the multi-zone combustion model in GT-Power, a commercial 0D/1D engine simulation software. As an alternative method, the data from Cantera is also implemented as a lookup table within GT-Power. The following sections explain the methodologies in more detail.

3.1. Dataset Generation Using Chemical Kinetic Mechanisms and Cantera

First, the chemical kinetic mechanisms developed by Andrae et al. [40] and Liu et al. [41] are used in Cantera under various operating conditions to generate the datasets for methanol/diesel and hydrogen/diesel, respectively. It is worth mentioning that the ignition delay time is defined as the time at which the OH species concentration peaks. Four input parameters are considered: pressure (P), temperature (T), the overall equivalence ratio (φ), and the molar percentage of methanol or hydrogen in the mixture. Table 3 shows the range of input data and the number of data points used for methanol/diesel and hydrogen/diesel blends.

Table 3. Data range used for training.

	P (MPa)	T (K)	φ	Molar Percentage	Number of Datapoints
Methanol/diesel	7–13	625–1800	0.5–3.5	0–95	5240
Hydrogen/diesel	4–10	625–1100	0.5–3.0	0–95	3485

In the following section, two machine learning methods employed in this work are explained in more detail.

3.2. Machine Learning Method

In the present work, two different methods, namely Artificial Neural Network (ANN) and Support Vector Regression (SVR), are introduced and used to train models on the ignition delay data generated as discussed in the previous section. This is done in MATLAB. These two different methods are then compared to determine which one performs better in predicting the targets.

Before the training starts, it is important to preprocess the inputs. Four types of inputs are used: pressure (P), temperature (T), overall equivalence ratio (φ), and methanol or hydrogen molar percentage in total fuel mass. Each input must be normalized separately to prevent any single feature from dominating due to its scale, thereby improving the algorithm's performance. Normalization also facilitates faster convergence during the training process, especially in neural networks. In this study, the "mapminmax" method, defined by Equation (1), is employed to normalize input data. This technique, commonly used in the preprocessing stages of machine learning and neural networks, ensures equal contribution of all features to the training process and typically scales the data between −1 and 1 or 0 and 1 [42].

$$y_n = \frac{(y_{max} - y_{min})(x_i - x_{min})}{(x_{max} - x_{min})} + y_{min} \qquad y_{max} = 1, \qquad y_{min} = -1 \tag{1}$$

Here, x_{max} and x_{min} represent the maximum and minimum values of the actual data, respectively, while y_n shows the normalized parameter. x_i is the actual data to be scaled.

3.2.1. Artificial Neural Network (ANN)

Artificial Neural Networks (ANNs) are computational models that work similarly to the human brain. ANNs are used in a wide range of topics for classification, regression, pattern recognition, and more [43]. Figure 3 depicts the general structure of the ANN employed in the present work. The inputs to the ANN are pressure (P), temperature (T), overall equivalence ratio (φ), and the molar percentage of methanol or hydrogen in the mixture, and the output is the ignition delay. The ANN consists of neurons organized in layers: an input layer including ten neurons, one hidden layer including 12 neurons, and an output layer. The neurons of a layer are individually linked to each neuron of the next layer via a connection with a specific weight. When a neuron receives inputs from neurons of the prior layer, each input is multiplied by the related weight of the connection. The small random values are initially used for the weights. During training, these weights are revised using optimization algorithms to reach the minimum error in the network's predictions. Each neuron also has a related bias term. After adding the weighted inputs, the bias is added to this sum. This process leads to non-zero output values even when all inputs are zero. After the weighted sum and bias addition, the result is passed through a transfer function (like sigmoid, hyperbolic tangent sigmoid, or Rectified Linear Unit), which makes the network non-linear. The network requires this non-linearity to deal with complex data. Overall, the primary function of ANNs is to minimize the error between predicted and target values by modifying the constants within the transfer functions between layers. Thus, selecting the transfer function and the specific errors in the training process are very important.

In the present study, the Levenberg–Marquardt algorithm is employed as the training function, while the mean squared error (MSE) method is used to evaluate the network's performance. Data division is random. Random data division in ANNs refers to randomly splitting the dataset into different groups for training, validation, and testing. Consequently, diverse and representative data are used in the model's performance evaluation. It will decrease the possibility of overfitting and provide a more precise assessment of its generalization capability.

The data are divided into three randomly discretized sets, for training, test, and validation. Approximately 70% of the data are used to train the network, 15% for testing, and the final 15% for validation. Figures 4 and 5 present diagrams comparing the model output (y-axis) with the target values (x-axis) for these three datasets, as well as for the complete data set. In the diagrams, the more data points (shown as circles) lie on the "Fit" line, the better the model's performance. According to the figures, the results demonstrate that the prediction accuracy (R) for training, validation, and testing exceeds 99% for both methanol/diesel and hydrogen/diesel ignition delay time data, indicating excellent model performance. The mean square errors for the best performance are 0.00066 and 0.000042 for methanol/diesel and hydrogen/diesel, respectively.

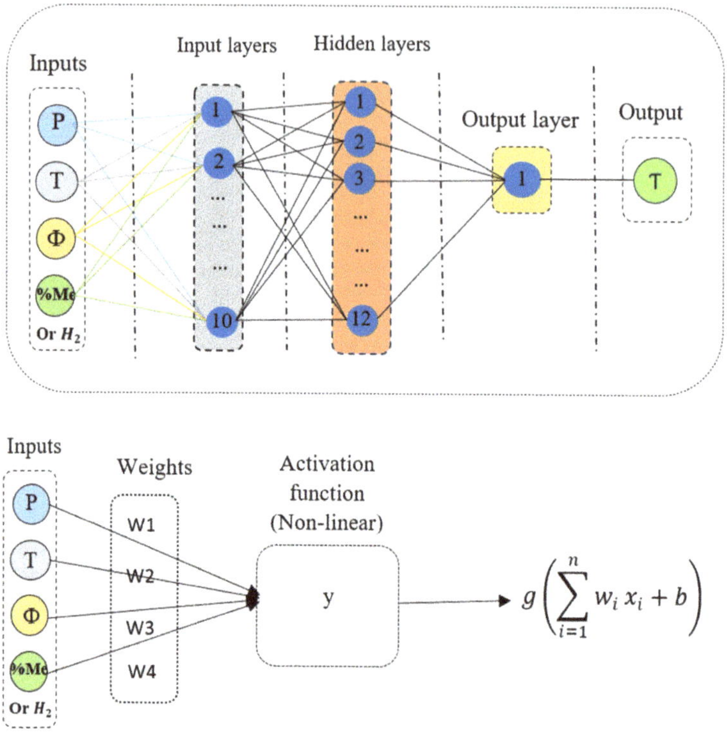

Figure 3. Artificial neural network architecture (w, x, b and g are weights, inputs, bias and activation function, respectively).

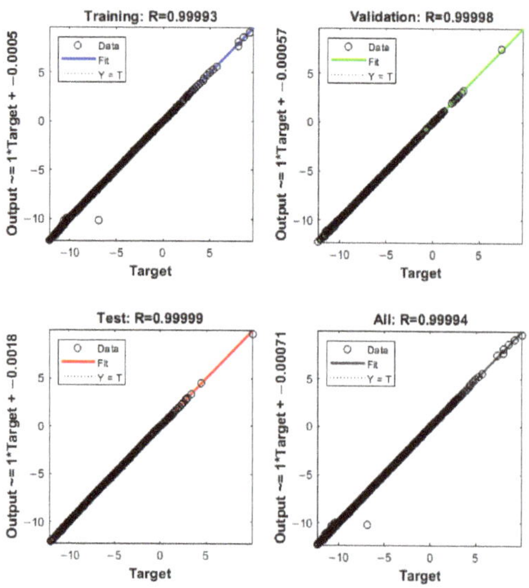

Figure 4. Regression results of the training for the methanol/diesel data set.

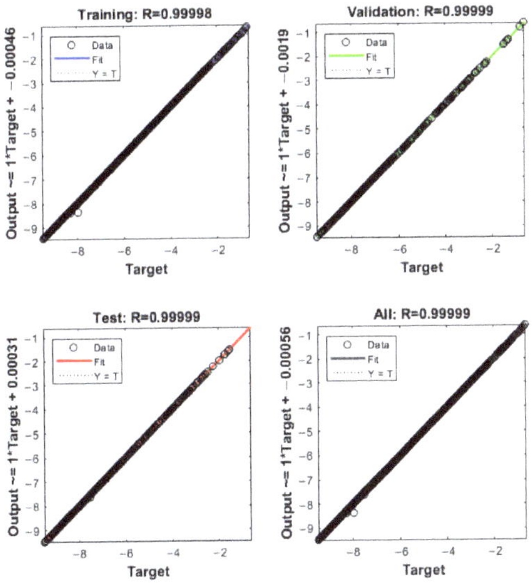

Figure 5. Regression results of ANN training for the hydrogen/diesel data set.

3.2.2. Support Vector Regression (SVR)

Support Vector Regression (SVR) is a machine learning algorithm based on so-called Support Vector Machines (SVM). It is widely used for regression tasks where a continuous target variable needs to be predicted [43]. SVR involves the principles of SVM, which is traditionally used for classification and regression problems. Traditional regression methods minimize the error directly, whereas SVR employs an ε-insensitive loss function. This means errors within a certain distance (ε) from the actual values are ignored, as illustrated in Figure 6. The model is only affected by data points that fall outside the ε margin (ξ and ξ^*). These crucial points, named support vectors, determine the position and direction of the regression line. A regularization parameter (C) is responsible for managing the compromise between the flatness of the regression function and the amount to which a deviation larger than ε is allowed. A larger C value leads to a more correct model that may overfit, while a smaller C allows for a more generalized model. SVR employs different kernels to deal with non-linear trends. Common kernels include linear kernel, polynomial kernel, and Gaussian kernel. This study trains the model using the Gaussian and polynomial kernels because the data show nonlinear trends.

Figures 7 and 8 show diagrams comparing the SVR-predicted values (y-axis) with the observed values (x-axis) using two different kernel functions: (a) Gaussian and (b) Polynomial. These diagrams illustrate how closely the predicted values match the observed data. The more blue points that lie on the red line ("Fit"), the more accurate the trained model is. Both kernels demonstrate good performance for both methanol/diesel and hydrogen/diesel. According to the results, error values for hydrogen/diesel models are 0.9895 for the Gaussian method and 0.9802 for the polynomial method. For methanol/diesel models, the error values are 0.9947 and 0.9939, respectively.

Both machine learning methods demonstrate high accuracy in predicting the targets. However, because the training time for the ANN method is approximately five times shorter, this method is selected for the remainder of this this work.

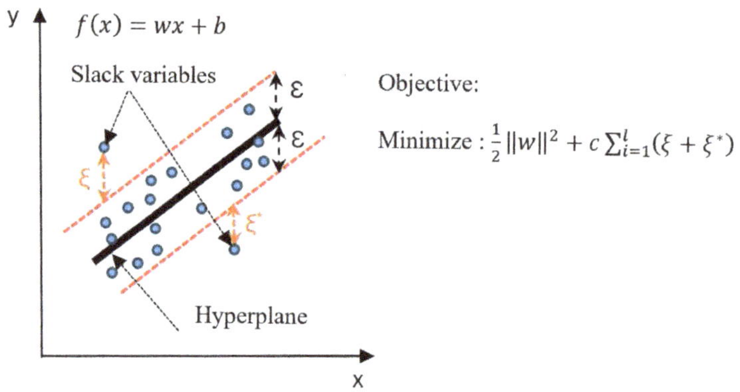

Figure 6. Linear SVR (w, b, c, ε and ξ, are weight, bias, regularization parameter, margin and slack variables, respectively). The red dotted lines indicate the boundaries of the ε-insensitive tube, within which no penalty is given for errors.

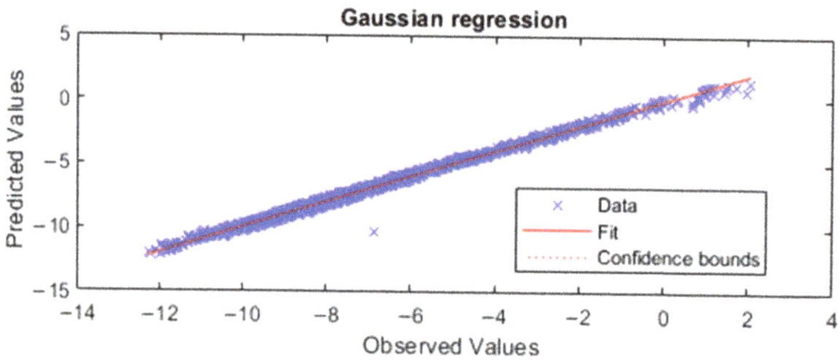

(a)

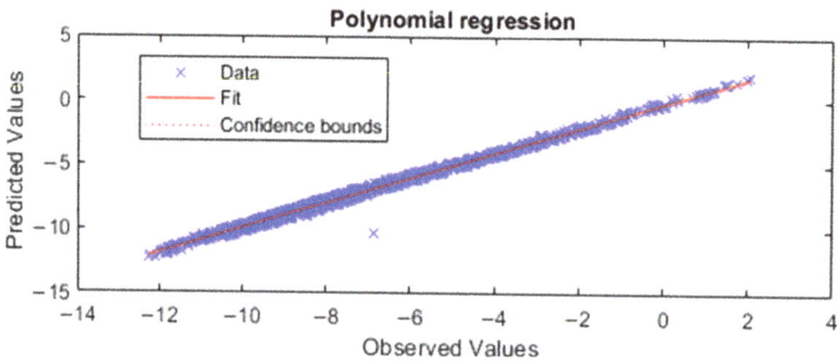

(b)

Figure 7. Regression results of SVR training for methanol/diesel dataset using two different kernels: (**a**) Gaussian, (**b**) polynomial.

It is worth mentioning that statistical processing can enhance the reliability of simulation-based studies by evaluating the impact of input parameters. Techniques like the Taguchi design improve efficiency by reducing the number of simulations needed. Similar methods have been successfully applied in engineering, such as in Milojević et al. [44], to optimize system performance.

Before explaining how the ANN method is integrated into the engine simulation software, the software used in this work is first introduced.

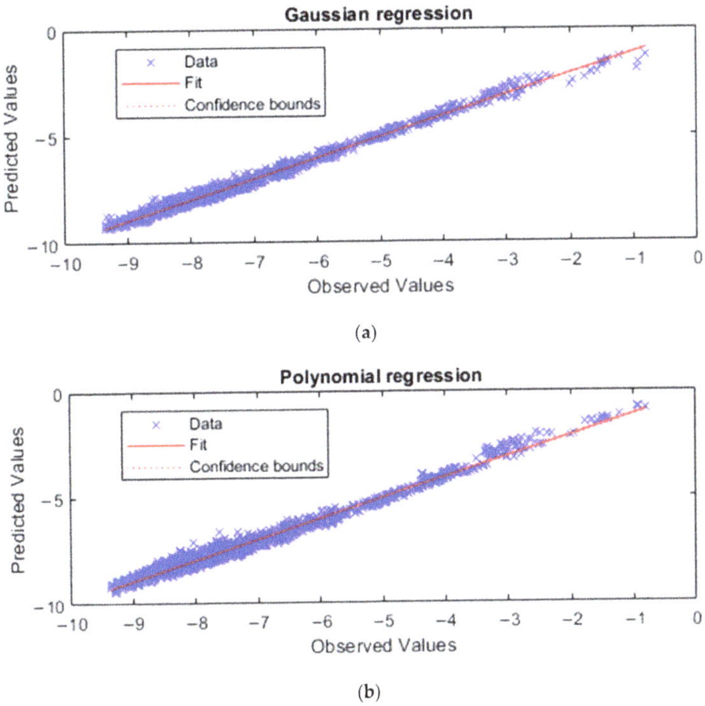

Figure 8. Regression results of SVR training for hydrogen/diesel dataset using two different kernels: (**a**) Gaussian, (**b**) polynomial.

3.3. *Multi-Zone Dual-Fuel Combustion Model in GT-Power*

This study employs GT-SUITE's predictive dual-fuel multi-zone combustion model to simulate a medium-speed dual-fuel engine. Researchers have previously used this model to simulate dual-fuel combustion of methane and diesel [45,46]. Three primary models are integrated into a multi-zone model to simulate dual-fuel combustion, to try to capture the physics of an autoigniting pilot spray that burns in a non-premixed mode and in turn ignites a premixed fuel-air mixture that burns in a flame propagation mode: a spray model (EngCylCombDIPulse), a transition function from spray to flame, and the flame model (SITurb). Different sub-models are employed by the DIPulse model to estimate spray entrainment, ignition delay, premixed combustion, and diffusion combustion, all referring to the pilot injection. The primary strategy involves tracking the fuel from injection, through evaporation, and mixing with the gas around the diesel spray, until it burns. The cylinder mixture is divided into three thermodynamic zones, each of which has its own species and temperature. The outer zone is called the main unburned zone, containing all cylinder content at intake valve closing time (IVC). The spray unburned zone is the inner zone, which includes injected fuel and entrained gas. There is a third zone between the mentioned zones called the burned zone, holding combustion products [46].

The Arrhenius equation presented in Equation (2) is used to estimate the ignition delay time of the blend in the pilot injection. The Ignition Delay Multiplier (C_{ign}) can be used to adjust the equation.

$$\tau_{\text{ign}} = C_{\text{ign}}\rho^{\ 1.5}\exp\left(\frac{3500}{T}\right)[O_2]^{-0.5} \quad (2)$$

The Livengood–Wu integral [47] is a commonly used method in ignition delay modeling. It relies on the conservation of delay principle, which states that the total ignition delay can be calculated by adding up the instantaneous ignition delays starting from the injection. Ignition, and thus the start of combustion, happens when the time integral equals one as is indicated in Equation (3).

$$\int_{t_{SOI}}^{t_{SOC}} \frac{1}{\tau_{ign}(t)} = 1 \tag{3}$$

Dierickx et al. [36] compared the GT-Power correlation and several other correlations for predicting the ignition delay of methanol–diesel and hydrogen–diesel co-combustion, using measured data from a medium-speed marine engine under real-world conditions. They found that the GT-Power methodology accurately predicts ignition delay in diesel-only applications but fails to do so when the methanol energy fraction increases.

The next section explains how the ANN was incorporated into this modeling framework.

3.4. Integration of the ANN into the Multi-Zone Dual-Fuel Combustion Model in GT-Power

To integrate the ANN from Section 3.2.1. into the dual-fuel multi-zone combustion model in GT-Power, it is first converted into a mathematical formula using its weights (w), biases (b), and transfer functions (f) [42].

The equation that relates the input and output parameters can be obtained as follows:

$$z = f_{out}\left\{\begin{pmatrix} w_{11} & w_{12}\cdots & w_{1h}\end{pmatrix} \times f_{hidden}\left\{\begin{pmatrix} w_{11} & \cdots & w_{1i} \\ \vdots & \ddots & \vdots \\ w_{h1} & \cdots & w_{hi}\end{pmatrix} \times f_{input}\left\{\begin{pmatrix} w_{11} & \cdots & w_{14} \\ \vdots & \ddots & \vdots \\ w_{i1} & \cdots & w_{i4}\end{pmatrix} \times \begin{pmatrix} y_1 \\ y_2 \\ y_3 \\ y_4 \end{pmatrix} + b_i\right\} + b_h\right\} + b_O\right\} \tag{4}$$

where f_{out}, f_{hidden}, f_{input} are the transfer functions of the output layer, hidden layer, and input layer, respectively, and b_o, b_h, b_i are the biases of the output layer, hidden layer and input layer. The scaled output (z) should be de-normalized using Equation (1) to obtain the actual output. This involves using the maximum and minimum values of the target data during the descaling process.

GT-Power software facilitates the development of new models for parameters such as ignition delay and flame speed through user-defined codes. The DIPulse user code, explained in the previous section, can be employed to incorporate user-defined models for estimating ignition delay time. It provides access to pulse conditions such as pulse temperature, cylinder pressure, and pulse composition.

In this study, the user code is used to implement a trained neural network for calculating τ_{ign} in Equation (3), using the pulse conditions (P, T, φ, and methanol or hydrogen molar percentage) as inputs.

3.5. Look-Up Table and Correlation Methods

Another approach to incorporate the ignition delay data from Section 3.1. into the multi-zone combustion model is through the use of a look-up table. Initially, a look-up table is generated to cover a wide range of operating conditions that covers all scenarios occurring during the injection pulse. Subsequently, specific commands are implemented in the DIPulse user code to retrieve $\tau_{ign}(t)$ as per Equation (3) from the look-up table.

As a final alternative, the phenomenological correlation proposed by Dierickx et al. [36] is also employed in this work to estimate the ignition delay. They developed a new correlation for dual-fuel operation with hydrogen or methanol in a medium speed single cylinder engine. Their approach emphasizes the importance of incorporating additional terms, such as the methanol equivalence ratio, to accurately describe methanol's inhibition effect. More detailed information about their correlation can be obtained from their paper [36].

4. Results and Discussions

The mentioned methods are now used to predict the ignition delay after being integrated in a multi-zone combustion model and the results are compared to the experimental data from the medium-speed dual-fuel engine as listed in Section 2. The final goal is to see which method most accurately predicts the ignition delay in real-world engine conditions.

4.1. *Methanol/Diesel*

Figures 9 and 10 compare the ignition delay predictions for the SCE1 and SCE2 methanol/diesel dual-fuel engine data points (as listed in Table 2) using the different methodologies. It should be mentioned that the measure ignition delay is derived from in-cylinder pressure data and is defined as the crank angle at which 2% of the total fuel mass has been burned. This point is commonly used as a reliable indicator of the start of combustion (SOC), as it captures the transition from fuel injection to the onset of significant heat release. To determine this, the apparent heat release rate is first calculated from the pressure trace using the first law of thermodynamics, and then integrated to obtain the cumulative heat release. The resulting mass fraction burned curve allows identification of the 2% burn point, which provides a consistent basis for comparing ignition delays under varying engine operating conditions. The gray area represents the measurement resolution of 1 crank angle degree. The methanol energy fraction varies from 50% to 75% for the SCE1 cases and from 46% to 56% for the SCE2 cases. It can be seen from both figures that the ignition delay of the pilot diesel is significantly influenced by increasing the methanol percentage in the premixed fuel under real engine conditions, which is in good agreement with the findings of previous works [32]. Ignition delay is represented in crank angle in the figures. It is worth mentioning that one crank angle degree at 1000 RPM is equal to 166.7 microseconds.

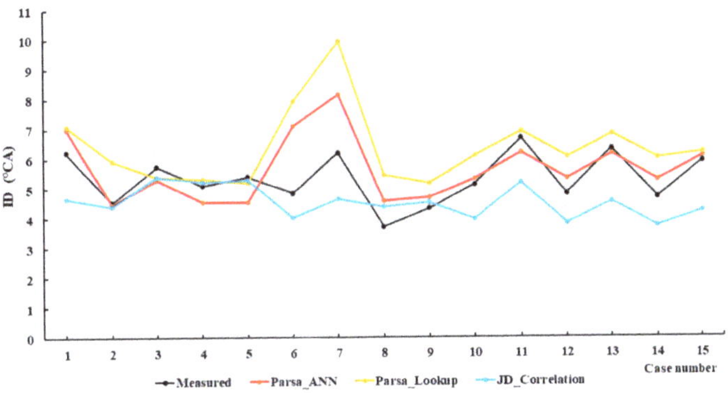

Figure 9. Comparison of results from different methodologies with measured data (SCE1). The gray area represents the measurement resolution of 1 crank angle degree.

According to Figure 9, the results from the ANN show more accurate predictions of ignition delay compared to other methods. The average relative errors of the ANN, correlation, and lookup table methods are 13%, 17%, and 23%, respectively. Furthermore, the ANN method has the lowest RMSE, approximately 0.91, compared to around 1.09 and 1.52 for the correlation and lookup table methods, respectively. In addition, as shown in Table 4, the ANN method has a lower mean squared relative error (MSRE) of 0.03. The MSRE is 0.04 for the correlation method and 0.09 for the lookup table.

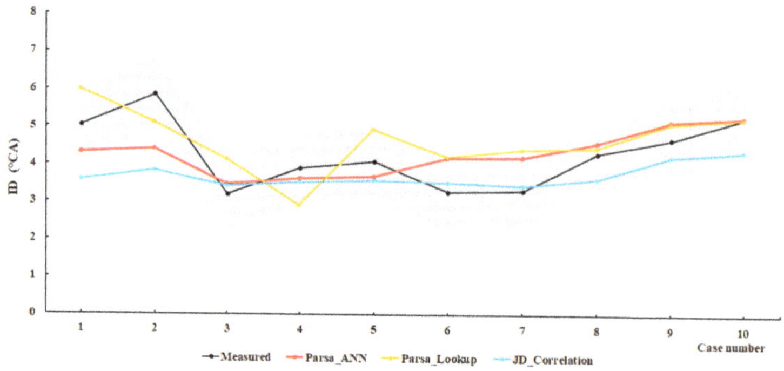

Figure 10. Comparison of results from different methodologies with measured data (SCE2). The gray area represents the measurement resolution of 1 crank angle degree.

Table 4. Comparison of RMSE and MSRE for the different methodologies.

Engine	Methodology	Root Mean Square Error (RMSE)	Mean Squared Relative Error (MSRE)
SCE1	Parsa_ANN	0.9127	0.0314
	Parsa_LookUP	1.5192	0.0894
	JD_Correlation	1.0909	0.0366
SCE2	Parsa_ANN	0.6939	0.1640
	Parsa_LookUP	0.7927	0.2115
	JD_Correlation	0.8996	0.1733

Figure 10 demonstrates the ignition delay predictions of the different methods for the SCE2 cases. It can be seen that here, too, that the ANN method predicts the ignition delay time more accurately compared to the other two methods. The average relative error of the ANN method is about 14%, which is lower than that of the correlation and lookup table methods, at 15% and 18.17%, respectively. Moreover, Table 4 shows that the ANN method has the lowest RMSE and MSRE, at approximately 0.7 and 0.2, respectively. Additionally, in terms of running time, the ANN and correlation methods are about 40% faster than using the lookup table.

The low accuracy of the lookup table method can be attributed to several factors: the four input data points required for prediction, the strongly non-linear trend of the ignition delay time, and the interpolation technique used. The interpolation method and the number of data points used to create the lookup table are critical factors affecting its accuracy in such situations. However, increasing the size of the lookup table and employing more complex interpolation methods will result in longer running times.

It can be concluded that the ANN methodology is the most accurate for predicting the ignition delay time of a methanol/diesel dual-fuel engine under real engine conditions. Since the data is derived from chemical kinetics mechanisms, developing a more accurate mechanism can potentially further enhance this method's accuracy.

4.2. Hydrogen/Diesel

Figure 11 compares the ignition delay predictions for the hydrogen/diesel operation using two different methods, ANN and correlation, against the experimental data for SCE1. There are 12 cases, and the hydrogen energy fraction (HEF) changes between 25% and 50%. The results from both methods indicate that the ignition delay of the diesel pilot is not considerably affected by the percentage of hydrogen in the premixed fuel under real-world engine conditions. This finding aligns with results from other literature [32].

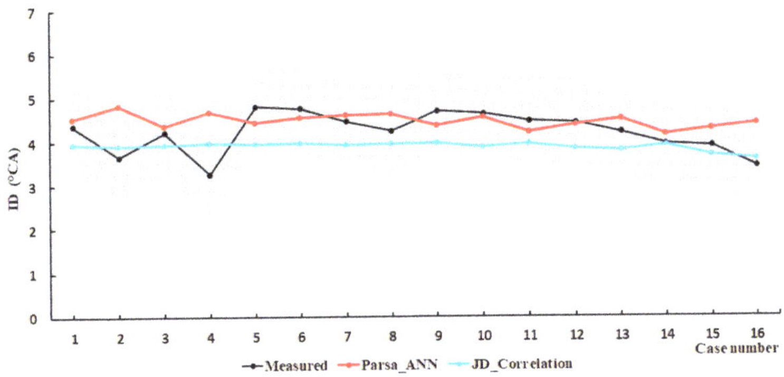

Figure 11. Comparison of results from ANN and correlation methods with measured data (SCE1). The gray area represents the measurement resolution of 1 crank angle degree.

The average relative error of both methods is about 10.9%. Furthermore, the RMSE is 0.57 for the ANN method and 0.53 for the correlation method. Both methods show good accuracy in predicting the ignition delay time of the diesel pilot in the presence of a hydrogen/air mixture. The ANN and correlation methods both show the same running time.

5. Conclusions

Accurate estimation of the ignition delay time of the pilot fuel is essential for simulating methanol/diesel and hydrogen/diesel dual-fuel combustion. There has been insufficient investigation into calculating this ignition delay under real engine conditions. This study aimed to determine the most accurate method for estimating ignition delay times in methanol/diesel and hydrogen/diesel dual-fuel engines under real world conditions. Data were used from a medium speed single cylinder setup running in relevant conditions, to serve as validation data. Unlike traditional correlations that require calibration based on specific engine parameters to produce accurate results, the proposed methodologies in this work can principally be used to calculate ignition delay in any dual-fuel engine. For the SCE1 dataset, the ANN achieved an RMSE of 0.91, outperforming the correlation method (RMSE = 1.09) and the lookup table method (RMSE = 1.52). Similarly, for the SCE2 dataset, the ANN method achieved the lowest RMSE of 0.7. For hydrogen/diesel operation, both ANN and correlation methods achieved good accuracy, with average relative errors of ~10.9% and RMSE values of 0.57 and 0.53, respectively. Additionally, the ANN method is approximately 40% faster than the lookup table methodology in terms of running time.

Author Contributions: S.P.: Writing—original draft, Visualization, Validation, Methodology, Investigation, Conceptualization. S.V.: Writing—review and editing, Supervision, Project administration, Funding acquisition. All authors have read and agreed to the published version of the manuscript.

Funding: This research was funded by the Flemish Agency for Innovation (VLAIO) through the inter cluster Strategic Basic Research project (cSBO) known as CHyPS (Clean Hydrogen-based Propulsion of Ships). The authors gratefully acknowledge this financial support.

Data Availability Statement: Some of the data presented in this study are available on request from the authors.

Conflicts of Interest: The authors declare no conflict of interest.

References

1. Faber, J.; Hanayama, S.; Zhang, S.; Pereda, P.; Comer, B.; Hauerhof, E.; van der Loeff, W.S.; Smith, T.; Zhang, Y.; Kosaka, H. Reduction of GHG emissions from ships—Fourth IMO GHG study 2020—Final report. *IMO MEPC* **2020**, *75*, 15.
2. Dnv, G. *Maritime Forecast to 2050, Energy Transition Outlook 2019*; DNV GL: Høvik, Norway, 2019.
3. Fetting, C. *The European Green Deal*; ESDN Report; ESDN Office: Vienna, Austria, 2020; p. 2.
4. Hedberg, A. *The European Green Deal: How to Turn Ambition into Action*; EPC Discussion Paper 04 November 2021; EPC: Brussels, Belgium, 2021.
5. Verhelst, S.; Turner, J.W.; Sileghem, L.; Vancoillie, J. Methanol as a fuel for internal combustion engines. *Prog. Energy Combust. Sci.* **2019**, *70*, 43–88. [CrossRef]
6. Huang, Z. Fuel blend combustion for decarbonization. *Proc. Combust. Inst.* **2024**, *40*, 105776. [CrossRef]
7. Hosseini, S.H.; Tsolakis, A.; Alagumalai, A.; Mahian, O.; Lam, S.S.; Pan, J.; Peng, W.; Tabatabaei, M.; Aghbashlo, M. Use of hydrogen in dual-fuel diesel engines. *Prog. Energy Combust. Sci.* **2023**, *98*, 101100. [CrossRef]
8. Verhelst, S. Recent progress in the use of hydrogen as a fuel for internal combustion engines. *Int. J. Hydrogen Energy* **2014**, *39*, 1071–1085. [CrossRef]
9. Olah, G.A. After oil and gas: Methanol economy. *Catal. Lett.* **2004**, *93*, 1–2. [CrossRef]
10. Tian, Z.; Wang, Y.; Zhen, X.; Liu, Z. The effect of methanol production and application in internal combustion engines on emissions in the context of carbon neutrality: A review. *Fuel* **2022**, *320*, 123902. [CrossRef]
11. Gautam, P.; Upadhyay, S.N.; Dubey, S. Bio-methanol as a renewable fuel from waste biomass: Current trends and future perspective. *Fuel* **2020**, *273*, 117783. [CrossRef]
12. Lonis, F.; Tola, V.; Cau, G. Assessment of integrated energy systems for the production and use of renewable methanol by water electrolysis and CO_2 hydrogenation. *Fuel* **2021**, *285*, 119160. [CrossRef]
13. Geo, V.E.; Nagarajan, G.; Nagalingam, B. Experimental investigations to improve the performance of rubber seed oil–fueled diesel engine by dual fueling with hydrogen. *Int. J. Green Energy* **2009**, *6*, 343–358. [CrossRef]
14. Parsa, S.; Jafarmadar, S.; Neshat, E. Application of waste heat in a novel trigeneration system integrated with an HCCI engine for power, heat and hydrogen production. *Int. J. Hydrogen Energy* **2022**, *47*, 26303–26315. [CrossRef]
15. Basile, A.; Dalena, F. *Methanol: Science and Engineering*; Elsevier: Amsterdam, The Netherlands, 2017.
16. Agarwal, A.K.; Valera, H.; Pexa, M.; Čedík, J. *Methanol: A Sustainable Transport Fuel for CI Engines*; Springer Nature: Berlin/Heidelberg, Germany, 2021. [CrossRef]
17. Jiang, M.; Sun, W.; Guo, L.; Zhang, H.; Jia, Z.; Qin, Z.; Zhu, G.; Yu, C.; Zhang, J. Numerical optimization of injector hole arrangement for marine methanol/diesel direct dual fuel stratification engines. *Appl. Therm. Eng.* **2024**, *257*, 124456. [CrossRef]
18. Li, Z.; Wang, Y.; Yin, Z.; Geng, H.; Zhu, R.; Zhen, X. Effect of injection strategy on a diesel/methanol dual-fuel direct-injection engine. *Appl. Therm. Eng.* **2021**, *189*, 116691. [CrossRef]
19. Dierickx, J.; Verbiest, J.; Janvier, T.; Peeters, J.; Sileghem, L.; Verhelst, S. Retrofitting a high-speed marine engine to dual-fuel methanol-diesel operation: A comparison of multiple and single point methanol port injection. *Fuel Commun.* **2021**, *7*, 100010. [CrossRef]
20. Yao, C.; Pan, W.; Yao, A. Methanol fumigation in compression-ignition engines: A critical review of recent academic and technological developments. *Fuel* **2017**, *209*, 713–732. [CrossRef]
21. Saravanan, N.; Nagarajan, G. Performance and emission studies on port injection of hydrogen with varied flow rates with Diesel as an ignition source. *Appl. Energy* **2010**, *87*, 2218–2229. [CrossRef]
22. Fayaz, H.; Saidur, R.; Razali, N.; Anuar, F.S.; Saleman, A.; Islam, M. An overview of hydrogen as a vehicle fuel. *Renew. Sustain. Energy Rev.* **2012**, *16*, 5511–5528. [CrossRef]
23. Havemann, H.; Rao, M.; Natarajan, A.; Narasimhan, T. Alcohol in diesel engines. *J. Indian Inst. Sci.* **2008**, *88*, 820.
24. Zhang, Z.; Tsang, K.; Cheung, C.S.; Chan, T.L.; Yao, C. Effect of fumigation methanol and ethanol on the gaseous and particulate emissions of a direct-injection diesel engine. *Atmos. Environ.* **2011**, *45*, 2001–2008. [CrossRef]
25. Shi, M.; Wu, B.; Wang, J.; Jin, S.; Chen, T. Optimization of methanol/diesel dual-fuel engines at low load condition for heavy-duty vehicles to operated at high substitution ratio by using single-hole injector for direct injection of methanol. *Appl. Therm. Eng.* **2024**, *246*, 122854. [CrossRef]
26. Imran, A.; Varman, M.; Masjuki, H.H.; Kalam, M.A. Review on alcohol fumigation on diesel engine: A viable alternative dual fuel technology for satisfactory engine performance and reduction of environment concerning emission. *Renew. Sustain. Energy Rev.* **2013**, *26*, 739–751. [CrossRef]
27. Sun, Z.; Lian, Z.; Ma, J.; Wang, C.; Li, W.; Pan, J. Combustion and Emissions Optimization of Diesel–Methanol Dual-Fuel Engine: Emphasis on Valve Phasing and Injection Parameters. *Processes* **2025**, *13*, 1183. [CrossRef]
28. Verhelst, S.; Wallner, T. Hydrogen-fueled internal combustion engines. *Prog. Energy Combust. Sci.* **2009**, *35*, 490–527. [CrossRef]
29. Liew, C.; Li, H.; Nuszkowski, J.; Liu, S.; Gatts, T.; Atkinson, R.; Clark, N. An experimental investigation of the combustion process of a heavy-duty diesel engine enriched with H2. *Int. J. Hydrogen Energy* **2010**, *35*, 11357–11365. [CrossRef]

30. Chintala, V.; Subramanian, K. A comprehensive review on utilization of hydrogen in a compression ignition engine under dual fuel mode. *Renew. Sustain. Energy Rev.* **2017**, *70*, 472–491. [CrossRef]
31. Karagöz, Y.; Sandalcı, T.; Yüksek, L.; Dalkılıç, A.S.; Wongwises, S. Effect of hydrogen–diesel dual-fuel usage on performance, emissions and diesel combustion in diesel engines. *Adv. Mech. Eng.* **2016**, *8*, 1687814016664458. [CrossRef]
32. Parsa, S.; Verhelst, S. *Numerical Investigation of the Ignition Delay and Laminar Flame Speed for Pilot-Ignited Dual Fuel Engine Operation with Hydrogen or Methanol*; 0148-7191; SAE Technical Paper; SAE: Warrendale, PA, USA, 2023.
33. Coulier, J.; Verhelst, S. Using alcohol fuels in dual fuel operation of compression ignition engines: A review. In Proceedings of the 28th CIMAC World Congress on Combustion Engine, Helsinki, Finland, 6–10 June 2016; pp. 1–12.
34. Yin, Z.; Yao, C.; Geng, P.; Hu, J. Visualization of combustion characteristic of diesel in premixed methanol–air mixture atmosphere of different ambient temperature in a constant volume chamber. *Fuel* **2016**, *174*, 242–250. [CrossRef]
35. Dhole, A.; Yarasu, R.; Lata, D. Investigations on the combustion duration and ignition delay period of a dual fuel diesel engine with hydrogen and producer gas as secondary fuels. *Appl. Therm. Eng.* **2016**, *107*, 524–532. [CrossRef]
36. Dierickx, J.; Mattheeuws, L.; Christianen, K.; Stenzel, K.; Verhelst, S. Evaluation and extension of ignition delay correlations for dual-fuel operation with hydrogen or methanol in a medium speed single cylinder engine. *Fuel* **2023**, *345*, 128254. [CrossRef]
37. Zang, R.; Yao, C. Numerical study of combustion and emission characteristics of a diesel/methanol dual fuel (DMDF) engine. *Energy Fuels* **2015**, *29*, 3963–3971. [CrossRef]
38. Xu, H.; Yao, C.; Xu, G. Chemical kinetic mechanism and a skeletal model for oxidation of n-heptane/methanol fuel blends. *Fuel* **2012**, *93*, 625–631. [CrossRef]
39. Decan, G.; Lucchini, T.; D'Errico, G.; Verhelst, S. A novel technique for detailed and time-efficient combustion modeling of fumigated dual-fuel internal combustion engines. *Appl. Therm. Eng.* **2020**, *174*, 115224. [CrossRef]
40. Andrae, J.C.; Head, R. HCCI experiments with gasoline surrogate fuels modeled by a semidetailed chemical kinetic model. *Combust. Flame* **2009**, *156*, 842–851. [CrossRef]
41. Liu, S.; Sun, T.; Zhou, L.; Jia, M.; Zhao, W.; Wei, H. A new skeletal kinetic model for methanol/n-heptane dual fuels under engine-like conditions. *Energy* **2023**, *263*, 125648. [CrossRef]
42. Lawal, A.I. An artificial neural network-based mathematical model for the prediction of blast-induced ground vibration in granite quarries in Ibadan, Oyo State, Nigeria. *Sci. Afr.* **2020**, *8*, e00413. [CrossRef]
43. Shahpouri, S.; Norouzi, A.; Hayduk, C.; Fandakov, A.; Rezaei, R.; Koch, C.R.; Shahbakhti, M. Laminar flame speed modeling for low carbon fuels using methods of machine learning. *Fuel* **2023**, *333*, 126187. [CrossRef]
44. Milojević, S.; Savić, S.; Mitrović, S.; Marić, D.; Krstić, B.; Stojanović, B.; Popović, V. Solving the problem of friction and wear in auxiliary devices of internal combustion engines on the example of reciprocating air compressor for vehicles. *Teh. Vjesn.* **2023**, *30*, 122–130.
45. Wenig, M.; Roggendorf, K.; Fogla, N. Towards predictive dual-fuel combustion and prechamber modeling for large two-stroke engines in the scope of 0D/1D simulation. In Proceedings of the 29th CIMAC World Congress on Combustion Engine Technology, Vancouver, BC, Canada, 10–14 June 2019; pp. 10–14.
46. Millo, F.; Accurso, F.; Piano, A.; Caputo, G.; Cafari, A.; Hyvönen, J. Experimental and numerical investigation of the ignition process in a large bore dual fuel engine. *Fuel* **2021**, *290*, 120073. [CrossRef]
47. Livengood, J.; PC, W. Correlation of autoignition phenomena in internal combustion engines and rapid compression machines. *Symp. (Int.) Combust.* **1955**, *5*, 347–356. [CrossRef]

energies

MDPI

Article

Determination of the Diffusion Coefficients of Binary CH_4 and C_2H_6 in a Supercritical CO_2 Environment (500–2000 K and 100–1000 atm) by Molecular Dynamics Simulations

Chun-Hung Wang [1,2], K. R. V. Manikantachari (Raghu) [3], Artëm E. Masunov [2,4] and Subith S. Vasu [3,*]

1 Department of Science, Northland Pioneer College, Little Colorado Campus, Winslow, AZ 86047, USA; chun-hung.wang@npc.edu
2 NanoScience Technology Center, University of Central Florida, Orlando, FL 32826, USA; amasunov@ucf.edu
3 Center for Advanced Turbomachinery and Energy Research (CATER), Department of Mechanical and Aerospace Engineering, University of Central Florida, Orlando, FL 32816, USA
4 School of Modeling, Simulation, and Training, University of Central Florida, Orlando, FL 32826, USA
* Correspondence: subith@ucf.edu

Abstract: The self-diffusion coefficients of carbonaceous fuels in a supercritical CO_2 environment provide transport information that can help us understand the Allam Cycle mechanism at a high pressure of 300 atm. The diffusion coefficients of pure CO_2 and binary CO_2/CH_4 and CO_2/C_2H_6 at high temperatures (500 K~2000 K) and high pressures (100 atm~1000 atm) are determined by molecular dynamics simulations in this study. Increasing the temperature leads to an increase in the diffusion coefficient, and increasing the pressure leads to a decrease in the diffusion coefficients for both methane and ethane. The diffusion coefficient of methane at 300 atm is approximately 0.012 cm^2/s at 1000 K and 0.032 cm^2/s at 1500 K. The diffusion coefficient of ethane at 300 atm is approximately 0.016 cm^2/s at 1000 K and 0.045 cm^2/s at 1500 K. The understanding of diffusion coefficients potentially leads to the reduction in fuel consumption and minimization of greenhouse gas emissions in the Allam Cycle.

Keywords: diffusion coefficient; supercritical CO_2; combustion; methane; ethane; molecular dynamics; ideal gas kinetic theory

1. Introduction

Diffusion is a fundamental process driven by the random thermal motion of molecules, which significantly impacts chemical and biological reaction rates [1]. It manifests differently across gas, liquid, and solid phases, and governs the overall rate of these processes. This study explores diffusion within supercritical fluid (SCF), which is a unique state where liquid and gas properties coexist. Such unique properties pose experimental and computational challenges [2]. SCF mass transfer processes hold immense potential in various fields: (i) the selective extraction of valuable components from food and environmental samples; and (ii) design and optimization of SCF reactors with precise control of temperature and pressure parameters to ensure proper mixing for efficient reactions. Supercritical carbon dioxide (sCO_2) is a promising tool in materials science and biological applications. Notably, the Allam Power Cycle, a recent breakthrough technology, utilizes sCO_2 for thermal energy conversion from carbon fuels [3]. The cycle transforms greenhouse gas CO_2 in traditional power plants into a reusable working fluid for spin turboexpanders and then generates electricity. The power generation cycle was developed in the last 10 years and has turned the air pollution problem into a solution [3]. However, further research is necessary to fully unlock its potential. Understanding diffusion is crucial for optimizing several key aspects of the technology of the Allam Cycle. First, diffusion dictates the mixing of fuels, like natural gas, with CO_2 at the molecular level. Improved mixing enhances combustion completeness. Second, efficient heat transfer from hot combustion products to CO_2 working fluid relies on diffusion within the gas stream. Third, diffusion influences the separation mechanism of CO_2 from flue gas after combustion. This study investigates the diffusion of hydrocarbon in sCO_2 due to its unique liquid-like and gaseous-like properties, which influences the diffusion coefficients of various hydrocarbons. Understanding the behavior of hydrocarbons in high-temperature and -pressure supercritical environments is vital [4].

The direct-fired sCO_2 cycle that utilizes CO_2 as the working fluid directly is an alternative to current electricity production. The Allam Cycle is a novel power generation technology and it converts carbon fuels into thermal energy and electricity while capturing the generated CO_2 and O_2. The cycle involves four stages. First, the cycle begins with the burning of fuels and oxygen. Fuels and pure O_2 combust within high-pressure and high-temperature sCO_2 in a combustor. The purpose of using pure O_2 is to replace air for cleaner combustion and to avoid nitrogen dilution. Second, the hot combustion products, CO_2 and water vapor, transfer heat to sCO_2 in a heat exchanger. Third, the heated CO_2 expands through a turbine and generates electricity. Lastly, cooled-down sCO_2 is collected in a compressor to complete the cycle. The relatively low critical temperature (37 °C) and critical pressure (8.4 MPa) of sCO_2 make it an attractive choice as an SCF solvent in the Allam Cycle. By elucidating the diffusion coefficients, we aim to gain a deeper understanding of the molecular transport phenomena within this cycle, ultimately paving the way for further power cycle design improvements. The Allam Cycle operates at high pressures (30–300 atm) and high temperatures (500–1150 °C) [3,4].

SCFs offer significant advantages due to their low viscosity and high solute diffusivity, leading to a multitude of industrial applications. The combination of CO_2 with hydrocarbons, like methane and ethane, in supercritical environments holds particular interest in various fields. For example, efficient carbon capture and storage are crucial for mitigating climate change. sCO_2 can effectively capture CO_2 emissions from industrial processes. sCO_2 can also be used to selectively separate and purify desired components from mixtures to avoid the usage of hazardous organic solvents. Furthermore, sCO_2 is a valuable solvent for extracting materials from various sources. These characteristics show the importance of understanding diffusion properties in the sCO_2 environment to optimize these industrial procedures. With its near-perfect CO_2 capture capability, the Allam Cycle requires stable diffusion in the combustor. This is the main reason for investigating the diffusion coefficients of methane and ethane in an sCO_2 environment to see how stable the diffusion of hydrocarbon is in such gaseous-like and liquid-like conditions. The Allam Cycle operates at a pressure range of 30–300 atm. At 300 atm, CO_2 exhibits optimal thermodynamic properties and translates to high thermal efficiency and reduced material selection challenge. In addition, the turbine inlet temperature can reach as high as 1150 °C, which is comparable to modern natural gas combined cycle plants. The broad pressure and temperature windows play significant roles in the Allam Cycle's performance and design [3,4], and this is the goal of our study.

This paper describes the results of the determination of the binary diffusion coefficients CO_2/CH_4 and CO_2/C_2H_6 by classical molecular dynamics (MD) simulations. CH_4 and C_2H_6 are prevalent hydrocarbon fuels employed in power cycles. In the past, Stubbs comprehensively reviewed MD and Monte Carlo (MC) simulations of supercritical H_2O as well as sCO_2 systems in detail, encompassing crucial aspects of the selection of force fields, the size of the simulation boxes, and the pH at different temperatures, pressures, and densities [5]. The thermophysical properties in the sCO_2 environment have been summarized in his review article. Several studied have employed various force fields to evaluate diffusion coefficients in systems relevant to this work. Aimoli et al. [6] and Moultos et al. [7] utilized the *tra*nsferable *p*otentials for *p*hase *e*quilibria (TraPPE) force field to assess the diffusion coefficients of pure CO_2, CH_4, and CO_2-H_2O mixtures across a broad range of temperatures (273–623 K) and pressures (0.1–100 MPa). Zhu et al. [8] employed the elementary physical model (EPM2) force field for a Gibbs ensemble MC simulation to determine diffusion coefficients at 304 K. The radial distribution function of CO_2 was analyzed to evaluate the validity of their simulations that the first-neighbor C-C distance is around 4.0 Å. Abbaspour and Nameni used a two-body Hartree–Fock dispersion-like potential to determine the self-diffusion coefficients of CO_2 and CO_2-CH_4 mixtures at approximately 300 K [9].

Several studies have investigated diffusion in an sCO_2 environment relevant to our research. Guevara-Carrion et al. performed CH_4 diffusion experiments with the Taylor dispersion technique and MD simulations [10] at temperatures in the range of 293–333 K and pressures in the range of 9.0–14.7 MPa, which is typical of Allam Cycle conditions. This group determined the self-diffusion, Fick, and Maxwell–Stefan coefficients of CH_4 diluted in an sCO_2 environment. Feng et al. collected a series of diffusion coefficients of *n*-hydrocarbon (C1–C14) in near-critical and sCO_2 environments (308 K and 323 K, respectively, at 10.5 MPa) from experiments and simulations at an infinite dilution. The ratios of carbon dioxide to hydrocarbon in their simulations are CO_2:CH_4 = 4000:110 and CO_2:C_2H_6 = 4000:58 to mimic the infinite dilution of hydrocarbon molecules [11]. Furthermore, in recent years, Asadov et al. experimentally studied the diffusion coefficients of CO_2-C_2H_6-heavy oil and CO_2-C_3H_8-heavy oil in an sCO_2 environment at temperatures in the range of 320–355 K and pressures in the range of 2–15 MPa [12]. It is important to note that these examples of previous studies focused on conditions below 623 K and 100 MPa.

More MD and MC simulations were applied to study various physical and chemical properties of hydrocarbons in an sCO_2 environment. Other MD studies included of the following: (1) the thermodynamic properties of CH_4 in an sCO_2 environment (CO_2:CH_4 = 400:100, 323 K at 9.94 MPa), such as potential energy and pressure, mean square force, and torque, were studied by Skarmoutsos et al. [13]; and (2) Gong et al. studied the evaporation mode transition of hydrocarbon fuels in subcritical and supercritical fluids (750–3600 K and 4–36 MPa) to gain insights into air–fuel mixing and combustion processes [14]. Other MC studies included the following: (1) the chemical potential of non-polar hydrocarbon in an sCO_2 environment (300–350 K at 10–500 bar) was examined by Chang [15]; and (2) the free energy of the solvation and structural properties of CH_4 in an sCO_2 environment (304 K at 80 and 200 atm) were studied by Tafazzoli et al. [16] These studies show the importance of providing thermodynamic parameters to understand hydrocarbon reactions in an sCO_2 environment.

Here, we focus on the self-diffusion coefficients of CO_2/CH_4 and CO_2/C_2H_6 mixtures in extreme temperatures (500–2000 K) and pressures (100–1000 atm) that were not easily studied experimentally and/or computationally in the past to mimic the condition of the Allam Cycle. Studying diffusion under such extreme conditions presents significant challenges for both experimental and theoretical approaches. From the perspective of the experimental challenges, specialized equipment is needed to handle such conditions safely and accurately. The high mobility of molecules and potential for chemical reactions at these extremes introduce noise to the measurements. Some materials used for the equipment might be unstable and potentially lead to contamination [3,4,17]. From the perspective of computer simulation challenges, the accuracy of MD simulations relies heavily on the selection of force fields for high-temperature and -pressure systems. In addition, diffusion is a relatively slow process. Simulating realistic timescales at high temperatures and pressures can be computationally prohibitive [18]. To the best of our knowledge, it is the first time extremely high temperatures (~2000 K) and pressures (~1000 atm) have been achieved by computer simulations, and this is systematically discussed under such conditions. Our simulations explore diffusion behavior at high temperatures and pressures relevant to the Allam Cycle. While these conditions might be challenging to achieve experimentally, our simulations provide valuable data for the understanding the system's behavior at its operating limits.

2. Computational Methods

The initial pure CO_2 diffusion coefficients at various temperatures and pressures were determined by ChemKin II Fortran version for comparison [19]. All the molecular dynamics simulations were performed using the LAMMPS program [20]. The force fields of CO_2 [21,22], CH_4 [23], and C_2H_6 [23,24] we selected were united-atom TraPPE because of its broad applicability and transferability to carbon dioxide and hydrocarbons. TraPPE has been successfully validated across various diffusion coefficient evaluations involving sCO_2 by comparison with available computational CO_2 self-diffusion coefficient results [7]. The cutoff radius of Coulomb and Lennard–Jones potentials was set to 14.0 Å. A time step of 1.0 fs was adopted. Simulations were performed under periodic boundary conditions in a cubic box measuring $35.0 \times 35.0 \times 35.0$ Å^3 and $40.0 \times 40.0 \times 40.0$ Å^3, with a particle–particle particle–mesh calculation in the k-space to better estimate long-range interactions. The initial geometry of all the simulations was generated by Packmol v20.15.0 [25]. A total of 77 CO_2 molecules and binary CO_2:CH_4 = 16:16, 32:32, or 42:42, and CO_2:C_2H_6 = 16:16 or 32:32 molecules were placed in the simulation box. Different numbers of molecules in simulation boxes of different sizes will lead to different pressures. The reason for using a one-to-one ratio of carbon dioxide versus hydrocarbon is to achieve high pressure in the simulations. The selected simulation temperatures are 750, 1000, 1250, 1500, 1750, and 2000 K for pure CO_2, and 300, 500, 1000, 1500, and 2000 K for both CO_2/CH_4 and CO_2/C_2H_6 mixtures. Energy minimization was the first step to stabilize the simulation box, which was generated originally using Packmol software. The criteria for stopping minimization were: a tolerance for energy of 10^{-5}, a tolerance for force of 10^{-7} kcal/mol/Å, max iterations of the minimizer of 5 million steps, and a max number of force/energy evaluations of 10 million steps. After energy minimization, a constant volume and temperature ensemble (NVT) were applied for the first 5 million steps (5 ns), followed by a constant volume and energy ensemble (NVE) for 5 million steps (5 ns) as the production run. A Nosé–Hoover thermostat was applied for the NVT [26,27]. The damping parameter was 200.0, meaning we relaxed the temperature at a timespan of 200 femtoseconds for the NVT. At least 5 to 10 independent test runs were averaged (with different initial geometries) to reduce fluctuations and then receive statistically reasonable results.

The calculations of the self-diffusion coefficients of the Einstein relation are based upon the mean squared displacement (MSD):

$$MSD \equiv \langle |r(t) - r(0)|^2 \rangle. \quad (1)$$

The equation describes the rate at which individual molecules move around due to random thermal motion within a medium. The slope of the MSD versus time is proportional to the diffusion coefficient of the diffusing atoms. The displacement of an atom is from its reference position, which is the original position at the time the simulation was started. MSD reflects the average squared distance a molecule travels over time, indicating how far it explores its surroundings. The MSD was collected in the NVE production runs. The unit of diffusion is cm^2/s.

3. Results and Discussion

Pure CO_2. In order to validate our simulation methods, we compared our 16, 49, 148, 288, 411, and 538 atm MD simulation results for diffusion coefficient D_{ij} to the ideal gas kinetic theory (IGKT, the selected pressures are 1, 49, 99, 147, and 296 atm) predictions by ChemKin II in Figure 1. In the MD simulation, the NVT and NVE are not able to assign the same pressure as the IGKT results. Therefore, the selection of pressures in the MD simulation should be as close to the IGKT pressure as possible. This comparison serves a two-fold purpose. Firstly, it ensures consistency with established principles of the IGKT. Secondly, it provides a reference point for interpreting the behavior observed in the MD simulations. D_{ij} of pure CO_2 from the IGKT creates a gas where particles are point masses with negligible interactions. The equation for calculating D_{ij} by the ideal gas kinetic theory is presented as:

$$D_{ij} = \frac{3}{16}\frac{\sqrt{\frac{2\pi k_B^3 T^3}{m_{ij}}}}{P\pi\sigma_{ij}^2\Omega^{(1,1)*}}, \quad (2)$$

where k_B is the Boltzmann constant and m_{ij} is the reduced molecular mass for the (i,j) species pair:

$$m_{ij} = \frac{m_i m_j}{m_i + m_j} \quad (3)$$

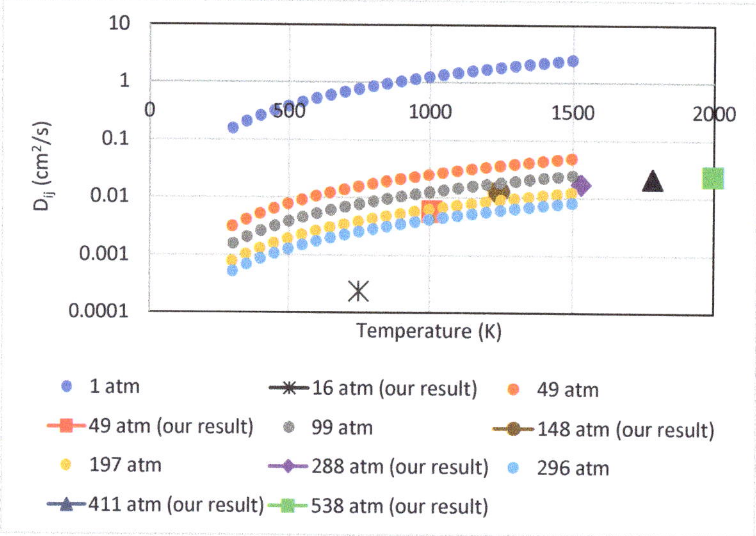

Figure 1. Self-diffusion coefficients (D_{ij}) of CO_2 computed by the ideal gas kinetic theory (1, 49, 99, 197, and 296 atm) and MD simulations (16, 49, 148, 288, 411, and 538 atm).

Σ_{ij} is the reduced collision diameter, and $\Omega^{(1,1)*}$ is the collision integral based on the Stockmayer potential [28]. Figure 1 shows the similarities between our MD simulation results and fitted ideal gas theory results, except at a low pressure of 16 atm at 750 K. The increase in temperature leads to the increase in D_{ij}. A higher temperature leads to faster molecular motion and a higher diffusion coefficient. The increased density in SCFs due to the pressure increase results in more frequent collisions and more efficient shuffling between molecules. At a high pressure (above 100 atm), the MD simulation results are closer to those of the ideal gas kinetic theory (within a factor of 10 of each other). At a very high pressure, potentially, sCO_2 can start to resemble a solid and diffusion slows down. This causes the slope of D_{ij} to decline as the temperature increases. An underestimate of D_{ij} at a lower temperature (750 K) aligns with our previous study of the chemical kinetics of combustion reactions in an sCO_2 environment well [29,30]. Table 1 shows the average diffusion coefficients of pure CO_2 under various temperatures and pressures by MD simulations and how spread out the data are. The results support the validity of using the chosen simulation conditions, such as force field TraPPE, to account for intermolecular interactions for the binary systems containing carbon dioxide and hydrocarbon, particularly at higher pressures above 50 atm, for accurate predictions compared to ideal gas assumptions.

Table 1. Diffusion coefficients (D_{ij}) of CO_2 under various temperatures (748–2040 K) and pressures (16–538 atm).

T (K)	P (atm)	D_{ij} (10^{-4} cm^2/s)
748	16 ± 3	2 ± 1
1012	49 ± 32	58 ± 24
1243	148 ± 7	128 ± 11
1530	288 ± 14	171 ± 9
1783	411 ± 23	215 ± 13
2040	538 ± 21	245 ± 16

Binary CO_2 and CH_4 mixtures. To understand the combustion of methane in an sCO_2 environment, it is necessary to study the diffusion of methane in CO_2/CH_4 mixtures. Figure 2 presents the diffusion coefficient of binary mixtures at various temperatures and pressures. The general trend of diffusion coefficient D_{ij} is that, as the temperature increases, the molecules in the mixture gain kinetic energy and result in the increase in D_{ij}. Both CO_2 and CH_4 move faster and collide with each other more frequently. A higher temperature provides enough thermal energy for molecules to diffuse. As the pressure increases at a constant temperature, D_{ij} decreases. A higher pressure depresses the diffusion of molecules as the mean free path of the fluid decreases and starts to resemble a solid so that a more frequent collision between molecules occurs and then hinders diffusion. Our simulation results reach as high as 930 atm at 2000 K. D_{ij} at near 300 atm is 0.012 at 1000 K and 0.032 cm^2/s at 1500 K. Interestingly, compared with the results of D_{ij} at 300 K from different groups, our results are ten-times higher than those of Guevara-Carrion et al.'s result of ~0.0002 cm^2/s at 300 K, 9 MPa [10], and Feng et al.'s result of 0.000272 cm^2/s at 299 K, 10.5 MPa [11]. It is important to remember that D_{ij} at low temperatures (300 and 500 K) in our simulation results is systematically tenfold higher than those the other groups reported. However, when we compare the diffusion coefficients in terms of absolute values, they are quite similar. These differences may be from the relatively low mole fraction of CH_4 in their study, while our study has a 1:1 carbon dioxide and hydrocarbon ratio. Such a discrepancy at lower temperatures aligns with our previous study of the chemical kinetics of combustion reactions in an sCO_2 environment well [29,30]. We are currently looking into the additional reasons behind this discrepancy to better understand the implications for our research.

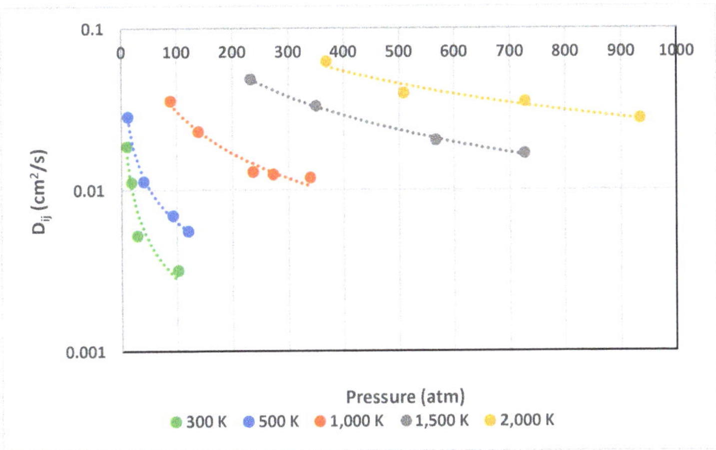

Figure 2. Self-diffusion coefficients (D_{ij}) of CH_4 in binary CO_2/CH_4 mixtures at various temperatures (300, 500, 1000, 1500, and 2000 K) and pressures below 1000 atm.

Binary CO_2 and C_2H_6 mixtures. To understand the combustion of ethane in an sCO_2 environment, it is necessary to study the diffusion of ethane in CO_2/C_2H_6 mixtures. Figure 3 presents the trend of the diffusion coefficient (D_{ij}) of CO_2/C_2H_6 that is similar to that of CO_2/CH_4 mixtures. We successfully obtained D_{ij} for all studied temperatures (500, 1000, 1500, and 2000 K), except 300 K and pressures below 600 atm. As the temperature increases, D_{ij} increases as a higher thermal energy for molecules. As the pressure increases at a constant temperature, D_{ij} decreases as more frequent collisions hinder diffusion. Our simulation can reach as high as 560 atm at 2000 K. D_{ij} at almost 300 atm is 0.016 cm^2/s at 1000 K and 0.045 cm^2/s at 1500 K. Our results can be compared with Feng et al.'s results, 0.000597 cm^2/s at 323 K, 10.5 MPa as well [11], though we are not able to achieve the simulation condition of 300 K at a mole fraction of 0.50 for C_2H_6.

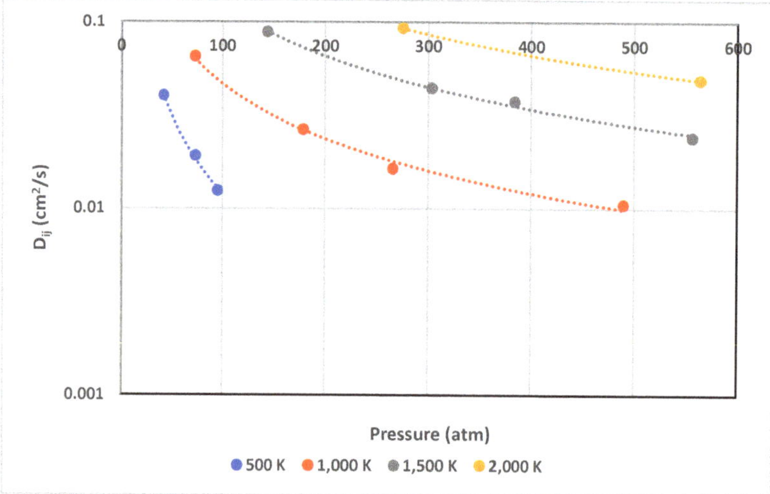

Figure 3. Self-diffusion coefficients (D_{ij}) of C_2H_6 in binary CO_2/C_2H_6 mixtures at various temperatures (500, 1000, 1500, and 2000 K) and pressures below 600 atm.

Overall, compared with the CO_2/CH_4 and CO_2/C_2H_6 results in Figures 2 and 3, respectively, the diffusion coefficient is inversely proportional to the molar volume of the solute and decreases quickly with the increasing carbon chain for short-chain *n*-alkanes. A smaller molar volume indicates a smaller molecule that experiences less friction as it moves through a solvent molecule. In general, both methane and ethane will experience similar increases in D_{ij} with the increasing temperature. The pressure influences the D_{ij} values of methane and ethane differently due to their subtle size difference. Table 2 shows the value of the diffusion coefficients of CO_2/CH_4 and CO_2/C_2H_6 under various temperatures and pressures. It is expected that the diffusion coefficient decreases as the mass and size of the molecule increase, while the difference is limited. This table also shows, in general, a greater fluctuation in D_{ij} as the pressure reduces. Feng et al. showed the limited difference of D_{ij} by experiments and MD simulations from CH_4 to $C_{14}H_{30}$ at 10.5 MPa [11]. For example, D_{ij} ranges from 0.000691 to 0.000220 cm^2/s at 323 K. Our simulations show the importance of the diffusion coefficients of binary CO_2/CH_4 and CO_2/C_2H_6 in an sCO_2 environment at high temperatures and pressures, which are challenging to achieve experimentally [2].

Table 2. Diffusion coefficients (D_{ij}) of hydrocarbon in CO_2/CH_4 and CO_2/C_2H_6 mixtures under various temperatures (300–2000 K) and pressures (10–1000 atm).

	CO_2/CH_4		CO_2/C_2H_6	
T (K)	P (atm)	D_{ij} (10^{-4} cm²/s)	P (atm)	D_{ij} (10^{-4} cm²/s)
300	12 ± 4	185 ± 29		
	20 ± 11	111 ± 24		
	31 ± 11	52 ± 8		
	104 ± 38	32 ± 3		
500	13 ± 7	283 ± 58	41 ± 3	404 ± 54
	42 ± 11	113 ± 27	74 ± 5	192 ± 34
	94 ± 4	69 ± 9	95 ± 14	124 ± 18
	121 ± 21	56 ± 17		
1000	91 ± 7	356 ± 40	73 ± 5	653 ± 88
	140 ± 9	229 ± 20	179 ± 14	266 ± 20
	238 ± 18	129 ± 21	266 ± 16	163 ± 15
	274 ± 20	125 ± 16	490 ± 31	105 ± 11
	341 ± 31	119 ± 15		
1500	232 ± 17	484 ± 69	144 ± 12	884 ± 146
	351 ± 27	330 ± 51	304 ± 22	444 ± 43
	567 ± 54	202 ± 16	384 ± 29	376 ± 37
	726 ± 74	167 ± 20	557 ± 37	242 ± 19
2000	371 ± 32	624 ± 68	275 ± 14	928 ± 109
	510 ± 54	397 ± 53	564 ± 25	490 ± 31
	729 ± 51	353 ± 34		
	936 ± 104	276 ± 45		

Table 3 shows the fitted equations under various temperatures in the range of 300–2000 K. The temperature is around 1100 °C (1423 K) and the pressure is around 300 bar (296 atm) of the Allam Power Cycle. The fitted D_{ij} for methane at 1500 K is 0.038 cm²/s and for ethane at 1500 K, it is 0.045 cm²/s.

Table 3. Fitted equations of D_{ij} (y) of binary CO_2/CH_4 and CO_2/C_2H_6 mixtures under various temperatures (300–2000 K) and pressures (x).

	CO_2/CH_4		CO_2/C_2H_6	
T (K)	Fitted Equation	R^2 Value	Fitted Equation	R^2 Value
300	$y = 0.1200x^{-0.813}$	0.9487		
500	$y = 0.1786x^{-0.723}$	0.9979	$y = 7.3009x^{-1.392}$	0.9979
1000	$y = 1.8117x^{-0.882}$	0.9860	$y = 4.1986x^{-0.977}$	0.9981
1500	$y = 8.5061x^{-0.949}$	0.9993	$y = 9.5610x^{-0.940}$	0.9978
2000	$y = 7.3284x^{-0.817}$	0.9339	$y = 13.7450x^{-0.890}$	1.0000

4. Conclusions

We determined the self-diffusion coefficients of pure CO_2 and simplest hydrocarbons in binary CO_2/CH_4 and CO_2/C_2H_6 mixtures at temperatures in the range of 300–2000 K and pressures in the range of 10–1000 atm. For the comparison to other experimental and computational results, we tested lower temperatures (as low as ~300 K) and pressures (as low as ~10 atm) as well. Our simulations at higher temperatures (as high as ~2000 K) and pressures (as high as ~900 atm) provide the data to better understand the Allam Power Cycle at 300 atm and make improvements in the future. The diffusion coefficients of methane at 300 atm are approximately 0.012 cm²/s at 1000 K and 0.032 cm²/s at 1500 K. The diffusion coefficients of ethane at 300 atm are approximately 0.016 cm²/s at 1000 K and 0.045 cm²/s at 1500 K. Our simulation results are far from a perfect fit and still need improvements, especially at lower temperatures and pressures. Nonetheless, this study provides valuable information for the improvement of the Allam Cycle in the future. The technology of the Allam Cycle is still under development. Besides the understanding of diffusion coefficients, the Allam Cycle requires further research and experiments to improve its overall performance and effectiveness to achieve commercial viability.

Author Contributions: Design of molecular dynamics simulations: C.-H.W.; ideal gas model and concept of equation of state: K.R.V.M.; simulation planning: A.E.M.; funding acquisition and manuscript review: S.S.V. All authors have read and agreed to the published version of the manuscript.

Funding: This work was mainly supported by the Department of Energy (DOE) under the grant number DE-FE0025260. Some aspects of the research were made possible with support from UCF.

Data Availability Statement: In the research and the article, only materials from publicly available sources of information are used. The simulation results obtained with LAMMPS software 2 August 2023 version are presented.

Acknowledgments: This study was prepared as an account of work sponsored by the United States Government agency. Neither the United States Government nor any agency thereof, nor any of their employees, makes any warranty, express or implied, or assumes any legal liability or responsibility for the accuracy, completeness, or usefulness of any information, apparatus, product, or process disclosed or represented that its use would not infringe privately owned rights. Referring to any specific commercial product, process, or service by trade name, trademark, manufacturer, or otherwise does not necessarily constitute or imply its endorsement, recommendation, or favoring by the United States Government or any agency thereof. The views and opinions of the authors expressed here do not necessarily state or reflect those of the United States Government or any agency thereof.

Conflicts of Interest: The authors declare no conflicts of interest.

References

1. Cussler, E.L. *Diffusion: Mass Transfer in Fluid Systems*; Cambridge University Press: Cambridge, UK, 2009.
2. Medina, I. Determination of diffusion coefficients for supercritical fluids. *J. Chromatogr. A* **2012**, *1250*, 124–140. [CrossRef] [PubMed]
3. Allam, R.; Martin, S.; Forrest, B.; Fetvedt, J.; Lu, X.; Freed, D.; Brown, G.W., Jr.; Sasaki, T.; Itoh, M.; Manning, J. Demonstration of the Allam Cycle: An update on the development status of a high efficiency supercritical carbon dioxide power process employing full carbon capture. *Energy Procedia* **2017**, *114*, 5948–5966. [CrossRef]
4. Chan, W.; Morosuk, T.; Li, X.; Li, H. Allam cycle: Review of research and development. *Energy Convers. Manag.* **2023**, *294*, 117607. [CrossRef]
5. Stubbs, J.M. Molecular simulations of supercritical fluid systems. *J. Supercrit. Fluids* **2016**, *108*, 104–122. [CrossRef]
6. Aimoli, C.; Maginn, E.; Abreu, C. Transport properties of carbon dioxide and methane from molecular dynamics simulations. *J. Chem. Phys.* **2014**, *141*, 134101. [CrossRef] [PubMed]
7. Moultos, O.A.; Orozco, G.A.; Tsimpanogiannis, I.N.; Panagiotopoulos, A.Z. Atomistic molecular dynamics simulations of H_2O diffusivity in liquid and supercritical CO_2. *Mol. Phys.* **2015**, *113*, 2805–2814. [CrossRef]
8. Zhu, A.; Zhang, X.; Liu, Q.; Zhang, Q. A fully flexible potential model for carbon dioxide. *Chin. J. Chem. Eng.* **2009**, *17*, 268–272. [CrossRef]
9. Abbaspour, M.; Nameni, E. Many-body effects in some thermodynamic properties of supercritical CO_2, CO_2-Ar, and CO_2-CH_4 using HFD-like potentials from molecular dynamics simulation. *J. Supercrit. Fluids* **2013**, *74*, 61–69. [CrossRef]
10. Guevara-Carrion, G.; Ancherbak, S.; Mialdun, A.; Vrabec, J.; Shevtsova, V. Diffusion of methane in supercritical carbon dioxide across the Widom line. *Sci. Rep.* **2019**, *9*, 8466. [CrossRef] [PubMed]
11. Feng, H.; Gao, W.; Sun, Z.; Lei, B.; Li, G.; Chen, L. Molecular dynamics simulation of diffusion and structure of some n-alkanes in near critical and supercritical carbon dioxide at infinite dilution. *J. Phys. Chem. B* **2013**, *117*, 12525–12534. [CrossRef] [PubMed]
12. Asadov, M.M.; Ramazanova, E.E.; Aliyev, E.N. Solubility, diffusion of components, and equilibrium in CO_2-ethane-heavy oil and CO_2-propane-heavy oil system. *Russ. J. Phys. Chem. A* **2021**, *95*, 868–874. [CrossRef]
13. Skarmoutsos, I.; Samios, J. Local intermolecular structure and dynamics in binary supercritical solutions. A molecular dynamics simulation study of methane in carbon dioxide. *J. Mol. Liq.* **2006**, *125*, 181–186. [CrossRef]
14. Gong, Y.; Ma, X.; Luo, K.H.; Xu, H.; Shuai, S. A molecular dynamics study of evaporation mode transition of hydrocarbon fuels under supercritical conditions. *Combust. Flame* **2022**, *246*, 112397. [CrossRef]
15. Chang, J. Expanded ensemble Monte Carlo simulations for the chemical potentials of supercritical carbon dioxide and hydrocarbon studies. *Korean J. Chem. Eng.* **2011**, *28*, 597–601. [CrossRef]
16. Tafazzoli, M.; Khanlarkhani, A. Investigation of the enhanced solubility of fluorinated methanes in CO_2 by Monte Carlo simulation: Absolute free energy of solvation and structural properties of solution. *J. Supercrit. Fluids* **2007**, *40*, 40–49. [CrossRef]

17. Fernandes, D.; Wang, S.; Xu, Q.; Buss, R.; Chen, D. Process and Carbon Footprint Analyses of the Allam Cycle Power Plant Integrated with an Air Separation Unit. *Clean Technol.* **2019**, *1*, 325–340. [CrossRef]
18. Frenkel, D.; Smit, B. *Understanding Molecular Simulation: From Algorithms to Applications*, 2nd ed.; Academic Press: Cambridge, MA, USA, 2002.
19. Kee, R.J.; Rupley, F.M.; Miller, J.A. Chemkin-II: A Fortran Chemical Kinetics Package for the Analysis of Gas-Phase Chemical Kinetics. United States: N. p. 1989. Web. Available online: https://www.osti.gov/biblio/5681118 (accessed on 10 August 2024).
20. Thompson, A.P.; Aktulga, H.M.; Berger, R.; Bolintineanu, D.S.; Brown, W.M.; Crozier, P.S.; In't Veld, P.J.; Kohlmeyer, A.; Moore, S.G.; Nguyen, T.D.; et al. LAMMPS–A flexible simulation tool for particle-based materials modeling at the atomic, meso, and continuum scales. *Comput. Phys. Commun.* **2022**, *271*, 108171. [CrossRef]
21. Potoff, J.J.; Siepmann, J.I. Vapor-liquid equilibria of mixtures containing alkanes, carbon dioxide, and nitrogen. *AIChE J.* **2001**, *47*, 1676–1682. [CrossRef]
22. Eggimann, B.L.; Sun, Y.; DeJaco, R.F.; Singh, R.; Ahsan, M.; Josephson, T.R.; Siepmann, J.I. Assessing the quality of molecular simulations for vapor-liquid equilibria: An analysis of the TraPPE database. *J. Chem. Eng. Data* **2020**, *65*, 1330–1334. [CrossRef]
23. Martin, M.G.; Siepmann, J.I. Transferable potentials for phase equilibria. 1. United-atom description on n-alkanes. *J. Phys. Chem. B* **1998**, *102*, 2569–2577. [CrossRef]
24. Shah, M.S.; Tsapatsis, M.; Siepmann, J.I. Transferable potentials for phase equilibria. Improved united-atom description of ethane and ethylene. *AIChE J.* **2017**, *63*, 5098–5110. [CrossRef]
25. Martinez, L.; Andrade, R.; Birgin, E.G.; Martinez, J.M. Packmol: A package for building initial configurations for molecular dynamics simulations. *J. Comput. Chem.* **2009**, *30*, 2157–2164. [CrossRef] [PubMed]
26. Nosé, S. A unified formulation of the constant temperature molecular dynamics methods. *J. Chem. Phys.* **1984**, *81*, 511–519. [CrossRef]
27. Hoover, W.G. Canonical dynamics: Equilibrium phase-space distributions. *Phys. Rev. A* **1985**, *31*, 1695–1697. [CrossRef] [PubMed]
28. Monchick, L.; Mason, E.A. Transport properties of polar gases. *J. Chem. Phys.* **1961**, *35*, 1676–1697. [CrossRef]
29. Wang, C.-H.; Panteleev, S.V.; Masunov, A.E.; Allison, T.C.; Chang, S.; Lim, C.; Jin, Y.; Vasu, S.S. Molecular dynamics of combustion reactions in supercritical carbon dioxide. Part 5: Computational study of ethane dissociation and recombination reactions $C_2H_6 \rightleftharpoons CH_3 + CH_3$. *J. Phys. Chem. A* **2019**, *123*, 4776–4784. [CrossRef] [PubMed]
30. Wang, C.-H.; Masunov, A.E.; Allison, T.C.; Chang, S.; Lim, C.; Jin, Y.; Vasu, S.S. Molecular dynamics of combustion reactions in supercritical carbon dioxide. 6. Computational kinetics of reactions between hydrogen atom and oxygen molecule $H + O_2 \rightleftharpoons HO + O$ and $H + O_2 \rightleftharpoons HO_2$. *J. Phys. Chem. A* **2019**, *123*, 10772–10781. [CrossRef] [PubMed]

energies

MDPI

Article

A Numerical Investigation of the Flame Characteristics of a CH_4/NH_3 Blend Under Different Swirl Intensity and Diffusion Models

Ahmed Adam [1,2], Ayman Elbaz [3], Reo Kai [1] and Hiroaki Watanabe [1,*]

[1] Interdisciplinary Graduate School of Engineering Sciences (IGSES), Kyushu University, Kasuga City 816-8580, Fukuoka Prefecture, Japan

[2] Mechanical Power Engineering Department, Faculty of Engineering—Mataria, Helwan University, Cairo 11795, Egypt

[3] Clean Energy Research Platform (CERP), King Abdullah University of Science and Technology, Thuwal 23955-6900, Saudi Arabia; ayman.elhagrasy@kaust.edu.sa

* Correspondence: whiroaki@tse.kyushu-u.ac.jp

Abstract: This study investigates the effects of diffusion modeling and swirl intensity on flow fields and NO emissions in CH_4/NH_3 non-premixed swirling flames using large eddy simulations (LESs). Simulations are performed for a 50/50 ammonia–methane blend at three global equivalence ratios of 0.77, 0.54, and 0.46 and two swirl numbers of 8 and 12, comparing the unity Lewis number (ULN) and mixture-averaged diffusion (MAD) models against the experimental data includes OH-PLIF and ON-PLIF reported in a prior study by the KAUST group. Both models produce similar flow fields, but the MAD model alters the flame structure and species distributions due to differential diffusion (DD) and limitations in its Flamelet library. Notably, the MAD library lacks unstable flame branch solutions, leading to extensive interpolation between extinction and stable branches. This results in overpredicted progress variable source terms and reactive scalars, both within and beyond the flame zone. The ULN model better reproduces experimental OH profiles and localizes NO formation near the flame front, whereas the MAD model predicts broader NO distributions due to nitrogen species diffusion. Higher swirl intensities shorten the flame and shift NO production upstream. While a low equivalence ratio provides enough air for good mixing, lower ammonia and higher NO contents in exhaust gases, respectively.

Keywords: turbulent non-premixed combustion; differential diffusion; flame structure; Flamelet progress variable

1. Introduction

The power sector forms a cornerstone of modern civilization, evolving to meet the growing global energy demand [1]. Among various energy sources, fuels remain the most energy-dense medium, capable of being burned to produce thermal energy for power generation in heat engines such as gas turbines and internal combustion engines.

Despite the advantages of fuels, their combustion generates chemical pollutants such as CO_2, CO, and NO_X, which contribute to greenhouse gas emissions and global warming. This has prompted the adoption of energy policies prioritizing carbon-neutral and carbon-free power generation to mitigate environmental impacts. In this context, hydrogen and ammonia have emerged as promising alternatives to fossil fuels. Being carbon-free species, their blends can offer a comparable performance to fossil fuels [2]. However, challenges related to their storage, transportation, and combustion processes remain significant [3].

Ammonia is a promising fuel because it can be transported and stored using existing infrastructure with minimal modifications, unlike hydrogen. Additionally, ammonia can serve as a hydrogen carrier, decomposing into hydrogen and nitrogen when needed [3]. However, ammonia faces challenges due to its low combustion performance, including a high ignition temperature (930 K) and a low laminar flame speed (0.072 m/s) [4]. To address these limitations, ammonia is often blended with hydrocarbons such as methane [5] or with hydrogen [2] to enhance its combustion characteristics.

In the industrial and power sectors, most flames are turbulent to meet power requirements while ensuring flame stability. This is often achieved using swirling combustion, where the air flow creates a recirculation zone near the burner rim. This recirculation increases the fuel's residence time in the highly reactive zone, enhancing combustion. As a result, the swirl number plays a critical role in determining the flame characteristics [6].

The LES/FPV (large eddy simulation/Flamelet progress variable) approach is a widely adopted modeling framework for turbulent combustion due to its ability to combine detailed chemistry with the resolution of large-scale turbulent structures. In this method, combustion is represented using precomputed Flamelet libraries parameterized by variables such as the mixture fraction and a progress variable, while LES resolves the energy-containing eddies that strongly influence flame dynamics [7,8]. A major advantage of LES/FPV is its computational efficiency compared to direct numerical simulation (DNS), enabling the study of complex, realistic configurations, such as gas turbines and engines, while capturing key turbulence–chemistry interactions [9]. It also facilitates a direct comparison with experiments by allowing the computation of filtered scalar quantities like LIF and Rayleigh signals [10]. However, the approach relies on assumptions such as the Flamelet regime being valid and the use of presumed probability density functions (PDFs) to reconstruct subgrid scalar statistics, which may reduce accuracy in regimes with ignition, extinction, or strong stratification [11]. Despite these limitations, LES/FPV remains a popular tool due to its balance between physical fidelity and computational feasibility, making it attractive for both academic and industrial combustion research [12].

The LES/FPV approach is used in the previous works to investigate the flame characteristics. For example, the flame structure of a piloted partially premixed dimethyl ether (DME) was investigated using LES/FPV. The numerical results were compared with experimental data, including temperature and species measurements (Raman/Rayleigh, LIF) and velocity fields (PIV), which showed good agreement [13]. The diffusion models of the Flamelet libraries were investigated for a turbulent non-premixed oxy-fuel flame using experimental data and LES simulations [14]. A species-weighted Flamelet model accounting for DD was developed and applied in the LES of oxy-fuel flames, showing improved predictions of flame structure and extinction behavior compared to traditional unity and species-constant Lewis number models [15]. LESs were performed for turbulent CH_4/H_2 flames with and without fuel stratification, using Flamelet libraries based on ULN and MAD models. Two approaches were used to access the libraries: one solving species transport equations and the other solving trajectory variables directly. The results showed that accounting for heat loss and using the trajectory-based approach improved agreement with experiments, and that DD effects played a key role in downstream flame behavior [16]. Building on these studies, the present work incorporates heat loss and mixture fraction variance in the Flamelet library to model turbulent non-premixed flames, and systematically compares two diffusion models: ULN and MAD.

Recent studies have advanced the understanding of ammonia and hydrogen combustion. Su et al. [17] analyzed chemiluminescence in NH_3–CH_4 laminar flames and identified CH^*/NH_2^* as a potential marker for ammonia content, but their work was limited to laminar conditions. Ostrov and Haas [18] developed a Wiebe-based model for hydrogen combustion in spark-ignition engines, focusing on mixture stratification and ignition control, though their study did not consider ammonia or swirling diffusion flames. Lan et al. [19] showed that hydrogen enrichment in natural gas flames reduces CO and CO_2 emissions but increases NO_x, emphasizing trade-offs in fuel composition; however, turbulence and diffusion modeling were not the focus. Liu et al. [20] linked pressure fluctuations in swirl combustors to heat release, swirl, and flame detachment using EMD, FFT, and POD, but without exploring diffusion effects. Wu et al. [21] showed the efficiency and emission benefits of diesel/ammonia dual-fuel engines but did not investigate the turbulent flame structure or DD effects. Wang et al. [22] investigated the DD effects in transcritical LO_2/CH_4 flames using large-eddy simulations and the FPV model, where Flamelets were solved in mixture fraction space with both unity and non-unity Lewis numbers. In the present study, DD is treated differently by solving the Flamelet equations in a physical space using the MAD model, which account for the different values of the Lewis number for the same species at different conditions, while a constant Lewis number considers only one value for the Lewis number per species regardless of its different conditions. Obando Vega et al. [23] focused on modeling the subgrid-scale effects in the LES of turbulent non-premixed flames using filtered tabulated chemistry to capture flame front wrinkling and strain-rate influences, while the DD was also considered by solving the Flamelet equation using a constant Lewis number in mixture fraction space.

To address the limitations highlighted in previous studies, such as the lack of turbulence, swirl effects, and advanced diffusion modeling, this work investigates CH_4/NH_3 non-premixed turbulent swirling flames using high-fidelity numerical simulations. It compares two widely used diffusion models in Flamelet equations, the unity Lewis number (ULN) and mixture-averaged diffusion (MAD) approaches, solved in mixture fraction and physical space, respectively. Their predictions of key scalar fields, including OH and NO, are validated against experimental data to evaluate each model's ability to capture the flame structure properly. The combined impact of swirl intensity and diffusion treatment on flame structure is further analyzed using Flamelet libraries. By benchmarking against experimental results from [6], the study provides new insights into the predictive performance of diffusion models under realistic turbulent combustion conditions.

2. Numerical Methods and Modeling

This study investigates two sets of simulation cases. The first set employs a ULN Flamelet library, assuming equal diffusivity for all species, while the second set incorporates DD effects using a mixture-averaged diffusion (MAD) model. The key configurations and parameters for each case are summarized in Table 1, where S_g denotes the swirl number defined in [6] and ϕ is the global equivalence ratio. The selected values are based on the experimental cases reported in [6], where ϕ = 0.54 produced the highest NO emissions at $X_{NH3} = 0.5$ and $S_g = 8$, while the other two equivalence ratios, although still lean, produced lower NO levels. These points were selected to explore the sensitivity of NO formation to slight variations in fuel–air equivalence under lean conditions.

Table 1. Investigation program.

	Case 1	Case 2	Case 3	Case 4
ULN	$S_g = 8$ $\phi = 0.46$	$S_g = 8$ $\phi = 0.54$	$S_g = 8$ $\phi = 0.77$	$S_g = 12$ $\phi = 0.54$
MAD	$S_g = 8$ $\phi = 0.46$	$S_g = 8$ $\phi = 0.54$	$S_g = 8$ $\phi = 0.77$	$S_g = 12$ $\phi = 0.54$

2.1. Geometry and Grid Configurations

The LESs are conducted on the same burner and combustor configuration mentioned in [6]. Figure 1 shows the rendered view of the metallic burner fixed with the fiber glass combustor: the burner consists of two concentric pipes, the central pipe supplies fuel to the combustion chamber at constant velocity equal to 13 m/s for all investigated cases, while the annular pipe has five inlets, one is axial and four are tangential, the flow rates of air inlets are manipulated to achieve different global equivalence ratios and swirl numbers. The swirl number is calculated using $S_g = \frac{\pi r_o d_a}{2A_t} \times \left(\frac{\dot{m}_{air,t}}{\dot{m}_{air,t}+\dot{m}_{air,a}}\right)^2$, where $r_o = d_a - d_f$, d_a is the annular large diameter, i.e., 27 mm and d_f is the fuel pipe inner diameter, i.e., 4.4 mm, A_t is the cross sectional area of all 4 tangential air inlets, while $\dot{m}_{air,t}$ and $\dot{m}_{air,a}$ are the total tangential and axial air mass flow rates, respectively. Based on this design, the structured grid is created.

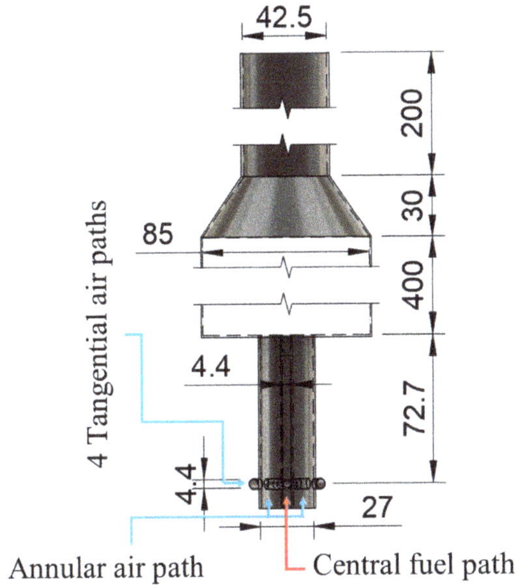

Figure 1. Setup configuration with dimensions in mm, the burner and exhaust system are metallic while the combustor is fiber glass.

2.2. Governing Equations

The governing equations in this LES non-adiabatic Flamelet generated manifold (NA-FGM) study are solved by front flow red (FFR) solver, they include the density weighted averaged conservation equations for mass, momentum, enthalpy $\tilde{h}$, NO mass fraction $\tilde{Y}_{NO}$, mixture fraction $\tilde{Z}$, and progress variable $\tilde{C}$ as in Equations (1)–(6).

$$\frac{\partial \bar{\rho}}{\partial t} + \nabla \cdot \left(\bar{\rho}\tilde{\boldsymbol{u}}\right) = 0 \tag{1}$$

$$\frac{\partial \bar{\rho}\tilde{\boldsymbol{u}}}{\partial t} + \nabla \cdot \left(\bar{\rho}\tilde{\boldsymbol{u}}\tilde{\boldsymbol{u}}\right) = -\nabla \overline{\mathrm{P}} + \nabla \cdot \bar{\sigma} + \nabla \cdot \tau \tag{2}$$

$$\frac{\partial \bar{\rho}\tilde{h}}{\partial t} + \nabla \cdot \left(\bar{\rho}\tilde{\boldsymbol{u}}\tilde{h}\right) = \nabla \cdot \left(\bar{\rho}\tilde{D}_h \nabla \tilde{h}\right) + \nabla \cdot q_h + Q_{rad} \tag{3}$$

$$\frac{\partial \bar{\rho} \widetilde{Y_{NO}}}{\partial t} + \nabla \cdot \left(\bar{\rho} \tilde{\boldsymbol{u}} \widetilde{Y_{NO}} \right) = \nabla \cdot \left(\bar{\rho} \widetilde{D_{Y_{NO}}} \nabla \widetilde{Y_{NO}} \right) + \nabla \cdot q_{Y_{NO}} + \widetilde{\dot{\omega}_{Y_{NO}}} \quad (4)$$

$$\frac{\partial \bar{\rho} \tilde{Z}}{\partial t} + \nabla \cdot \left(\bar{\rho} \tilde{\boldsymbol{u}} \tilde{Z} \right) = \nabla \cdot \left(\bar{\rho} \widetilde{D_Z} \nabla \tilde{Z} \right) + \nabla \cdot q_Z \quad (5)$$

$$\frac{\partial \bar{\rho} \tilde{C}}{\partial t} + \nabla \cdot \left(\bar{\rho} \tilde{\boldsymbol{u}} \tilde{C} \right) = \nabla \cdot \left(\bar{\rho} \widetilde{D_C} \nabla \tilde{C} \right) + \nabla \cdot q_C + \widetilde{\dot{\omega}_C} \quad (6)$$

where τ is the subgrid stress derived from the tubulence model, P is the pressure, q_ϕ denotes the subgrid-scale terms for the scalar ϕ, and $\tilde{\boldsymbol{u}}$ represents velocity in the three spatial directions, while D_ϕ is the diffusion coefficient of the scalar ϕ, which is calculated using Equation (7).

$$D_\phi = \frac{\mu}{Pr_\phi} \quad (7)$$

2.3. Chemistry Modeling

This study adopts the Flamelet progress variable (FPV) approach, where equilibrium thermo-chemical states of laminar Flamelets are precomputed. The Flamelet equations in [24] are solved with the Okafor mechanism [25], which includes 59 species and 356 reactions. For non-premixed combustion, a wide range of scalar dissipation rates and stain rates up to extinction are computed for ULN and MAD models, respectively. These calculations are repeated for the different percentages of the total heat in the adiabatic case. The resulting data are then used to generate the Flamelet library, where the enthalpy difference Δh from the adiabatic case is computed. The maximum Δh is estimated based on the obtained Flamelet solutions. The library conditions thermo-chemical states on $\tilde{Z}, \widetilde{Z''^2}, \tilde{C}$, and $\widetilde{\Delta h}$. The mixture fraction variance $\widetilde{Z''^2}$ follows an exponential distribution from 0 to 0.25, and the corresponding states $\tilde{\phi}$ are computed using Equations (8)–(12) in [8].

$$\tilde{\phi} = \int_0^1 \phi(Z)\, \tilde{P}(Z) dZ \quad (8)$$

$$\tilde{P} = \frac{Z^{\alpha-1}\,(1-Z)^{\beta-1}}{\Gamma(\alpha)\Gamma(\beta)} \Gamma(\alpha+\beta) \quad (9)$$

$$\alpha = \tilde{Z}\gamma \quad (10)$$

$$\beta = \left(1 - \tilde{Z}\right)\gamma \quad (11)$$

$$\gamma = \frac{\tilde{Z}\left(1-\tilde{Z}\right)}{\widetilde{Z''^2}} - 1 \geq 0 \quad (12)$$

Furthermore, during the LES, the total enthalpy transport equation is solved to determine the physical enthalpy at each grid point. The difference between this enthalpy and the adiabatic enthalpy from the library represents the local heat loss. This enthalpy difference is then used to retrieve the appropriate thermo-chemical state from the Flamelet library, ensuring an accurate representation of non-adiabatic effects in the simulation.

The Flamelet libraries contain equilibrium solutions of the Flamelet equations. Since NO has a larger timescale than most major species, particularly those forming the progress variable, the Flamelet model inevitably overpredicts NO [26]. To address this, the transport Equation (4) is solved while using the Flamelet library, with its source term modeled by Equation (13). This approach mitigates the high NO destruction rate, which would otherwise result from its mass fraction overprediction in equilibrium state.

$$\tilde{\dot{\omega}}_{NO} = \tilde{\dot{\omega}}_{NO}^{+flm} + \tilde{Y}_{NO} \frac{\tilde{\dot{\omega}}_{NO}^{-flm}}{\tilde{Y}_{NO}^{flm}} \tag{13}$$

The flm superscript denotes the data retrieved from the Flamelet library, while $+$ and $-$ indicate production and destruction rates, respectively. $\tilde{Y}_{NO}$ is obtained from the solution of the transport Equation (4), and $\tilde{\dot{\omega}}_{NO}$ represents the updated source term for the next time step calculation.

2.4. Flamelet Library

In this study, two approaches for treating molecular transport are considered, which are the MAD model and the ULN model. The MAD model accounts for species-dependent diffusivities by incorporating the effects of DD based on molecular properties and local composition. In contrast, the ULN model simplifies the diffusion process by assuming equal thermal and mass diffusivities (i.e., Lewis number of unity) for all species, which reduces the computational complexity but may lead to inaccuracies in predicting species distributions and flame structure under conditions where DD is significant.

From a numerical standpoint, the ULN Flamelet model offers significant advantages in terms of computational implementation and flexibility. ULN Flamelets are typically solved in the mixture fraction space, allowing for the pre-tabulation of the entire S-curve, including both stable and unstable branches. This capability is particularly useful for studying ignition, extinction, and flame stabilization phenomena. In contrast, the MAD model must be solved in physical space due to its dependence on local species gradients and diffusivities. However, the MAD model in physical space is limited to stable flame regime analyses, leaving the unstable branch unexplored and highlighting opportunities for further development.

Additionally, practical observations from this study show that ULN Flamelets exhibit greater numerical robustness under heat loss conditions. Stable solutions could be obtained with up to 25% heat loss relative to the total energy. In comparison, MAD Flamelet calculations became unstable beyond 12.5% heat loss. This demonstrates a practical advantage of the ULN model when simulating flames with radiative or conductive losses. On the other hand, the MAD diffusion model emphasizes the significant impact of DD on flame stability under non-adiabatic conditions. The higher diffusivity shortens the residence time of reactive species, resulting in earlier quenching and flame extinction at lower heat loss levels compared to the ULN Flamelet model.

Figure 2 (left) presents the scatter plot of the progress variable against the mixture fraction, specifically for the adiabatic flame without variance in the mixture fraction, where each point is colored according to the temperature. The ULN Flamelet solution (top) shows a more concentrated distribution near the stoichiometric mixture fraction compared to the MAD model (bottom). This behavior is attributed to the ULN approach's capability to capture the unstable branch of the S-curve, which improves solution continuity and resolution in the flame stabilization region. However, the obtained Flamelet library interpolates any missing data, resulting in both models appearing similar despite the differences in the original solutions, as demonstrated in Figure 2 (right). This interpolation introduces greater uncertainty in the MAD model, as a larger portion of its library is not directly solved but inferred. Such interpolation may be particularly critical when predicting the progress variable source term, which is extracted from the library during the LES, as illustrated in Figure 3 (right).

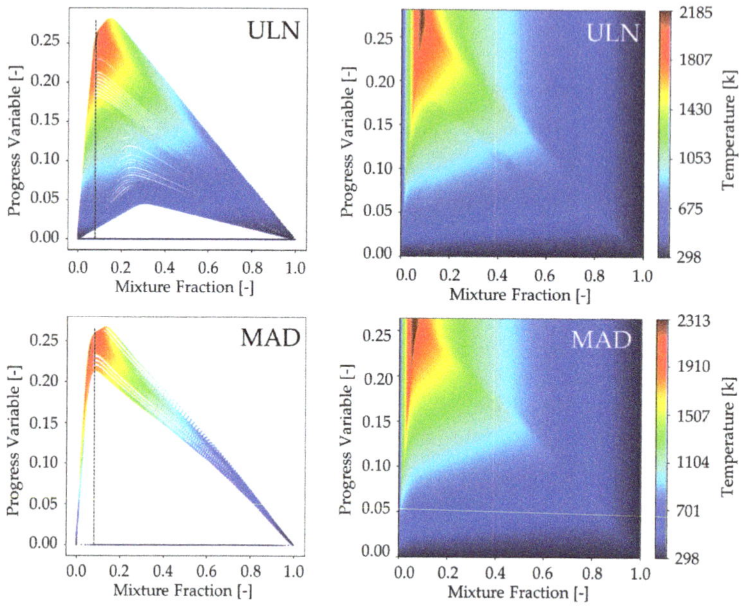

Figure 2. Temperature distribution in Flamelet solutions and corresponding Flamelet libraries for the ULN model (**top**) and MAD model (**bottom**). Flamelet solutions (**left**), and the Flamelet libraries are generated using an in-house code (**right**).

Figure 4 compares the distributions of Y_{OH} and Y_{NO} in the ULN and MAD Flamelet libraries, highlighting how DD affects their spatial behavior. Figure 3 further illustrates the source terms of the progress variable (left) and NO (right), using a seismic colormap in which red indicates production and blue indicates destruction. The green alpha shape marks the region where the Flamelet solution exists; outside this boundary, the Flamelet library is either interpolated or copied. This visualization highlights the dynamic balance between production and destruction in different flame regions and reveals the limitations of each model in covering the Flamelet library. It also provides insight into how diffusion models influence the key species responsible for pollutant formation and heat release.

In addition, the MAD library shows a narrower solved region compared to the ULN library. This gives an advantage to the ULN model, as during LES, the source terms for the progress variable and NO are retrieved from the library and substituted into the governing equations. A limited solved region in the MAD library may lead to inaccurate results due to the absence of unstable branch solutions.

In this study, special emphasis is placed on nitric oxide (NO) and hydroxyl radical (OH) due to their crucial roles in ammonia combustion chemistry. NO is a primary pollutant and one of the most harmful nitrogen-containing emissions, which is especially relevant in ammonia-based fuel systems where nitrogen content is inherently high. Its formation is governed by complex mechanisms, including prompt NO, thermal NO, and fuel NO pathways, all of which are influenced by flame temperature, local equivalence ratio, and radical pool composition.

OH, on the other hand, serves as a critical intermediate in the oxidation process and plays a key role in NO formation through reactions such as $NH + OH \rightarrow NO + H_2$. It is also an indicator of high-temperature reaction zones and is often used to characterize flame front structures and heat release regions. Monitoring the OH distribution thus provides insight into the location and intensity of the reaction zones, while also indirectly informing about the potential for NO formation.

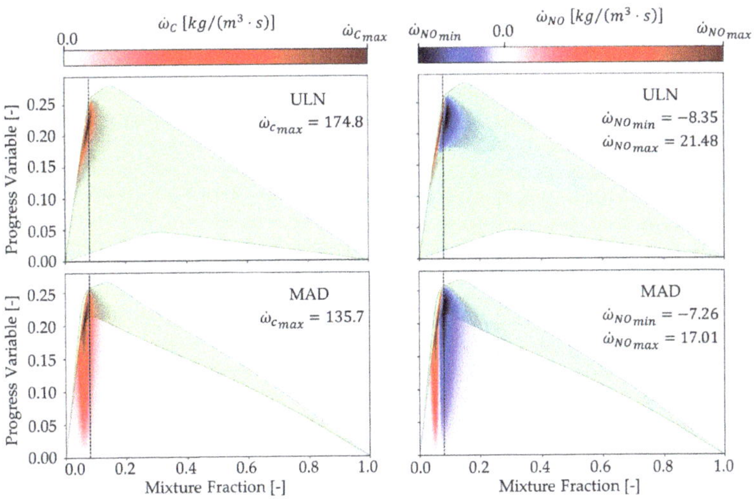

Figure 3. Distribution of source terms $\dot{\omega}_C$ (**left**) and $\dot{\omega}_{NO}$ (**right**) in the ULN (**top**) and MAD (**bottom**) Flamelet libraries. The seismic colormap is used, with white corresponding to the zero source term values. The green contour shows the Flamelet solution region.

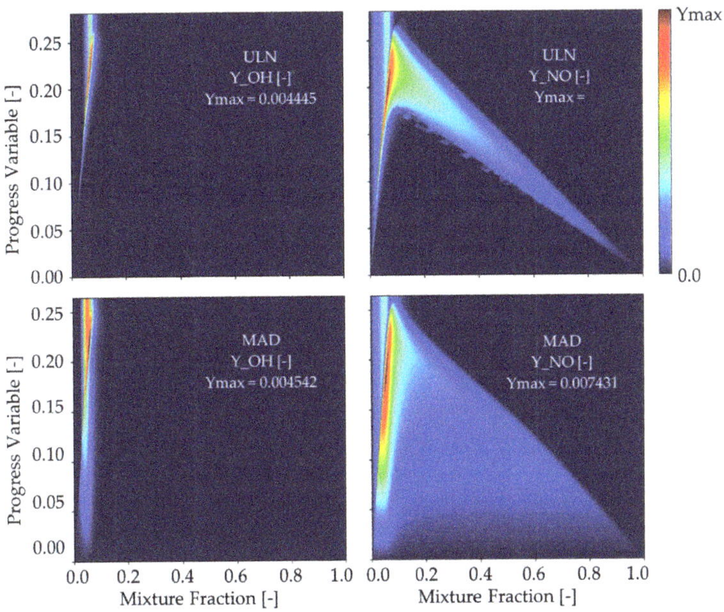

Figure 4. Distribution of Y_{OH} (**right**) and Y_{NO} (**left**) in the ULN (**top**) and MAD (**bottom**) Flamelet libraries.

3. Results and Discussion

The presented results are time-averaged over 200,000 time steps, corresponding to an averaging period of approximately 30 ms. Since the experimental measurements include OH and NO distributions, the validation is carried out in Section 3.2.

3.1. Temperature Distribution and Flow Field

The temperature distribution in all MAD cases is noticeably higher than in the ULN cases, with unrealistically elevated temperatures observed downstream of the flame and away from the flame front, as shown in Figure 5. This raises concerns about the MAD model, which is expected to be more accurate due to its consideration of DD effects, unlike the ULN model, which suppresses these effects by assuming equal diffusivities for all species. However, the MAD model mispredicts scalar fields such as temperature and OH concentration, as will be further demonstrated in Section 3.2. This discrepancy is discussed in detail in that section.

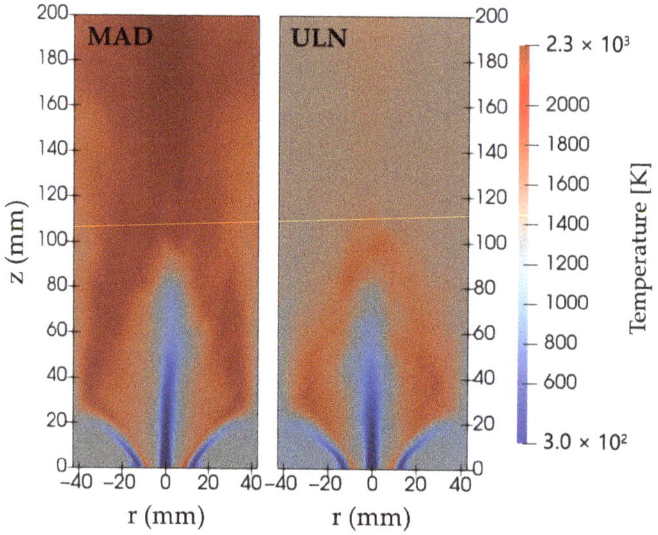

Figure 5. Averaged temperature distribution for Case 2, (left) MAD model, and (right) ULN model.

The flow field is primarily governed by the flow rates of the main streams and the geometry of the burner and combustor. Figure 6 shows the velocity magnitude distributions for the ULN model (top row, a–d) and the MAD model (bottom row, A–D). For the same swirl number (a–c and A–C), reducing the equivalence ratio increases the air flow rate, enhancing the swirling motion. However, increasing the swirl number has a more pronounced effect on swirl strength as in (b and d) or (B and D).

While the main flow structures are largely preserved, noticeable differences in velocity magnitude appear across the entire domain when comparing the two diffusion models. The MAD model shows higher velocity magnitudes over broader regions both inside and downstream of the flame, suggesting that enhanced diffusion contributes to stronger momentum transfer and a more sustained swirling motion.

The velocity magnitude distribution shown in Figure 6 reveals that stronger swirling motion results in higher negative axial velocity in the central region of the combustor, as shown in Figure 7, both inside and downstream of the flame. This behavior is more pronounced in cases with higher swirl numbers, where the enhanced centrifugal force generates a stronger low-pressure zone along the axis. As a result, the central recirculation zone becomes more intense, drawing hot combustion products and reactants back toward the burner.

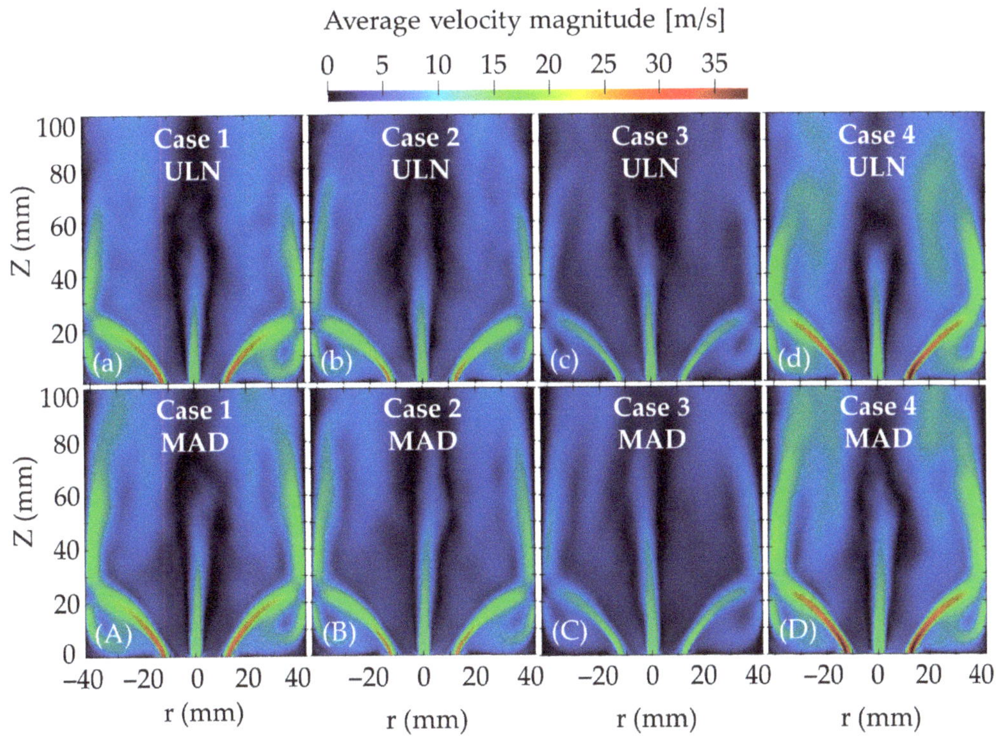

Figure 6. Averaged velocity magnitude distribution for Cases 1, 2, 3, and 4: the top row corresponds to the ULN model (**a**–**d**), and the bottom row to the MAD model (**A**–**D**).

This recirculation not only contributes to flame stabilization but also enhances mixing between fuel and oxidizer, promoting more complete combustion. However, for low-reactivity fuels such as ammonia, CH_4/NH_3 blends do not always exhibit this behavior. As reported in [6], increasing the ammonia content reduces the influence of swirl on flame stability. Beyond a certain blend ratio, higher swirl numbers may even worsen flame stability. This is attributed to the strong shear regions created by intense recirculation, where the variance of the mixture fraction is also intense, which negatively affects the slow chemical kinetics of ammonia oxidation.

Moreover, the MAD model consistently exhibits stronger negative axial velocities compared to the ULN model, suggesting that the diffusion formulation in MAD enhances the momentum transfer and supports a more vigorous swirling motion. These differences indicate that both the swirl number and diffusion model play a significant role in shaping the flow field and, consequently, combustion characteristics.

3.2. *Effect of Diffusion Model on NO/OH Distribution and Kinetics*

Figure 8 compares the time-averaged OH mass fraction distributions predicted by the ULN and MAD models with experimental PLIF data [6], while the NO mass fraction is shown in Figure 9. The OH distribution predicted by the ULN model shows better agreement with the experimental measurements, while the MAD model exhibits a noticeably narrower OH region near the burner, indicating the stronger confinement of radical species due to DD. For NO, both models capture the general distribution reasonably well, although differences exist: the MAD model predicts a slightly narrower NO region compared to ULN, yet both remain in an acceptable agreement with the experimental profile. Downstream of 65 mm from the burner rim, the MAD model shows a more extended tail in NO concentration as shown in Figure 10a; however, due to the lack of experimental data in this region, the accuracy of this prediction cannot be confirmed. In addition, the OH distribution in this region is high as well for MAD cases.

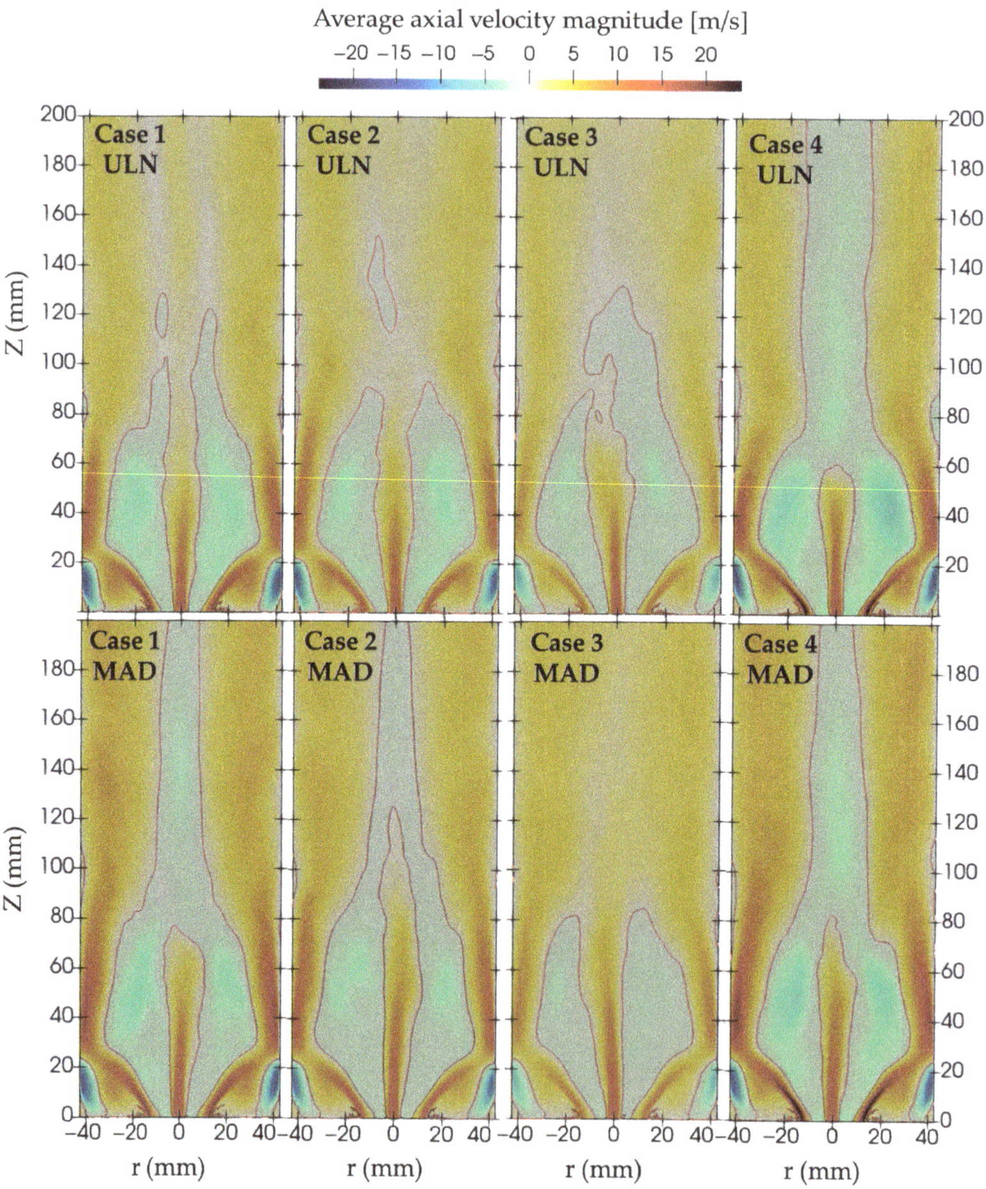

Figure 7. Averaged axial velocity magnitude distribution for Cases 1, 2, 3, and 4: the (**top**) row corresponds to the ULN model, and the (**bottom**) row to the MAD model. Isolines represent zero axial velocity to show recirculation zone boundaries.

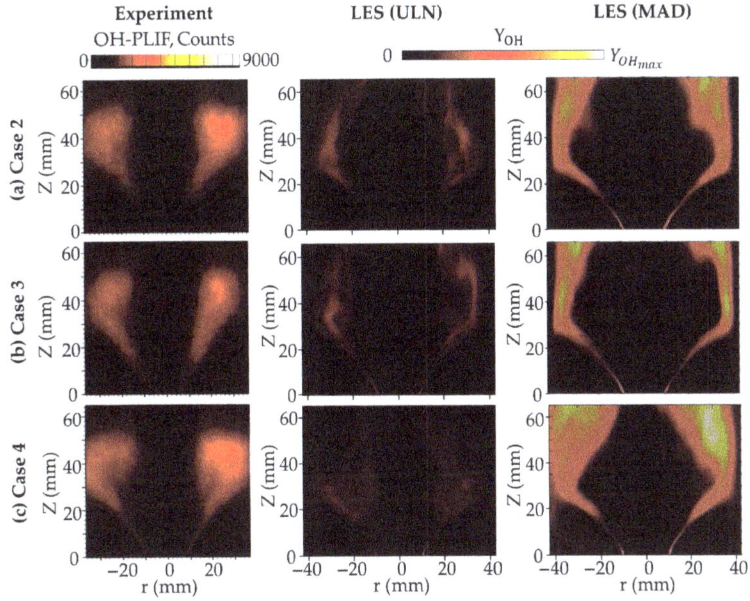

Figure 8. Time-averaged OH mass fractions predicted by LES using the ULN and MAD models, compared with OH PLIF signal intensities from the experimental data in [6].

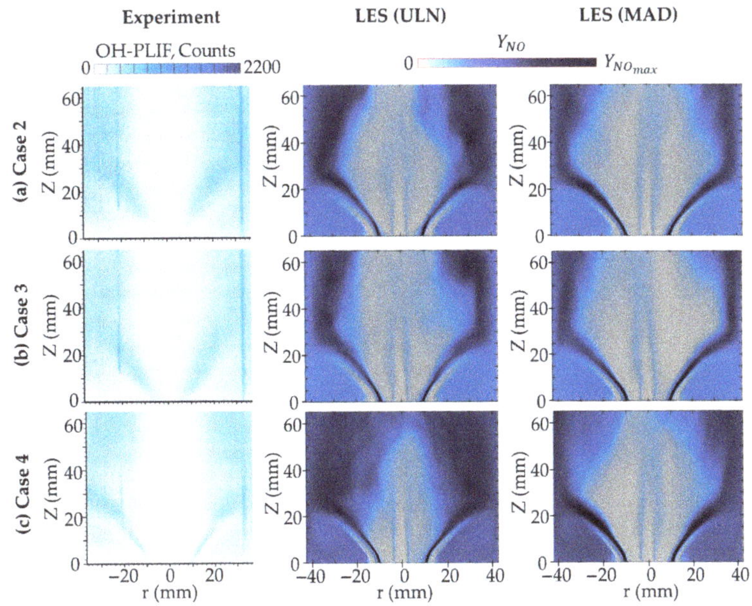

Figure 9. Time-averaged NO mass fractions predicted by LES using the ULN and MAD models, compared with NO PLIF signal intensities from the experimental data in [6].

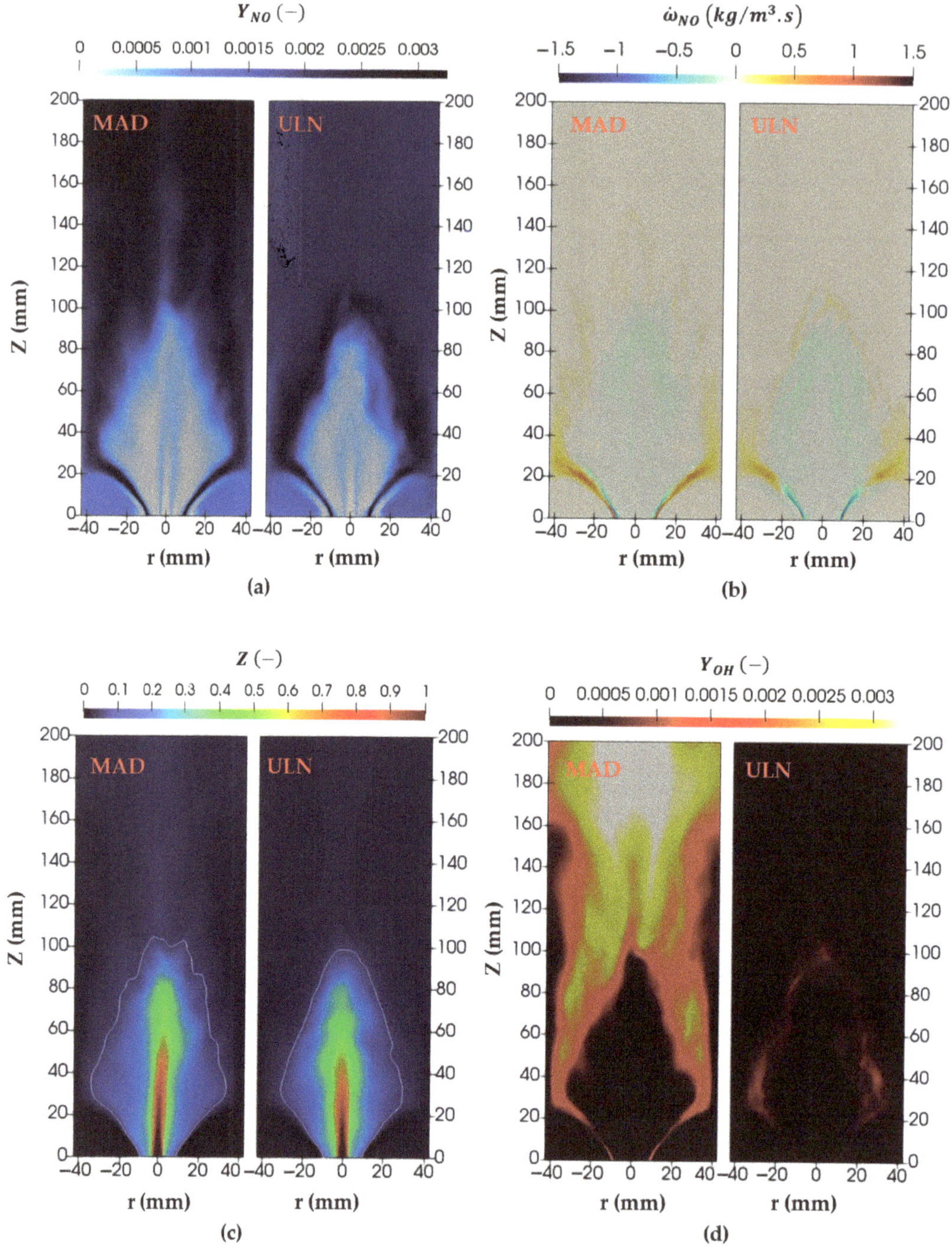

Figure 10. (**a**) Averaged values comparison between MAD (**left**) and ULN (**right**) models for case 2: (**a**) NO mass fraction distribution, (**b**) NO source term distribution, (**c**) Mixture fraction distribution with Z_{st} white isoline, and (**d**) OH mass fraction distribution.

Although the diffusion model only has a minor impact on the overall flow field, it significantly affects the NO distribution and its source term. Figure 10a shows the NO mass fraction distribution for Case 2 using both the ULN and MAD models. In the MAD case, higher NO concentrations are observed further downstream of the flame, while in the ULN case, the peak NO concentrations are localized near the flame front.

The differences in NO distribution around the flame front between the ULN and MAD models arise primarily from how species diffusion is treated. In the ULN model, the assumption of equal diffusivity for all species leads to a more localized flame front, where peak temperatures and reaction rates align, resulting in an NO formation concentrated near the front. In contrast, the MAD model accounts for species-specific diffusivities, allowing the faster diffusion of reactive radicals such as OH and O. This broadens the flame structure and promotes NO formation farther downstream. However, the high NO concentrations observed downstream, despite mixing with excess air, are not realistic and are attributed to the artificial increase in the progress variable, as will also be discussed later in this section for OH.

This is evident in the NO source term distribution shown in Figure 10b. In the ULN case, the source term is predominantly negative in the early flame stages near the burner rim, indicating NO consumption. In contrast, the MAD case shows both positive and negative source terms in the same region, reflecting ongoing NO production. This behavior is attributed to the interpolated NO source term shown in Figure 3 (left and bottom), which appears artificially high due to interpolation. This occurs because NO is not directly solved in the Flamelet model within the unstable branch region, which lies between the lowest stable branch and the quenching branch.

The NO source term is retrieved from the Flamelet library during the LES calculations. As a result, the artificially high source term leads to excessive NO production, which is then transported downstream with unrealistically high values.

The OH mass fraction distribution shown in Figure 10d for case 2 indicates that the MAD model predicts elevated OH concentrations downstream of the flame front, in the flue gas region where it mixes with excess air. This behavior is consistent across all other cases as well. Consequently, the progress variable also exhibits higher values in these regions for MAD cases, as illustrated in Figure 11b. This trend can be traced back to the distribution of the progress variable source term, which is concentrated near the burner rim in the early stages of the flame. Since the LES framework solves the transport equation for the progress variable by sampling its source term from the precomputed Flamelet library, any interpolation artifacts in the library are directly reflected in the LES results.

In the case of the MAD model, a substantial portion of the Flamelet library—particularly between the extinction curve (where the progress variable equals zero) and the lowest stable computed Flamelet—is not explicitly solved but interpolated. This is shown in Figure 2 (left) and Figure 3 (left) for the MAD library. This unsolved region is significantly larger in MAD than in the ULN model, primarily due to the absence of unstable branch solutions in MAD. Furthermore, source term values above the highest stable curve are often extrapolated or copied from the highest computed solution, which adds to the inaccuracy.

These limitations in the MAD library contribute to an overprediction of the progress variable and OH mass fraction, both downstream of the flame and within the flame itself. The broader region of positive source term values in the MAD library—caused by interpolation across the high-progress-variable zone where unstable branches typically exist—leads to a more rapid increase in the progress variable throughout the combustion domain.

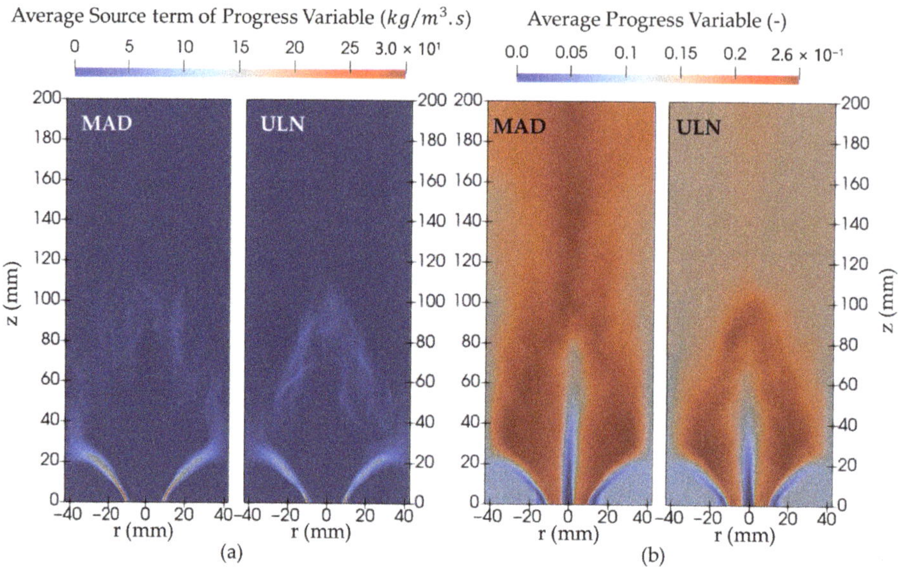

Figure 11. Averaged values comparison between MAD (**left**) and ULN (**right**) models for case 2: (**a**) averaged source term of progress variable distribution, (**b**) averaged progress variable distribution.

3.3. *Effect of Swirl Number on NO Emissions*

As discussed earlier in Figure 6, increasing the swirl number alters the shape of the flow field as expected. Figure 12 further illustrates the impact of swirl numbers 8 and 12 on the NO mass fraction and source term distributions in the ULN model. A higher swirl number shortens the reaction region while intensifying the NO source terms near the burner rim. As a result, NO production becomes more concentrated in the early stages of the flame, whereas NO consumption is more broadly distributed downstream with lower intensity. As observed in [6], increasing the swirl number leads to similar NO emissions across different NH_3/CH_4 blends. This trend is attributed to the enhanced central recirculation and flame stabilization at high swirl, which promotes intense combustion and NO production near the burner. Under these conditions, the flame is shorter, and most NO is produced in the early stages via the thermal and fuel-bound nitrogen pathways, with limited residence time for downstream reduction. In contrast, lower swirl numbers produce a more extended flame with longer residence times and distributed heat release, allowing blend-specific differences in NO production and post-flame reduction pathways to become more pronounced. Consequently, the NO emissions become more dependent on the NH_3/CH_4 blend ratio under lower swirl conditions.

3.4. *Effect of Global Equivalence Ratio on Flame Structure and NO Emissions*

In this study, the fuel is supplied at a fixed velocity of 13 m/s for all cases. Therefore, changes in the global equivalence ratio are achieved solely by adjusting the air mass flow rate, which is inherently higher than that of the fuel. As a result, this method can also introduce momentum effects into the flow field. The flame shape changes accordingly, as a lower global equivalence ratio has a similar effect to increasing the swirl number but not the same. It enhances the velocity magnitudes and strengthens the swirling motion, as shown in Figure 6; on the other hand, the swirl number controls tangential and axial momentum ratios. Consequently, varying the global equivalence ratio significantly influences the flame structure and its emissions, including NO.

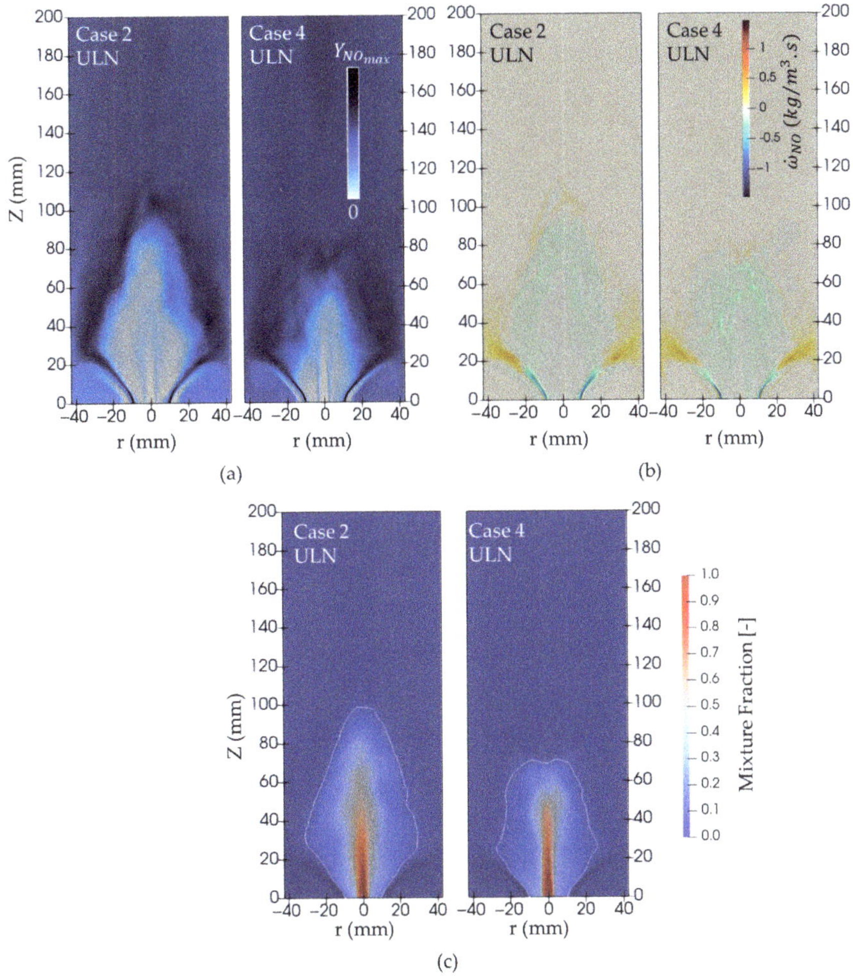

Figure 12. Comparisons between cases with different swirl numbers 8 (**left**) and 12 (**right**) on each panel for the ULN model, panel (**a**): averaged NO mass fraction, panel (**b**): averaged NO source term, and panel (**c**): averaged mixture fraction with isoline at the stoichiometric value.

Figure 13 shows the mixture fraction fields for cases 1, 2, and 3 using the ULN model. At a high equivalence ratio (0.77), the air momentum is relatively low compared to the fuel jet, preventing proper interaction between the swirling air and the fuel jet. This results in poor combustion. In contrast, lower global equivalence ratios (0.54 and 0.46) introduce a greater amount of air and higher air momentum, enhancing the air–fuel mixing. This leads to a reduction in the mixture fraction below the stoichiometric value throughout the domain, around 100 mm downstream of the burner rim, indicating improved combustion. Furthermore, although the mixing process improves at lower equivalence ratios, the flame length remains nearly identical for both cases. This is because, in non-premixed combustion, once sufficient mixing is achieved, the diffusion process becomes the dominant factor governing combustion, leading to similar flame lengths. However, as mentioned in Section 3.3, increasing the swirl number has the strongest effect on the flow field, resulting in shorter flames, and therefore increasing the swirl number can improve the combustion process for a high equivalence ratio like 0.77, as mentioned in [6].

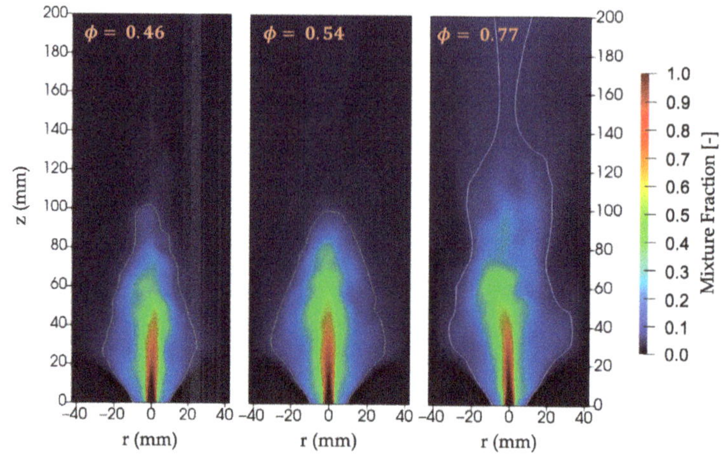

Figure 13. Mixture fraction fields for ULN cases 1, 2, and 3, corresponding to equivalence ratios of 0.46, 0.54, and 0.77, respectively, shown from left to right. Isolines at the stoichiometric mixture fraction are also included.

Figure 14 shows the normalized NO mass fractions at 15% O_2 mass fractions, this normalization process serves to eliminate the effect of different amounts of air associated with different equivalence ratios on the NO mass fraction. The high global equivalence ratio shows the lowest maximum NO mass fractions among the maxima of the three cases, which is expected due to the poor combustion process, which exhibits lower NO and higher NH_3 emissions, as mentioned in [6]

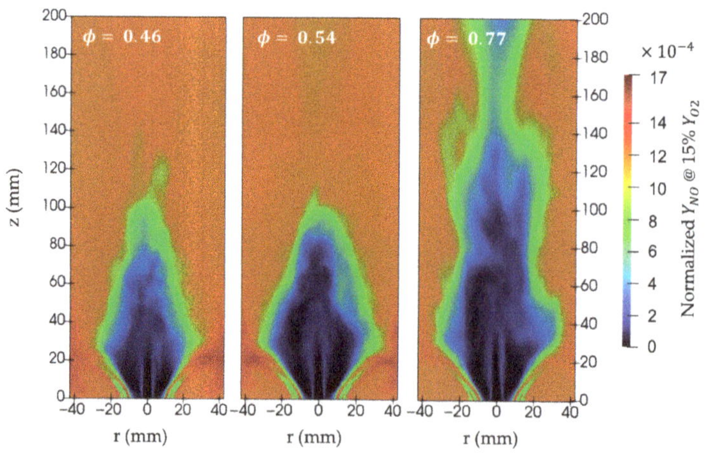

Figure 14. Normalized Y_{NO} @ 15% Y_{O2} fields for different equivalence ratios of 0.46, 0.54, and 0.77 for ULN cases 1, 2, and 3 from left to right, respectively.

4. Conclusions

This study investigated the effects of the swirl number and diffusion modeling on the flow field and NO emissions in CH_4/NH_3 flames. Large eddy simulations (LES) were conducted for a 50% ammonia/50% methane mixture by volume, across various equivalence ratios and swirl numbers. Two diffusion models were considered: the unity Lewis number (ULN) model and the mixture-averaged diffusion (MAD) model. Several cases were selected to match those in [6] for comparison.

The key findings are as follows:

- While the choice of diffusion model has a minor effect on the overall flow field, it significantly influences flame structure, and the distribution of reactive species.
- The MAD model predicts higher OH concentrations and broader reaction zones compared to the ULN model, which better matches experimental OH profiles with more localized flame structures. This overprediction in the MAD case arises from artificially elevated progress variable source terms in the Flamelet library, particularly in regions where an unstable branch was not resolved, unlike the ULN model, which successfully recovers this branch. As a result, the MAD model exhibits exaggerated progress variable growth throughout the flame.
- For NO emissions, the ULN model confines NO formation and reduction to a narrow region near the flame front, whereas the MAD model shows more spatially distributed NO profiles, driven by differential diffusion (DD) of nitrogen-based intermediates.
- Increasing the swirl number leads to shorter flames and upstream-shifted NO formation zones. This limits the role of post-flame NO chemistry and suppresses NO reduction pathways associated with intermediate species such as NH and NH_2.
- Reducing the global equivalence ratio to a certain level ensures good air–fuel mixing near the burner rim, leading to improved combustion. This is associated with reduced fuel slip to the exhaust and increased NO emissions.
- Future work includes the development of unstable branch calculation for the MAD model to enable accurate resolution for it and reduce interpolation artifacts in the Flamelet library. These findings enhance the understanding of how diffusion models and turbulent flow conditions influence NO chemistry in ammonia–methane flames and can inform strategies for low-emission combustion design.

Author Contributions: Conceptualization, A.A.; Methodology, A.A.; Validation, A.E.; Formal analysis, A.A. and R.K.; Investigation, A.A.; Resources, A.E.; Data curation, A.A.; Writing—original draft, A.A.; Writing—review & editing, A.E. and R.K.; Visualization, A.A.; Supervision, R.K. and H.W.; Project administration, H.W.; Funding acquisition, H.W. All authors have read and agreed to the published version of the manuscript.

Funding: This work was supported by MEXT as "Program for Promoting Researches on the Supercomputer Fugaku" and used computational resources from the supercomputer Fugaku provided by the RIKEN Center for Computational Science) (Project ID: hp230193, hp240205, hp250232, hp220241). Part of this work was supported by JSPS KAKENHI Grant Number 23K22686.

Data Availability Statement: The original contributions presented in this study are included in the article. Further inquiries can be directed to the corresponding author.

Conflicts of Interest: The authors declare no conflict of interest.

Abbreviations

The following abbreviations are used in this manuscript:

ULN	Unity Lewis Number
MAD	Mixture Averaged Diffusion
DD	Differential Diffusion
LES	Large Eddy Simulation
FPV	Flamelet Progress Variable

References

1. IEA. Efficiency and Demands. Available online: https://www.iea.org/world/efficiency-demand (accessed on 12 June 2025).
2. Wiseman, S.; Rieth, M.; Gruber, A.; Dawson, J.R.; Chen, J.H. A comparison of the blow-out behavior of turbulent premixed ammonia/hydrogen/nitrogen-air and methane–air flames. *Proc. Combust. Inst.* **2021**, *38*, 2869–2876. [CrossRef]

3. Elbaz, A.M.; Wang, S.; Guiberti, T.F.; Roberts, W.L. Review on the recent advances on ammonia combustion from the fundamentals to the applications. *Fuel Commun.* **2022**, *10*, 100053. [CrossRef]
4. Valera-Medina, A.; Xiao, H.; Owen-Jones, M.; David, W.I.F.; Bowen, P.J. Ammonia for power. *Prog. Energy Combust. Sci.* **2018**, *69*, 63–102. [CrossRef]
5. Okafor, E.C.; Somarathne, K.K.A.; Ratthanan, R.; Hayakawa, A.; Kudo, T.; Kurata, O.; Iki, N.; Tsujimura, T.; Furutani, H.; Kobayashi, H. Control of NO_x and other emissions in micro gas turbine combustors fuelled with mixtures of methane and ammonia. *Combust. Flame* **2020**, *211*, 406–416. [CrossRef]
6. Elbaz, A.M.; Hassan, Z.O.; Roberts, W.L. Exploring the influence of swirl intensity on stability, emissions, and flame structure in non-premixed NH_3/CH_4 swirling flames. *Proc. Combust. Inst.* **2024**, *40*, 105644. [CrossRef]
7. Pitsch, H. Large-eddy simulation of turbulent combustion. *Annu. Rev. Fluid Mech.* **2006**, *38*, 453–482. [CrossRef]
8. Peters, N. Turbulent Combustion. In *Cambridge Monographs on Mechanics*; Cambridge University Press: Cambridge, UK, 2000.
9. Franzelli, B.; Riber, E.; Gicquel, L.Y.M.; Poinsot, T. Large Eddy Simulation of combustion instabilities in a lean partially premixed swirled flame. *Combust. Flame* **2012**, *159*, 621–637. [CrossRef]
10. Schulz, C.; Dreizler, A.; Ebert, V.; Wolfrum, J. Combustion Diagnostics. In *Springer Handbook of Experimental Fluid Mechanics*; Tropea, C., Yarin, A.L., Foss, J.F., Eds.; Springer: Berlin/Heidelberg, Germany, 2007; pp. 1241–1315. [CrossRef]
11. Knudsen, E.; Pitsch, H. A general flamelet transformation useful for distinguishing between premixed and non-premixed modes of combustion. *Combust. Flame* **2009**, *156*, 678–696. [CrossRef]
12. Gicquel, L.Y.M.; Staffelbach, G.; Poinsot, T. Large Eddy Simulations of gaseous flames in gas turbine combustion chambers. *Prog. Energy Combust. Sci.* **2012**, *38*, 782–817. [CrossRef]
13. Popp, S.; Hunger, F.; Hartl, S.; Messig, D.; Coriton, B.; Frank, J.H.; Fuest, F.; Hasse, C. LES flamelet-progress variable modeling and measurements of a turbulent partially-premixed dimethyl ether jet flame. *Combust. Flame* **2015**, *162*, 3016–3029. [CrossRef]
14. Gierth, S.; Hunger, F.; Popp, S.; Wu, H.; Ihme, M.; Hasse, C. Assessment of differential diffusion effects in flamelet modeling of oxy-fuel flames. *Combust. Flame* **2018**, *197*, 134–144. [CrossRef]
15. Jiang, X.; Guo, J.; Wei, Z.; Quadarella, E.; Im, H.G.; Liu, Z. A species-weighted flamelet/progress variable model with differential diffusion effects for oxy-fuel jet flames. *Combust. Flame* **2023**, *251*, 112674. [CrossRef]
16. Wen, X.; Dressler, L.; Dreizler, A.; Sadiki, A.; Janicka, J.; Hasse, C. Flamelet LES of turbulent premixed/stratified flames with H2 addition. *Combust. Flame* **2021**, *230*, 111428. [CrossRef]
17. Su, H.; Wu, Y.; Yan, J.; Jiang, L. Experimental Study on Chemiluminescence Properties of Ammonia-Methane Non-Premixed Laminar Flames. *Energies* **2025**, *18*, 402. [CrossRef]
18. Osetrov, O.; Haas, R. Modeling Homogeneous, Stratified, and Diffusion Combustion in Hydrogen SI Engines Using the Wiebe Approach. *Energies* **2025**, *18*, 3004. [CrossRef]
19. Lan, Y.; Wang, Z.; Xu, J.; Yi, W. The Impact of Hydrogen on Flame Characteristics and Pollutant Emissions in Natural Gas Industrial Combustion Systems. *Energies* **2024**, *17*, 4959. [CrossRef]
20. Liu, C.; Ge, X.; Zhang, X.; Yang, C.; Liu, Y. Research on the Characteristics of Oscillation Combustion Pulsation in Swirl Combustor. *Energies* **2024**, *17*, 4164. [CrossRef]
21. Wu, Q.; Liang, X.; Zhu, Z.; Cui, L.; Liu, T. Numerical Simulation Research on Combustion and Emission Characteristics of Diesel/Ammonia Dual-Fuel Low-Speed Marine Engine. *Energies* **2024**, *17*, 2960. [CrossRef]
22. Wang, S.; Wang, H.; Luo, K.; Fan, J. The Effects of Differential Diffusion on Turbulent Non-Premixed Flames LO_2/CH_4 under Transcritical Conditions Using Large-Eddy Simulation. *Energies* **2023**, *16*, 1065. [CrossRef]
23. Vega, P.J.O.; Coussement, A.; Sadiki, A.; Parente, A. Non-Premixed Filtered Tabulated Chemistry for LES: Evaluation on Sandia Flames D and E. *Fuels* **2022**, *3*, 486–508. [CrossRef]
24. Pitsch, H.; Peters, N. A consistent flamelet formulation for non-premixed combustion considering differential diffusion effects. *Combust. Flame* **1998**, *114*, 26–40. [CrossRef]
25. Okafor, E.C.; Naito, Y.; Colson, S.; Ichikawa, A.; Kudo, T.; Hayakawa, A.; Kobayashi, H. Experimental and numerical study of the laminar burning velocity of CH_4–NH_3–air premixed flames. *Combust. Flame* **2018**, *187*, 185–198. [CrossRef]
26. Ihme, M.; Pitsch, H. Modeling of radiation and nitric oxide formation in turbulent nonpremixed flames using a flamelet/progress variable formulation. *Phys. Fluids* **2008**, *20*, 055110. [CrossRef]

MDPI AG
Grosspeteranlage 5
4052 Basel
Switzerland
Tel.: +41 61 683 77 34

Energies Editorial Office
E-mail: energies@mdpi.com
www.mdpi.com/journal/energies

www.ingramcontent.com/pod-product-compliance
Lightning Source LLC
Chambersburg PA
CBHW061439210926
51850CB00008B/1558
9783725850099